Vegetation of the Arabian Peninsula

Geobotany 25

Series Editor

M.J.A. WERGER

The titles published in this series are listed at the end of this volume.

Vegetation of the Arabian Peninsula

edited by

Shahina A. Ghazanfar

and

Martin Fisher

Department of Biology,
Sultan Qaboos University,
Muscat, Oman

Springer-Science+Business Media, B.V.

A C.I.P. Catalogue record for this book is available from the Library of Congress.

ISBN 978-90-481-5020-5 ISBN 978-94-017-3637-4 (eBook)
DOI 10.1007/978-94-017-3637-4

Printed on acid-free paper

Contents

Contributors

Abdul-Nasser Al Gifri
Block 50, Blgd 13-201, Mansoora, Aden, Yemen.

Benno Böer
Ecology Department, National Avian Research Centre, PO Box 9903, Sweihan, United Arab Emirates

David Membery
Department of Meteorology, Directorate General of Civil Aviation and Meteorology, Seeb International Airport, PO Box 204 Seeb 113, Oman

E Fay Robertson
King Khaled Wildlife Research Center, National Commission for Wildlife Conservation and Development, PO Box 61681, Riyadh 11575, Saudi Arabia

Harald Kürschner
Institut für Systematische Botanik und Pflanzengeographie, Freie Universität Berlin, Altensteinstr. 6, 14195 Berlin, Germany

Ingeborg Guba
Department of Petroleum and Mining Engineering, Sultan Qaboos University, PO Box 33 Al Khod 123, Muscat, Oman

Jameel A Abbas
Environmental Research Centre, University of Bahrain, PO Box 32038, Bahrain

James P Mandaville
10150 N. Calle del Carnero, Tucson, Arizona 85738, USA

Ken Glennie
Department of Geology and Petroleum Geology, University of Aberdeen, Aberdeen AB24 3UE, Great Britain

Martin Fisher
Department of Biology, Sultan Qaboos University,
PO Box 36 Al Khod 123, Muscat, Oman

Philip J Seddon
National Wildlife Research Center, National Commission for Wildlife Conservation and Development, PO Box 1086, Taif, Saudi Arabia

Samira Omar
Kuwait Institute for Scientific Research, PO Box 24885,
13109 Safat, Kuwait

Shahina A Ghazanfar
Department of Biology, Sultan Qaboos University,
PO Box 36 Al Khod 123, Muscat, Oman

Shaukat A Chaudhary
Ministry of Agriculture and Water Resources, PO Box 17285,
Riyadh 11484, Saudi Arabia

Ulrich Deil
Biologisches Institut II/Geobotanik, Schänzlestr. 1, 79104 Freiburg, Germany

Preface

The inspiration for this book came from our ten years of journeys and wanderings through the varied landscapes of Arabia, and in particular through those of its hospitable southeastern corner, Oman. We owe a particular debt to Sultan Qaboos University, which during this time has provided us with both a stimulating working environment and a home.

Transliteration of Arabic place and other names into English script is a task fraught with difficulties. We have followed 'accepted' spellings wherever these were not contrary to our common sense, and in other cases we have rendered names into Roman English script using phonetic spellings. Our main task in this respect was to ensure conformity between the fifteen contributing authors. Diacritical signs have mostly been avoided, since their use is neither widely followed nor readily understood.

Arabic words which have been commonly taken into the English language, such as 'sabkha' for a salt flat and 'wadi' for a valley with a seasonal watercourse, are not italicised in usage. However, other Arabic terms which are occasionally used in English but not as widely known, such as ***harrah*** for a basaltic lava field and ***hima*** for a traditional grazing reserve, are italicised throughout the text.

We are deeply indebted to the following scholars and enthusiasts for their various and invaluable contributions to this book: Andrew S. Gardner for his critical comments on Chapters 2, 3 and 12, Bruno Mies for critical comments on Chapter 5, Vic Hichings for his help with Chapter 3, Marinus Werger, the series editor, for his critical and encouraging comments on all chapters, and Ralph H Daly, Andrew Spalton, Ali al Kiyumi, David Insall and Mani Grobler for their help with the Oman section of Chapter 12.

In addition, we thank the Darwin scholar Duncan Porter for tracking down the quote at the beginning of Chapter 1, Michael Gallagher for the use of the invaluable library at the Natural History Museum, Oman, Ross Hesketh for his immortal contribution to the appearance of this book by introducing us to the Galliard font, David Insall for advice on the transliteration of Arabic names into Roman English script, Lisa Garner for the preparation of the line figures in Chapter 8, Klaus Müller-Hohenstein for the use of unpublished data and maps in Chapter 6, and Gina Douglas, Librarian at the Linnean Society, London, for information on Ethelbert Blatter. For the provision of climatological data we thank the Directorate General of Civil Aviation and Meteorology, Seeb International Airport, Oman, the staff of the White Oryx Project, Oman, and the Meteorological and Environmental Protection Authority, Saudi Arabia. We are indebted to SPSS for supplying a copy of SYSTAT version 7 which was used to prepare a number of the figures, Compaq Computer corporation for generously mending a computer, and the USGS for providing the satellite photograph for the front cover. The final layout for this book was prepared by Martin Fisher.

Shahina A Ghazanfar & Martin Fisher, Sultanate of Oman, October 1997

Chapter 1

Introduction

Shahina A Ghazanfar & Martin Fisher

"Hence, a traveller should be a botanist, for in all views plants form the chief embellishment. Group masses of naked rock even in the wildest forms, and they may for a time afford a sublime spectacle, but they will soon grow monotonous. Paint them with bright and varied colours, they will become fantastic; clothe them with vegetation, they must form a decent, if not a most beautiful picture."

From *Narrative of the Surveying Voyages of His Majesty's Ships Adventure and Beagle*, Charles Darwin (1839)

For nearly two and a half millennia *al Jazirat al 'Arabiyah* (the Island of the Arabs) has attracted botanists, naturalists and travellers from far and wide. Some were intrigued by the fertility and agricultural wealth of *Arabia Felix* in the southwest, others by the harshness and romance of *Arabia Deserta*, the relatively barren central desert. This interest in the Arabian Peninsula and its plants began as early as 420 BC, when frankincense was traded from southern Arabia to Egypt and through to Europe(Groom 1981). Later on, botanical observations were made by the Arab explorers of the 9th to the 14th centuries, noteworthy amongst them being ash Sharif al Idrisi (1100-1165/66 AD), Abbas Annabati (*c.* 1216 AD), Ibn Baithar (d. 1248 AD) and Ibn Battuta (1304-1368/69).

During the 16th and 17th centuries Arabia was explored by European travellers seeking new trading routes, and by naturalists who collected plants and described the vegetation of the regions that they visited. It was not, however, until the 18th century that a scholarly expedition to the Arabic-speaking countries, supported by the King of Denmark, was planned by Johann David Michaëlis (1717-1791), a professor at the University of Göttingen, and Johann Hartwig Ernst Bernstoff (1712-1772) the foreign minister of the King of Denmark and Norway. The aim of the *Arabian Journey*, as it came to be called, was to study the history and languages of the region and to survey and record the natural history of southern Arabia and Egypt. On this ill-fated expedition, Pehr Forsskål (1732-1763), the expedition's botanist, collected 2,093 plant species, of which 693 were from northern Yemen, and made notes on the vegetation. Forsskål's floristic lists and descriptions of species, organized according to the Linnean system of classification, with additional information on local names and uses of plants, was published with an introduction by Carsten Niebuhr (1772), the leader and only survivor of the expedition. More recently Hepper and Friis (1994) published a catalogue of Forsskål's plant collections, with validations, synonymy and the present location of his herbarium material.

Perhaps the only professional plant collector to visit eastern Arabia in the 19th century was Pierre Rémi Martin Aucher-Éloy, an intrepid French plant collector

S. A. Ghazanfar and M. Fisher (eds.), Vegetation of the Arabian Peninsula, 1-4.

who visited Oman (then known as the Immamate of Muscat) in 1838. He collected over 250 species from northern Oman of which many were new to science (Ghazanfar 1996a). His extensive collections from Iran, Syria, Egypt, Sinai and Oman were used by Edmund Boissier (1810-1881) to produce the five volume flora of the orient, *Flora Orientalis* (Boissier 1867-1888). Though Aucher-Éloy's botanical notes were scanty, he gave accurate descriptions of the mountain habitats and made observations on climate and cultivation.

One of the greatest travellers and botanists to visit Arabia in modern times was Georg August Schweinfurth (1836-1925). His travels took him to Socotra (1881) and southern Arabia (1881, 1887, 1888, 1889) where he searched for the species collected by Forsskål. His published accounts of the plant life (Schweinfurth 1884, 1891, 1894, 1896, 1899, 1912) contributed greatly to the study of the botany of Arabia.

During the late 19th and early 20th centuries the Peninsula was visited by a number of notable travellers whose collections and botanical accounts provided the background knowledge that paved the way for the work of the modern wave of natural scientists. Sir Richard Francis Burton (1821-1890), Albert Deflers (1841-1921), Theodore Joseph Bent (1852-1897), Sir Isaac Bailey Balfour (1853-1922), Joseph Friederich Nicolaus Bornmüller (1862-1948), Sir Percy Zachariah Cox (1864-1937), William Lunt (1871-1904), Hermann von Wissmann (1895-1979) Ethelbert Blatter (1877-1934) and others (see Wickens 1982 for a list of those who collected on the Arabian Peninsula) made invaluable contributions to the study of Arabian botany. Over the last fifty years, following on from these early explorations, the study of the flora and vegetation of the Peninsula has blossomed, and in addition to the studies cited in the present book, references to other work can be found in the bibliographies of Miller, Hedge and King (1982) and Fisher (1995a).

Many travellers, both ancient and modern, expressed the view that Arabia is a land of great contrasts. This was most elegantly expressed by Niebuhr (1772) in his memoirs:

> *"Intersected by sandy deserts and vast ranges of mountains it presents on one side nothing but desolation in its most frightful form, while the other is adorned with all beauties of the most fertile regions. Such is its position that it enjoys at once all the advantages of hot and of temperate climates. The peculiar productions of regions the most distant from one another are produced here in equal perfection."*

These diverse regions include the sand seas of the Rub' al Khali and the Great Nafud, the lofty mist-covered mountains of the Asir and Hijaz, the monotony of the hyperarid gravel plains, the lush tropical monsoonal landscape of Dhofar, the geologically unique mountains of northern Oman and the barren basalt fields of the north of the Peninsula. This diversity of landscapes and landforms is matched in equal part by the diversity of the vegetation; high altitude montane woodlands, succulent shrub communities, xerophytic psammophytes and *Acacia* parkland all have their place.

With the relatively recent establishment of universities and research institutes in the countries of the Peninsula, and with an enhanced understanding of the fragility of the biological diversity of the region, there has been an increased awareness over the last two decades of the need to document and study the vegetation. Whilst much work yet remains, especially in the ecological sphere, and some regions require further botanical exploration, we believe that it is time to take stock. In this book 15 authors provide for the first time a comprehensive overview of the phytogeography and vegetation of the Arabian Peninsula. The geographical coverage includes Bahrain, Kuwait, Oman, Qatar, Saudi Arabia, the United Arab Emirates and Yemen. In general we have not considered the vegetation of Socotra, except inasmuch as mention is made of its biogeography and relatively high species richness and endemism. We have organised the book into 11 chapters which describe the climate, geology, phytogeography, the vegetation of the major habitats and landforms, the biology of the bryophytes and lichens, the plants of economic importance, and the patterns of plant diversity and endemism and the status of plant conservation.

The chapters on climate and geology provide the background information necessary to understand the composition, dynamics and history of the Arabian vegetation. Chapter 2 examines the atmospheric and meteorological conditions that bring 'weather' to Arabia, and provides a detailed description and analysis of the climate, with details of both annual, seasonal and geographical variations. Chapter 3 provides an outline of the structure of the landforms of the Peninsula and summarises its geological history.

Chapter 4 examines the phytogeographical links of the Arabian Peninsula with Asia and Africa and provides an introductory overview of the main vegetation types and indicator species. The chapter describes the geographical and historical affinities of the major phytochoria which traverse the Peninsula, the role of Mediterranean and Irano-Turanian floristic intruders, the history of the vegetation and the occurrence of disjunct and relict taxa.

Chapter 5 deals with the bryophytes (mosses and liverworts) and lichens. The study of these groups on the Peninsula has been somewhat neglected in the past, and it is only in recent years that the composition of the bryophyte and lichen communities have been described and their ecological and morphological adaptations investigated.

Chapters 6 to 10 describe in detail the vegetation and plant ecology of the various landforms and habitats: the mountains and wadis (Chapter 6), the gravel plains (Chapter 7), the sand bodies (Chapter 8), the coasts and sabkhas (Chapter 9) and the fresh and brackish water bodies (Chapter 10). Each chapter describes the phytogeography and plant communities of a particular landform or habitat and the adaptations of its characteristic plants.

Chapter 11 deals with the plants of economic importance and the traditional uses of local plants for medicine, perfumes and fragrance, dyes, and for the manufacture of utilitarian objects.

Chapter 12 describes the diversity of the plants of the Peninsula and summarises the current efforts to preserve both vegetation and landscapes. Species diversity in the various countries of the Peninsula, the patterns of endemism of the various regions, current threats to plant diversity and vegetation cover are all described, and a detailed summary is given of the protected areas and relevant legislation currently in place in each country of the Peninsula.

The challenge now is to use the knowledge summarised in this book for the study of the historical and functional ecology of the vegetation of the Arabian Peninsula, and for the conservation of its unique flora.

Chapter 2

Climate

Martin Fisher & David A Membery

"I asked how much rain was required to produce grazing, and he answered, 'It is no use if it does not go into the sand this far,' and he indicated his elbow. 'How long does it have to rain to do that?' 'A heavy shower is enough. That would produce grazing that was better than nothing, but it would die within the year unless there were more rain. If we get really good rain, a whole day and night of rain, the grazing will remain green for three and even four years."

From *Arabian Sands*: Wilfred Thesiger conversing with Muhammad al Auf shortly before their crossing of the Empty Quarter in 1946 (Thesiger 1959).

2.1 Introduction

For an ecosystem in which productivity is primarily limited by the availability of water, an understanding of climate, and its attendant effects on vegetation, is indispensable. Muhammad al Auf understood this well, and Thesiger noted later how during their crossing of the Empty Quarter his guides would invariably stop wherever rain had fallen and check the depth of its penetration for later reference. Just as knowledge of the spatial and temporal distribution of rainfall over the previous few years was useful to Muhammad al Auf and other guides, in order that they could locate grazing for their camels during lengthy desert passages, information on climate patterns is required for a full understanding of the present composition and distribution of the vegetation of Arabia.

The frequently scarce, often erratic, and sometimes sudden and profuse occurrence of rainfall, combined with the generally high temperatures across the

S. A. Ghazanfar and M. Fisher (eds.), Vegetation of the Arabian Peninsula, 5-38.

subtropical desert belt, has sculpted the ecological, cultural and geomorphological landscapes of the Arabian Peninsula. Analogous to the way that Arabia is 'at the junction' from a phytogeographical point of view (see chapter 4), having components of the African, Mediterranean and southwest Asian floras, the meteorological events that bring rainfall to the Peninsula come from four different directions: the Mediterranean, central Asia, the tropical maritime regime of the Indian Ocean and tropical Africa. These four atmospheric influences largely operate at different times of the year, bringing a variable degree of seasonal 'weather', and hence rainfall, to the Peninsula. Given certain conditions, Arabia can also generate its own weather, largely due to the orographic influences of the southwestern mountains of Saudi Arabia and Yemen and the northern mountains of Oman. In addition, the influence of the high albedo of the central Arabian desert in the summer plays a fundamental role in the maintenance of the southwest monsoon, though the benefit to Arabia is small, generally bringing rainfall only to a relatively small area of the coastal regions at the border of Yemen and Oman, and to a lesser extent to the southwestern highlands.

The influence of rainfall and temperature regimes on the composition, distribution and abundance of the vegetation of Arabia is intertwined with both geomorphological and phytogeographical influences (see Chapters 3 and 4 respectively). Geomorphology, phytogeographical history and climate have acted together over time to shape the nature of the vegetation as we see it today, but it is largely to climate that we must turn for an understanding of the maintenance of the current species composition of the varied plant communities of the Peninsula.

In this chapter we first of all examine the atmospheric and meteorological conditions that bring weather to Arabia; this is the foundation for an understanding of the climatic patterns. We follow this with a description and analysis of the climate, in which we examine both annual and seasonal variations, and present a compilation of climagrams, data summaries and climate maps for the whole area. To our knowledge, this is the first time that a comprehensive meteorological and climatological analysis for the whole of the Arabian Peninsula has been compiled.

2.2 Meteorology

Arabia is a land of contrasts. From the searing summer heat of the Rub' al Khali to the moderate climate and occasional snowfall on the southwestern mountains and the Hajar mountains of Oman, one passes from the hyperarid (<100 mm rainfall) coasts and plains, through the arid (100-250 mm rainfall) higher plains and foothills and the semiarid (250-500 mm rainfall) mountain slopes and summits, to the temperate (>500 mm rainfall) mountains of the extreme southwest. Topographical influences on the meteorology of the Arabian Peninsula are strong. The western mountains influence rainfall production and surface wind flow all along the Red Sea from Sinai to Bab al Mandab and, although seemingly remote, the Zagros mountains of western Iran play an important role in the production of precipitation over the extreme east of the Peninsula. Kuwait, Bahrain, the United Arab Emirates and Musandam all benefit from the mass ascent of the westerly flow that affects the lower part of the atmosphere every winter, albeit with marked variations from year to year. Topographical influences also help to 'steer' mobile winter-time weather systems to the extent that they enter the Arabian Peninsula from the eastern Mediterranean and leave it via the Gulf of Oman.

In marked contrast to winter-time mobility, the meteorological situation in the summer is remarkably static. Throughout high summer a thermally driven low pressure circulation is to be found over southern Arabia approximately centred on 20°N 50°E, with its axis oriented southwest-northeast. Under clear skies, the loose sandy surface absorbs maximum solar radiation, leading to intense surface heating. Convective activity is vigorous but no clouds form because the ambient humidity is so low; however large quantities of dust are transported into the lower atmosphere through energetic thermal activity, leading to the widespread hazy conditions that characterize summer in the region.

2.2.1 Air-masses

It is customary in meteorology to describe the typical weather of a region in terms of the air-masses that dominate the area. There are two major air-mass types: Polar and Tropical, with two important subdivisions: Continental and Maritime. Each air-mass has distinct characteristics depending on the seasonal pressure distribution and the nature of the underlying surface. Air-mass characteristics are relatively conservative and so meteorological parameters such as temperature, humidity, dew-point and more particularly wet-bulb potential temperature can be used to identify an air-mass even when it is far from its source region. Two air-masses dominate the weather across Arabia, namely Polar Continental during the winter and Tropical Continental during the summer, with minor incursions of Polar Maritime and Tropical Maritime (Figure 2.1). In high summer conditions are such that the Arabian Peninsula is itself a source region of Tropical Continental air. Of the four air-mass types to affect Arabia, this is the most reliable, dominating the weather across the region from early June to late September.

2.2.1.1 Polar Continental

The Polar Continental air-mass that affects the Arabian Peninsula during winter originates over central Asia. It is characterized by very low temperatures at the surface due to strong radiational cooling on long winter nights, and as a result a temperature inversion develops; i.e. the normal decrease of temperature with height is inverted, and the temperature actually increases with height above the surface. As a result of warmer air above colder air, the lower atmosphere is regarded as stable and convective clouds do not occur. The characteristics of a Polar Continental air-mass are dry weather with generally clear skies and relatively

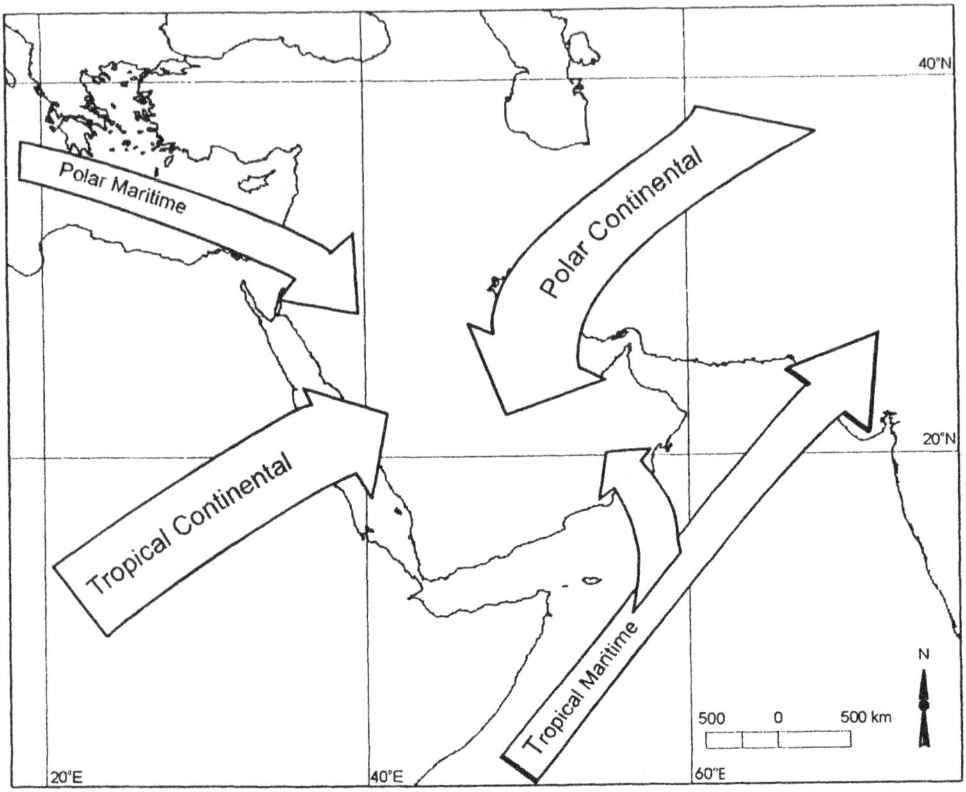

Figure 2.1 The directions of influence of the four air-masses that control the weather across Arabia. Note the disturbance to the Polar Continental air-mass as it crosses the Zagros mountains of Iran and the greater influence of the Polar and Tropical Continental air masses.

low temperatures. The Zagros mountains play an important role in protecting Arabia from the low temperatures experienced in central Asia. Occasionally a strong outbreak of Polar Continental air can lead to showers developing over the coastal areas of the Arabian Gulf and the Gulf of Oman because the cold air-mass has been heated from below by the warmer waters. Polar continental air does not affect the region during summer.

2.2.1.2 Polar Maritime

Occasionally the region is affected by Polar Maritime air that has the North Atlantic as its source region. This air-mass makes brief incursions during the winter months in connection with mobile 'weather systems' moving from west to east. These weather systems are a modification of mid-latitude depressions, with 'warm and cold fronts', although the warm front is usually very weak by the time the system has traversed North Africa and the Mediterranean. By contrast the cold front is frequently quite active, with thunderstorms developing over northern

Saudi Arabia and Kuwait ahead of the frontal zone, and with strong *shamal* winds along the Arabian Gulf in the much modified Polar Maritime air-mass behind the front.

2.2.1.3 Tropical Continental

Tropical Continental air-masses are formed over the hot and dry land surfaces of North Africa in late spring and early summer when surface heating is extreme. This air invades northwestern Saudi Arabia in advance of eastward moving low pressure systems over the Mediterranean; it advects hot and very dry air from Egypt and Sudan, producing some of the highest temperatures of the year in northern Arabia. In some locales the hot desert wind is known as the *khamsin* or the *sharav* (Winstanley 1972). During high summer the static conditions that exist over the interior of Arabia lead to the area being a source region of Tropical Continental air itself. This air-mass is characterized by cloudless skies, low humidity and very high temperatures (often >45°C), and the intense surface heating leads to vigorous dry convection and the transport of dust particles into the mid-troposphere (*c.* 4,500 m) so that hazy conditions invariably prevail.

2.2.1.4 Tropical Maritime

With the advent of the summer months the thermal equator marches north with the northward progress of the sun so that surface heating becomes intense over northern India, the Tibetan plateau and the Arabian Peninsula. A series of strong thermal low pressure centres develop, particularly in the Rub' al Khali, Baluchistan and Pakistan. Around these thermal low pressure centres a major cyclonic circulation develops known generally as the Indian monsoon system. Tropical Maritime air flowing over the northern Indian Ocean and the Arabian Sea in the Indian monsoon circulation affects parts of Yemen, southwestern Saudi Arabia and coastal Oman but its influence is greatly limited by the Tropical Continental air-mass that prevails over the bulk of the Arabian Peninsula.

2.2.2 The General Circulation over the Arabian Peninsula

2.2.2.1 Winter Circulation (December-February)

The true 'winter' season can be defined as the months of December, January and February, occasionally extending through to the middle of March, and during this period the Arabian Peninsula is dominated by a ridge of high pressure extending from the Siberian anticyclone over central Asia. The predominant air-flow around this system is from the northeast, and so the weather is typical of a Polar Continental air-mass, being relatively cool and dry with clear skies both day and night. However, this so-called 'northeast monsoon' is considerably modified by two effects before reaching the Peninsula. The first effect is due to the Zagros mountains, which act as an obstruction to the northeasterly air-flow, leading to the creation of a topographically induced pressure trough downwind of the mountains along the Iranian coast of both the Arabian Gulf and the Gulf of Oman. This trough prolongs the passage of the air over the water and involves a second important effect in that the water is very much warmer than the land mass to the northeast. Thus, by the time the air-flow reaches eastern Saudi Arabia, Qatar, the Emirates and northern Oman it is warmer than along the Iranian coast.

Usually the weather is cloud-free during this season, but occasional outbreaks of very cold air from the northeast can lead to unstable conditions developing, with the cold Polar Continental air being vigorously warmed from below. This can lead to the development of thunderstorms over the UAE and eastern Oman, though this is exceptional, since the main rainfall producing mechanism during the winter is associated with mobile 'upper troughs' moving from west to east.

These troughs develop over North Africa and the Mediterranean in the strong westerly flow that exists every winter in the northern hemisphere. Perturbations in this westerly flow lead to a succession of sinusoidal troughs and ridges steadily advancing from west to east, and the appearance of a low pressure circulation at the surface is evidence of a mobile trough in the upper troposphere. Mobile upper troughs and low pressure 'systems' are most active during the months of December, January and February but they occasionally extend into the transitional month of March. The plateau of Turkey forms a natural barrier to these 'systems' so that most cross the eastern Mediterranean travelling on to cross northern Saudi Arabia and the Arabian Gulf. On average there are four or five a month, although there is marked variation from year to year.

When a sharp upper trough extends its influence as far south as Yemen, strong southwesterly flow is to be expected ahead of the trough, leading to widespread ascent and the formation of extensive cloud in the middle and upper troposphere across much of central and eastern Arabia. If, at the same time, Tropical Maritime air is advected from the Arabian Sea by southerly winds in the lower troposphere and there is significant vertical motion through forced ascent over the Iranian mountains, then the stage is set for the production of widespread rain. Rain can fall by this mechanism up to 500 km upwind of the Zagros mountains, including Kuwait, Bahrain, Qatar, eastern districts of Saudi Arabia, the Emirates and northern Oman. Should the atmosphere be potentially unstable, forced ascent may well alter the temperature profile of the lower atmosphere, releasing the potential instability through pockets of deep convection and so thunderstorms with occasional heavy rain may develop.

Normally the sequence of events described above is transient, with the trough and low pressure system continuing to move away eastwards towards northern India. Occasionally, however, troughs 'stall' and then cloud and rain may well persist for several days. Even more occasionally the global circulation pattern will dictate the upper flow over the Arabian Peninsula such that 'troughing' persists even longer than a few days. During December 1995 for example, a 'blocking' situation over Europe and western Russia led to the formation of a persistent Ω pattern with the downstream trough affecting much of the Middle East. Heavy rains fell across southern Iraq, eastern Saudi Arabia, the Emirates and northern Oman, with heavy snows in the Zagros mountains. At Abu Dhabi it was the wettest and cloudiest December since records began in 1972, whilst in Fujairah flood damage was described as the worst natural disaster to affect the UAE since its inception. At Khasab in Musandam a total of 283 mm of rain fell during the month with more than 200 mm recorded in the eastern Hajar, and 113 mm at Seeb, more than in the average year (Membery 1997).

However, the conditions described above are the exception rather than the rule, and on many occasions the passage of an upper trough is identified only by patchy cloud, then a wind shift accompanied by increasing surface pressure. If the pressure rises significantly over the Peninsula strong to gale force dust-raising northerly winds will develop, particularly over eastern Saudi Arabia, Bahrain, Qatar and the Emirates and even occasionally extending as far south as Dhofar. These strong northerlies are known as the *shamal* and can blow for several days, briefly introducing Polar Maritime air to the Arabian Peninsula. As the upper trough moves eastwards across Pakistan, pressure rises over Iran lead to the development of *shamal* winds over northern Oman to the east of the Hajar mountains, also known as the *na'shi* along the Batinah and Sharqiyah coasts of Oman.

2.2.2.2 Spring Transitional Months (March-May)

The spring transitional months are so-called because the weather encountered at this time of the year is partly a result of weak westerly upper troughs in the mid-troposphere described earlier and partly a result of changes taking place over Asia and the Arabian Sea. These changes take place in a relatively short period of time due to the northward advance of the thermal equator. It is an important season for rainfall, especially over the southwestern mountains, but the rainfall producing mechanisms are not straightforward. Strong contrasts in temperature between land and sea and between mountains and valleys stimulate local circulations, leading to the production of convective rainfall.

In a continuation of the trough-ridge synopsis introduced in the previous section, the atmosphere is periodically unstable at a time when there are major daytime temperature differences between land and water surfaces, and these conditions are conducive to the formation of North African desert depressions which affect the northern part of the Peninsula at this time. There is a marked temperature contrast between the cool waters of the eastern Mediterranean (Polar Maritime and Polar Continental air) and the high temperatures existing over the desert interior (Tropical Continental air). Should such a situation occur underneath an upper trough then dynamic vertical motion encourages the production of precipitation.

Over northern Saudi Arabia and Kuwait, thunderstorm activity reaches a peak at this time of the year because Polar Maritime air is undercutting Tropical Continental air; on average six storms affect the region during April and May. Further south, Pedgley (1970) refers to the presence of the Oman Convergence

Zone across the interior regions of Oman with the axis of activity around 20°N. Along this zone, a northwesterly low-level flow from the Arabian Gulf meets a marine southeasterly flow from the Arabian Sea. Under the influence of surface convergence and assisted by vertical motion from any slow-moving upper trough, clouds develop deep enough to produce occasional rain over some of the driest territory in Oman. For example at Fahud, at the edge of the Rub' al Khali, *c*. 60% of the meagre annual total of *c*. 30 mm falls during March-May, whilst on Masirah Island more than 50% of the annual total falls during this period (Figure 2.11). Any late winter upper trough brings with it the potential for heavy rain, for as the seasonal temperature rises, so does the atmosphere's ability to hold moisture.

Exceptional rains occurred over eastern Yemen, the Rub' al Khali and western Oman during March and April 1997 when a series of slow-moving upper troughs advected potentially unstable moisture laden Tropical Maritime air from the Horn of Africa. Wadi flows were the heaviest in living memory, and the salt pan of Umm as Samim on the edge of the Rub' al Khali became a vast lake.

2.2.2.3 Summer Circulation (June-September)

The one season that can be counted upon each year across the Peninsula is the summer season, which begins during May and ends during September or early October. Summer is characterized by clear but hazy skies with high temperatures throughout, with night minimum temperatures amongst the hottest in the world. During summer a succession of thermal lows develop across the region, with centres over northwestern India, Pakistan, Baluchistan and the Rub' al Khali. These thermal low pressure centres are formed as a direct consequence of intense surface heating. Hot air above the surface expands and rises, leading to relatively low surface pressure values, and to the development of a cyclonic circulation around the thermal lows. Over northern Arabia another thermal low develops along the Euphrates-Tigris flood plain of southern Iraq, leading to the infamous dust storms of southern Iraq and Kuwait and the *shamal* winds of eastern Saudi Arabia. Peak months for this activity are June and July when the northwesterly wind frequently approaches gale force at Kuwait.

From a wider perspective, all of these thermal lows contribute to the dynamics of the southwesterly flow over the Indian Ocean and Arabian Sea, and are very closely associated with the development of the Asian monsoon system. Having travelled a long way over the Indian Ocean and the Arabian Sea, the Tropical Maritime air is moist and potentially unstable as it approaches the coast of Yemen and Oman, but the southwesterly flow is prevented from releasing its instability for two reasons.

Firstly, between one and two kilometres above sea level there is a notable shift in wind direction from the southwest to the northwest with a noticeable increase in temperature with height, as the northwesterly wind advects Tropical Continental air from the Rub' al Khali. This temperature increase is known as an 'inversion', since it is contrary to the normal decrease of temperature with height. Now, with hot air from the interior above cooler maritime air, instability is restricted to a shallow layer near the surface.

Secondly, the southwesterly wind blows over cold coastal waters that have formed through a process known as upwelling. The mechanics of upwelling result from a combination of meteorological and oceanographic effects, further explained below, but the consequence of the cold sea surface is to cool the lowest layers of the maritime air-mass. As elsewhere in meteorology, cooling an air-mass from below increases its stability, and again this stabilizing process limits vertical mixing and prevents the realization of possible precipitation. Ironically, it is the very aridity of the Arabian Peninsula that leads to the success of the Indian monsoon system. By confining the moisture to a shallow layer at the bottom of the atmosphere and preventing dilution through vertical mixing, the moist maritime air is channelled towards the Indian subcontinent. Gradually, the dominating effect of the temperature inversion dissipates and the potential instability is realized through forced orographic lifting over the western Ghats. Rainfall produced as a result of evaporation and condensation processes over the western Arabian Sea precipitates out over the Indian sub-continent (Smith 1986a, 1986b).

In any standard climatological atlas the Inter-Tropical Convergence Zone (the ITCZ) is positioned over southern Arabia, connecting tropical Africa with monsoonal India and the wet lands of southeast Asia. However, only on rare occasions during the summer months does the ITCZ ever exhibit its true characteristics of thunderstorms and heavy rain, usually in relation to either a tropical storm or a tropical cyclone. Both are agents of change which can destroy the ever-present temperature inversion, referred to above, by forcing Tropical Maritime air into the hyperarid Peninsula. Another opportunity to break through the obstructive temperature inversion exists over the southwestern mountains, and to a lesser extent over the Hajar mountains of northern Oman. These elevated platforms intercept maximum radiation from the overhead sun, and through conduction, the air above the mountains is warmer than the relatively cooler air of the free atmosphere. Local convection can take place, subject of course to available moisture.

There is satellite evidence that mid-tropospheric moisture streaming westwards (i.e. opposite to the low-level flow) from the Indian monsoon system supports convection once initiated by providing the necessary moist environment, but it is thought that local land and sea breeze circulations are primarily responsible for the formation of this local orographic convection. A wide area of montane Yemen and southwestern Saudi Arabia and a smaller area of Jebel al Akhdhar in the western Hajar mountains of northern Oman receive >300 mm of rain per year, a large part of which falls during the summer months of July and August (Figure 2.11).

As a result of the pressure gradient that exists between high pressure over the Arabian Sea and low pressure over the Rub' al Khali, strong south to southwesterly winds affect the coastal waters of eastern Yemen, Dhofar and al Wusta during the summer months. Through frictional affects these strong southwesterly winds impart stress to the ocean surface, forcing it forward in a northeasterly direction,

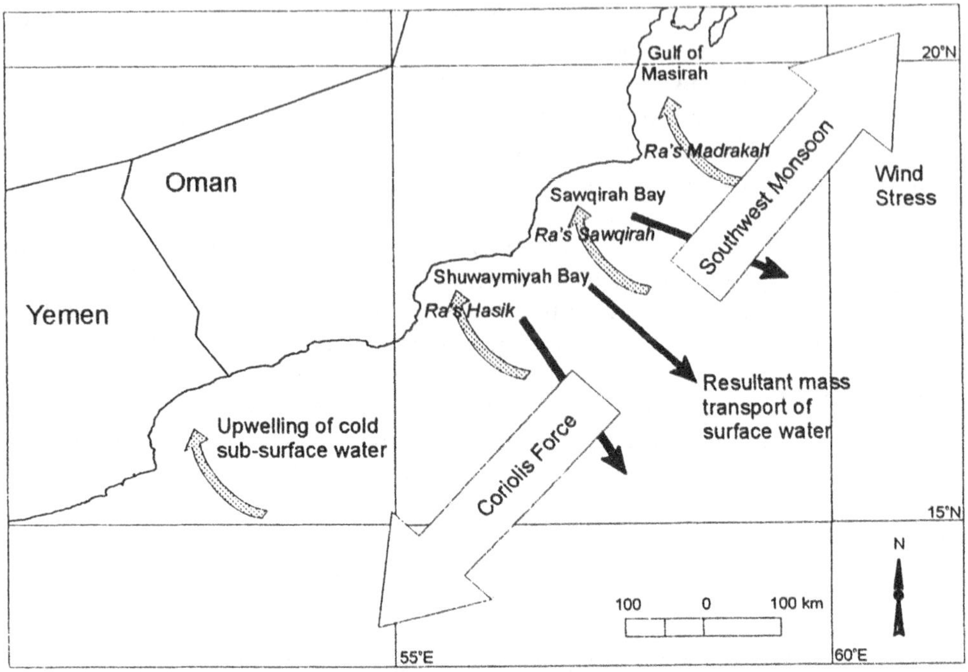

Figure 2.2 The mechanism of the upwelling that occurs off the southern coast of Arabia in the summer, illustrating the opposing influences of wind stress and the Coriolis force, the resultant easterly mass transport of surface water and the compensatory upwelling of cold sub-surface water (see text for further details).

but the coastal waters do not move in this direction because of the influence of the Coriolis force, which in the northern hemisphere opposes the wind stress. The resultant of these two forces is that the average motion of the current through the coastal waters is approximately 90° to the right of the surface wind, i.e. in an easterly direction (Figure 2.2). As a result of this mass transport of surface water towards the east, a compensatory upwelling of cold sub-surface water takes place such that minimum sea surface temperatures can be up to 5°C cooler than ambient offshore values. Satellite pictures suggest that the regions of intense upwelling are the bays downstream of coastal headlands, for example Shuwaymiyah Bay in the shelter of Ras Hasik, Sawqirah Bay downstream of Ras Sawqirah and the Gulf of Masirah in the lee of Ras Madrakah (Figure 2.2). Off the Dhofar coast the sea surface temperature is some 4-8°C cooler than the moist southerly winds flowing over it, and the maritime air-mass is cooled below its dew-point, leading to saturation and the formation of mist, fog and low cloud.

Driven by the southerly wind, the fog and low cloud move inland over the Salalah plain, but their progress is restricted by the crescent-like curvature of the Dhofar mountains and the fact that hot desert air is to be found around mountain-top level (on average just above 1,200 m). So, through a combination of topography and a strong temperature inversion, a stable situation develops with

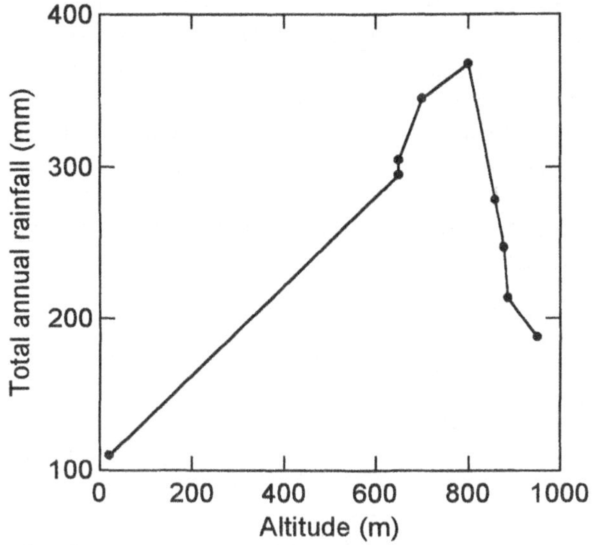

Figure 2.3 Variation in total annual precipitation with altitude on Jebel Dhofar in southern Oman.

persistent thick stratus clinging to the southern slopes of the mountains. The situation only allows for weak convection to take place within the marine layer, and a solid layer of stratocumulus develops above the fog and low stratus, with the tops of the cloud to be found at the base of the inversion. In addition to the fog and low cloud, light rain and drizzle are frequent during this period, especially overnight and during the early morning. Precipitation over the coastal areas of Salalah is usually in the form of fine drizzle with daily totals seldom exceeding 5 mm, whilst rainfall totals over the mountains are considerably greater. However, because of the limited depth of the cloud, maximum totals are not on the mountain tops as would be expected, but around 700-800 m (Figure 2.3).On the leeward slopes of the mountains, rainfall totals during the *khareef* diminish rapidly, with desert conditions returning within a few kilometres.

Wherever plants and shrubs intercept the moist airflow total 'precipitation' is very much greater than that observed in conventional raingauges. Experimental raingauges with foliage growing on a wire screen on top of the gauge have measured up to five times the precipitation recorded in a conventional raingauge during the monsoon in Dhofar (Anon. 1996). The experimental gauges collect both the vertical component (light rain and drizzle) and the horizontal component (occult precipitation intercepted by natural foliage).

2.2.2.4 Autumn Transitional Period (October-November)

October and November, apart from being the second tropical storm period, is also the season of change between the end of the southwest monsoon and the start of the winter-time northeast 'monsoon'. Apart from the threat of tropical storms, it is generally a quiet period; rain producing systems are not very active and even convection over the mountains is subdued. However, occasional showers can affect coastal areas of northern Oman during the autumn months when cool continental air crosses the relatively warm waters of the Gulf of Oman. The only other area to receive significant rainfall during the 'northeast monsoon' is the archipelago of Socotra. As November progresses into December, eastward moving troughs in the upper troposphere become more frequent and the yearly cycle begins to repeat itself again.

2.2.3 Tropical Storms and Cyclones

Tropical storms and cyclones are almost entirely confined to two cyclone seasons, namely the pre-monsoonal period (May-June) and post-monsoonal period (October-November). Most storms originate over the southeastern Arabian Sea in the vicinity of the Laccadive Islands, but some late season storms start over the southern Bay of Bengal and move westwards across southern India, regenerating as they cross over the warm waters of the Arabian Sea. A tropical storm develops from an initial disturbance which, under favourable environmental conditions, grows first into a tropical depression (winds <34 knots) and then a tropical storm, with winds around the cyclonic circulation of 34-63 knots. Under certain circumstances it may develop further into a tropical cyclone with winds >63 knots in the circulation around the centre.

For tropical storms or cyclones to develop, certain conditions are necessary to kick-start the initial perturbation, and a very important ingredient is the supply of energy into the system to create strong horizontal and vertical air motions. The main source of energy comes from the release of latent heat during the condensation-precipitation process, and for this reason tropical storms develop and continue to evolve into cyclones only over water with temperatures $>27^{\circ}C$. Storms and cyclones rarely form within $\pm 8^{\circ}$ of the equator where the Coriolis force is insufficient to maintain a circulation, or in zones of strong vertical wind shear (for example beneath a jet stream) as both factors inhibit the development of an organized vortex. Tropical depressions and storms can be described as moderately vigorous warm-cored cyclonic disturbances; warm-cored in the sense that temperatures at mid-tropospheric levels within the cyclonic circulation are warmer than those outside it. The warming process comes from the release of latent heat of condensation during the production of precipitation. If development is allowed to continue unhindered the warming process will lead to an expansion of the entire tropospheric column, ultimately leading to the creation of an upper level anticyclone (i.e. at *c.* 12,000 m). Outflow from the upper-level anticyclone creates a positive feedback effect which stimulates further low-level inflow of warmth and moisture, convective activity and an additional release of latent heat. Additionally, by invoking thermodynamic considerations, it can be shown that the warming process synchronously alters the distribution of potential vorticity throughout the tropospheric column, enhancing cyclonic vorticity in the lower troposphere and weakening it in the upper troposphere. The result is a fully developed self-perpetuating tropical cyclone.

Conventional wisdom has it that tropical storms and cyclones develop most frequently over the southeastern quadrant of the Arabian Sea, and once formed they move northwesterly towards the Arabian Peninsula, sometimes curving northeastwards towards Gujarat and Pakistan and sometimes curving westwards towards the Gulf of Aden. About one in three approaches the Arabian Peninsula, and storms and cyclones cross the coast of Oman and Yemen about once every three years, and of these only half are likely to be true cyclones with winds reaching at least force 12. (Table 2.1). Usually they cross the coast between Masirah Island and Salalah, but considering the lack of observations between these

Table 2.1 Number of tropical storms >34 knots and number of cyclones >64 knots affecting the coastline of Oman and Yemen during 1890-1996 (after Pedgley 1970, with modifications).

	J	F	M	A	M	J	J	A	S	O	N	D	Total
Storms + Cyclones	0	0	0	0	9	8	1	1	1	8	7	1	36
Cyclones	0	0	0	0	6	4	0	0	0	3	3	1	17

two points prior to satellite surveillance it is more than likely that some storms and cyclones went undetected.

Cyclones have been known to enter the Gulf of Aden and more rarely the Gulf of Oman. On 4th June 1890, a tropical cyclone brought 24 hours of torrential rain to the Batinah Coast and Muscat, with severe flooding and widespread damage to property. Close to 300 mm of rain fell on Muscat, and the ensuing destruction led directly to the deaths of at least 700 people. In June 1977, a severe cyclone crossed Masirah Island; maximum sustained winds were *c.* 90 knots with gusts to 120 knots, and the 24 h rainfall was 431 mm. In June 1996 a tropical storm crossed the Omani coast near Ras Madrakah and brought >200 mm of rain to the eastern Hajar mountains and >150 mm to the mountains of Dhofar (Membery in press). This tropical storm went on to bring widespread death and destruction to central and southern Yemen; more than 400 people were reported as drowned or missing and more than 40,000 hectares of arable land was devastated. 189 mm of rain fell at Ma'rib, 164 mm at San'a and 136 mm at Taiz.

Storms or cyclones are practically unknown at the height of the monsoon during the summer months of July-September because the sea surface temperature across most of the Arabian Sea falls below 27°C. Gale force southwesterly winds thoroughly mix the warm surface layers with colder waters from beneath, leading to a drop in temperature below the 27°C threshold, and indeed sometimes below 20°C along the coast. One rare tropical storm affected much of central Oman during August 1983 when a monsoon depression (with a mid-tropospheric circulation) tracked westwards across northern India and then developed into a tropical storm over the relatively warmer waters of the northern Arabian Sea. As it crossed Masirah Island surface winds of up to 47 knots were recorded; rainfall associated with the storm amounted to 46 mm (Membery 1985). Another monsoon depression crossed northern Oman during the last week of July 1995. Coastal areas received 50-100 mm, whilst there was more than 200 mm in the mountains. Jebel Shams in the western Hajar received its annual total of 300 mm during this week alone. The depression continued westwards to the UAE, giving a total of 41 mm to Al Ain and 18 mm to Abu Dhabi; rain was even recorded as far west as Qatar, Bahrain and eastern Saudi Arabia.

2.3 Climate

As the description of the meteorological influences has shown, the relatively great latitudinal and longitudinal extent of the Arabian Peninsula opens it to a number of atmospheric phenomena both external and local to the area. Whilst these day-to-day phenomena have direct and immediate local effects, such as that of heavy rainfall on seed germination, the influence of meteorological patterns over longer periods of time, i.e. the influence of climate, helps to shape the overall characteristics of the ecology and vegetation of the region. In this section we describe, analyse and map the temporal and spatial patterns of the climate of Arabia.

2.3.1 The Data

We describe the climate using data from the majority of climate stations currently operating in the Peninsula. We have in this respect faced four problems: firstly, the paucity of recording stations relative to the size of the area; secondly, the relative unevenness of station coverage, with few stations in the general area of the Rub' al Khali of central southern Arabia and little accessible data from Yemen; thirdly, the relatively short period of time that some of the more remote stations have been in operation; and, fourthly, difficulties in obtaining data for the same set of years for all stations, this being important for a comparative statistical analysis.

Whilst a number of meteorological stations in Arabia have been in operation since the seventies, especially on the coast and in some of the larger inland cities of Saudi Arabia, many stations in the drier interior only began operation in the eighties, and in the case of a few important stations in the drier interior of Oman, in 1985. The data set used here therefore covers 1985-1995, giving eleven years of mostly continuous data for 43 stations. The data covers the main climatological stations of Bahrain, Oman, Saudi Arabia and the United Arab Emirates. Data for a further 70 stations, available as either annual summarised data or as monthly summaries for years other than 1985-1995, were also used as an aid in preparing the climate maps. Locations of all stations are provided in Figure 2.6.

The number of years of records required for reliable estimation of climate parameters depends on the degree of inter-annual variability characteristic of the climate of the area concerned, with stations that exhibit relatively greater inter-annual variability requiring longer records for the estimation of long-term means.

In general, ten or more years of data provides a reasonable estimate of mean total annual rainfall and variability parameters (Fisher 1994), and fewer years are required for mean annual temperature due to its relatively lower inter-annual variability.

2.3.2 Climate Analysis

Statistical summaries of climate for ecological purposes are often restricted to the use of climagrams and long-term averages, although much more can be learnt by framing the analysis in the context of the components of a time series (Fisher 1994). These components are the level or mean value, a trend component due to any long term movements in level over time, a seasonal component, and an irregular component unique to each point in time. We use this system of analysis here, following an earlier exploratory analysis of the climate of Oman (Fisher 1994).

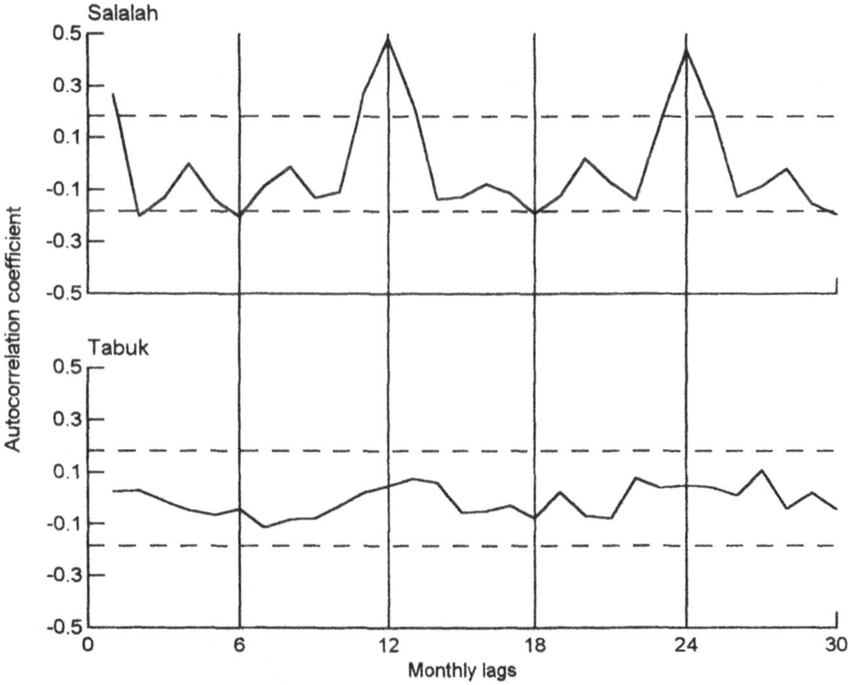

Figure 2.4 An illustration of the autocorrelation coefficient of total monthly rainfall using two extreme examples: Salalah, with the highest 12-month autocorrelation coefficient (0.48) of the 43 stations used here, and Tabuk, with one of the lowest (-0.05). Autocorrelation coefficients are given for 30 monthly lags using the whole of the 11-year data set (1985-1995) for both stations; i.e. 132 months in all. The dashed horizontal lines represent the approximate level of the autocorrelation coefficient which must be obtained for significance (± 0.183, see text for details) and the solid vertical lines indicate the 6-, 12-, 18- and 24-month autocorrelation coefficients for both stations.

There are two levels at which the serial properties of a climatic parameter are of interest here: at the monthly and at the annual. For the monthly series, we use the autocorrelation coefficient as a measure of seasonality, but we do not investigate the monthly mean, which is of less interest here than the annual mean. An autocorrelation coefficient is obtained from the correlation between the total rainfall or mean temperature of each month and the same data lagged by one or more months. A value of $\geq 2/\sqrt{n}$, where n = number of months in the data set, is an approximate guide to significant coefficients (Diggle 1990). For 11 years of data this value is 0.183. Peaks or troughs at six and 12 months, possibly with roll on effects at 18 and 24 months, indicate seasonality (Figure 2.4). The autocorrelation coefficient at a lag of 12 months is a useful measure of seasonality.

For the annual series we use the means of total rainfall and annual temperature as measures of level. The 'seasonality' of the annual series, which is equivalent to the identification of cycles >1 year long, would require a series longer than 11 years, and is therefore not examined here. We use the coefficient of variation (standard deviation/mean) of total annual rainfall and mean annual temperature as a measure of inter-annual variability.

Three further annual temperature parameters are also used here: absolute maximum, absolute minimum and mean monthly range, the latter calculated as the mean of the maximum monthly range of each year. We do not consider the trend component of any of the climatic time series, since this is more relevant to an examination of long-term climate change and would require series longer than 11 years.

The climagram is a standard method used for the description of climate (Walter & Lieth 1960). Both rainfall and temperature are plotted together, with the precipitation scale two times the temperature scale; a relatively humid period is presumed to prevail when the precipitation curve lies above that of temperature, and a relatively drier period when precipitation lies beneath temperature.

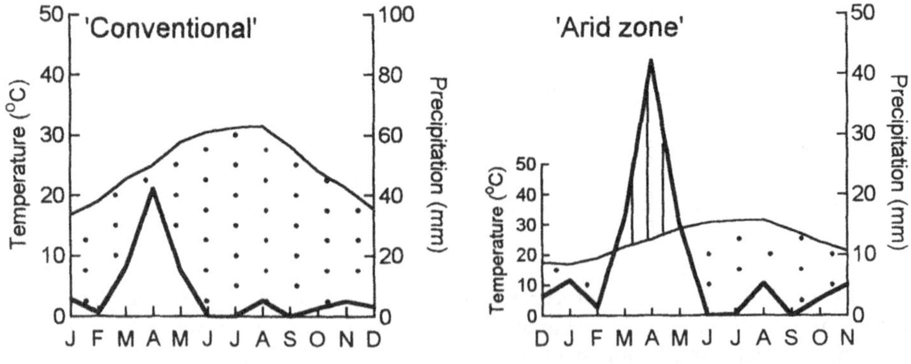

Figure 2.5 A comparison of 'conventional' and 'arid zone' climagrams using rainfall in Bisha (see Figure 2.6 for location) as an example. In a standard climagram the temperature axis is 0.5 x the rainfall axis and months are from January to December, whilst in an 'arid zone' climagram the temperature axis is 2 x the rainfall axis and the starting month can vary, being December (the beginning of our winter) in this example.

However, when used for arid areas the precipitation curve nearly always lies beneath that of temperature, and thus climagrams from regions with potentially subtle but important differences in climate appear rather similar, and do not provide a clear indication of when vegetation growth might be expected. McGinnies (1985) proposed a modified climagram for arid regions with the temperature scale twice that of precipitation (rather than *vice versa*), and with the freedom to liberate the time scale from the January to December orthodoxy. This suggestion was followed in an earlier analysis of the climate of Oman (Fisher 1994), and it is continued in the present analysis; the period when the precipitation curve lies above that of temperature being regarded as the time when vegetation growth is most likely. Since the meteorological analysis has shown that the annual weather regime of the Arabian Peninsula begins in December, so do our climagrams. Figure 2.5 illustrates the differences between the two types of climagrams. Climate statistics for the 43 stations used are given in Table 2.2 and climagrams in Figure 2.6.

2.3.3 Temperature Patterns

Temperature is markedly seasonal throughout Arabia, with highly significant 12-month autocorrelation coefficients (0.292-0.909) at all stations (Table 2.2). The lowest mean temperatures occur in the winter months of December-February and the highest during the summer months of June-September (Figure 2.6). The exceptions to this norm occur in those coastal areas affected by the southwest monsoon, in which there is a depression of summer temperature during the months of June and July. This can be seen in the climagrams for those stations within the monsoon-affected part of the southern coast (Qairoon Hariti and Salalah), and also to the north of the monsoon-affected southern mountains at Thumrait and on the island of Masirah further along the coast. There is even a small temperature depression in August as far north along the coast as Sur and Seeb (Figure 2.6). In contrast to the rainfall patterns (see section 2.3.4 below), inter-annual variability is very low, with coefficients of variation of mean annual temperature in the range 0-0.072 (Table 2.2).

Mean annual temperatures vary greatly across the Peninsula, from 18°C in the far northwest to 31°C on the southwestern coasts, with temperature falling with altitude in the usual manner in the southwestern mountains and in the northern mountains of Oman (Figure 2.7). In the northern central region there is a 1-3°C depression of mean temperature across approximately latitudes 24-28°N due to the high ground of the Nejd. The hottest regions lie in the south, i.e. in the Rub' al Khali, along the southern coasts of Yemen and Oman (excepting the coastal regions affected by the monsoon which have as a consequence a 1-2°C depression of mean temperature), along the Red Sea coast to about 25°N and all along the northern and eastern coasts of Oman and the Emirates. It should be noted that the contours of mean temperature for the Rub' al Khali are constructed using data from stations which are at its periphery.

Table 2.2 Altitude, World Meteorological Organization (WMO) number, mean total annual rainfall, maximum and minimum absolute temperatures, mean monthly maximum temperature range, mean annual temperature, 12-month autocorrelation coefficients (ACF) of total monthly rainfall and mean monthly temperature, coefficients of variation (CV) of total annual rainfall and mean annual temperature, and mean annual number of fog days for 43 meteorological stations (see Figure 2.6 for locations). Statistics calculated from monthly summaries for 1985-1995, except that fog observations only available for 1987-1995 for some stations (*), and for 1992-1993 at Ja'aluni. Mean number of fog days for the nearby Mina Raysut are given after that for Salalah; these two stations are similar meteorologically in all except observation of fog.

Station	Altitude (m)	WMO no.	Rain (mm)	Max temp (°C)	Min temp (°C)	Temp range (°C)	Mean temp (°C)	ACF of rain	ACF of temp	CV of rain	CV of temp	Mean fog days
Abha	2093	41112	253	34.1	0.0	16.0	18.6	0.22	0.88	0.42	0.03	13.7
Abu Dhabi	16	41217	80	47.6	5.4	15.9	27.1	-0.02	0.91	0.79	0.00	
Ahsa	178	40420	108	49.4	-2.3	17.5	26.5	0.01	0.89	0.80	0.02	10.5
Arar	549	40357	66	47.0	-5.6	16.5	21.8	0.21	0.89	0.46	0.03	8.4
Baha	1652	41055	150	38.6	0.0	13.6	22.7	0.16	0.85	0.32	0.03	4.0
Bahrain	2	41150	86	45.7	7.0	9.0	26.5	0.09	0.90	0.70	0.02	-
Bisha	1162	41084	97	42.8	-0.8	18.9	24.8	0.36	0.84	0.56	0.03	0.7
Buraimi	299	41244	104	50.8	2.5	18.2	27.9	0.08	0.79	1.23	0.06	-
Dhahran	17	40416	91	49.5	2.6	15.2	26.5	0.19	0.89	0.61	0.02	21.1
Dubai	8	41194	116	47.3	7.4	13.2	27.2	0.03	0.91	0.60	0.00	-
Fahud	170	41262	24	50.7	5.6	19.4	28.8	0.04	0.69	1.21	0.05	-
Gassim	647	40405	145	47.0	-4.0	17.7	24.2	0.17	0.89	0.39	0.03	7.8
Gizan	7	41141	129	45.3	11.8	11.6	30.6	-0.05	0.87	0.61	0.01	1.4
Guriat	504	40360	53	47.7	-8.0	19.9	19.8	0.11	0.87	0.49	0.03	6.0
Hail	1002	40394	116	43.5	-9.4	17.6	21.5	0.07	0.90	0.47	0.03	6.6
Ja'aluni	154	-	39	49.0	6.5	19.3	26.6	0.00	0.85	1.39	0.03	54.0
Jeddah	4	41024	47	49.0	9.8	13.8	28.4	0.17	0.86	0.96	0.02	4.8
Jouf	669	40361	63	46.0	-7.0	16.6	21.5	0.18	0.89	0.36	0.02	5.1
Khamis Mushait	2056	41114	215	36.0	-0.8	15.3	20.0	0.20	0.84	0.31	0.04	2.2
Khasab	3	41420	198	49.0	8.4	9.6	28.2	0.20	0.89	0.64	0.02	0.3*
Madinah	636	40430	78	47.5	3.0	15.1	27.9	0.16	0.89	0.53	0.02	0.2
Majis	4	41246	125	50.0	5.7	12.6	26.1	0.12	0.90	0.71	0.02	5.4*
Makkah	240	41030	96	49.8	10.0	16.0	30.8	0.24	0.87	0.68	0.02	0.0
Marmul	269	41304	41	49.0	5.4	17.3	27.7	0.01	0.72	1.82	0.07	-
Masirah	19	41288	51	45.2	12.3	10.4	26.6	-0.01	0.87	1.20	0.01	3.7*
Najran	1212	41128	50	42.0	-0.5	17.0	24.7	0.25	0.85	0.70	0.02	0.5
Qairoon Hariti	878	-	236	38.3	4.2	10.6	21.6	0.26	0.77	0.58	0.01	-
Qaysumah	358	40373	133	50.3	-4.0	17.1	25.0	0.06	0.90	0.36	0.02	10.3
Rafah	444	40362	100	48.5	-5.8	18.8	23.0	0.23	0.89	0.42	0.03	7.2
Ras al Khaymah	31	41184	131	48.2	4.4	16.9	27.4	0.14	0.91	0.61	0.00	-
Riyadh	614	40437	126	47.4	-4.4	18.5	24.8	0.13	0.88	0.63	0.02	5.1
Saiq	1755	41254	350	36.3	-3.6	12.6	18.1	0.12	0.89	0.33	0.03	3.4*
Salalah	20	41316	85	44.7	10.8	11.0	26.4	0.48	0.81	0.37	0.01	0.8* (16.2*)
Seeb	15	41256	86	49.2	10.0	11.0	28.7	0.09	0.88	0.78	0.01	2.0*
Sharjah	35	41196	115	49.2	2.5	16.8	26.8	0.00	0.91	0.66	0.00	-
Sharurah	725	41136	36	45.3	0.8	18.4	27.5	0.13	0.85	0.66	0.03	0.1
Sur	14	41268	92	48.9	10.6	12.7	29.3	0.01	0.88	0.69	0.02	0.0*
Tabuk	768	40375	46	44.4	-3.7	16.8	22.0	0.05	0.88	0.56	0.03	0.5
Taif	1453	41036	204	39.5	-1.2	14.7	22.9	0.10	0.87	0.42	0.03	12.1
Thumrait	467	41314	47	46.0	1.6	16.8	27.1	-0.04	0.29	1.50	0.07	0.1*
Turaif	852	40356	82	44.4	-8.0	17.3	18.7	0.10	0.88	0.43	0.03	12.5
Wejh	24	40400	26	44.1	5.1	11.4	24.9	0.06	0.85	0.74	0.02	6.7
Yenbo	10	40439	30	49.4	6.5	15.2	27.6	0.29	0.86	1.07	0.02	12.1

Mean temperatures in our four seasons follow a similar north to south pattern in the winter, spring and autumn, with temperatures of 10-26°C, 18-30°C and 20-30°C, respectively, from northwest to southeast (Figure 2.8). There is a marked change in the mean temperature pattern during June-September, with a rather uniform mean temperature of 30-34°C across central and eastern areas of the Peninsula. This summer pattern is due to the overriding influence of the thermal low pressure centre over the general region of the Rub' al Khali (see section 2.2.2.3).

For the eleven years of data used here, absolute maximum temperatures are 40-50°C over much of the Peninsula, with lower values in the mountains, and absolute minimum temperatures are (-9)-(+12)°C (Figure 2.9), though maximum and minimum absolute values will, unlike a mean, respectively increase and decrease as more data is gathered. The pattern of absolute minimum temperature across the Peninsula generally follows the winter pattern of an increase from the northwest to the southeast, with lower values on higher ground, though with a marked continental depression away from the moderating influence of the sea. Frosts are common in the north and at higher altitudes in the mountains. The geographical pattern of absolute maximum temperature is largely similar to that of mean summer temperature, being 46-50°C across latitudes 17-32° from central regions towards the east.

Mean maximum temperature range, as with absolute minimum temperature, shows the moderating affect of the oceans. In coastal stations the range is *c.* 10-12°C, whilst the continental interior has ranges of *c.* 16-20°C (Figure 2.10). Since the sea both gains and loses heat more slowly than the land, it moderates the temperature of the coastal regions, whilst away from this influence, the continental interior both heats up more during the day and undergoes relatively greater radiational cooling at night.

2.3.4 Rainfall Patterns

Whilst the combined patterns of rainfall and temperature determine overall climate, in a predominantly arid area such as the Arabian Peninsula primary productivity, the spectrum of plant life-forms and the seasonality of flowering are largely influenced by the amount and the seasonal distribution of rainfall, rather than by the seasonal march of temperature. This can be seen in the flowering periods of plants within a wadi in northern Oman over three years of differing rainfall pattern (Ghazanfar 1997, see Figures 7.3, 7.4). At a broader scale, the flowering period in the mountains, foothills and plains of northern Oman is in the late winter through to early spring, coincident with the rainfall peak that occurs at this time (see for example the climagrams for Seeb and Sur, Figure 2.6). In contrast, the main flowering period in the monsoon-affected escarpment mountains and coastal plains of Dhofar occurs immediately after the cessation of the southwest monsoon at the end of August (see the climagrams for Salalah and Qairoon Hariti, with rainfall peaks in July and August, Figure 2.6).

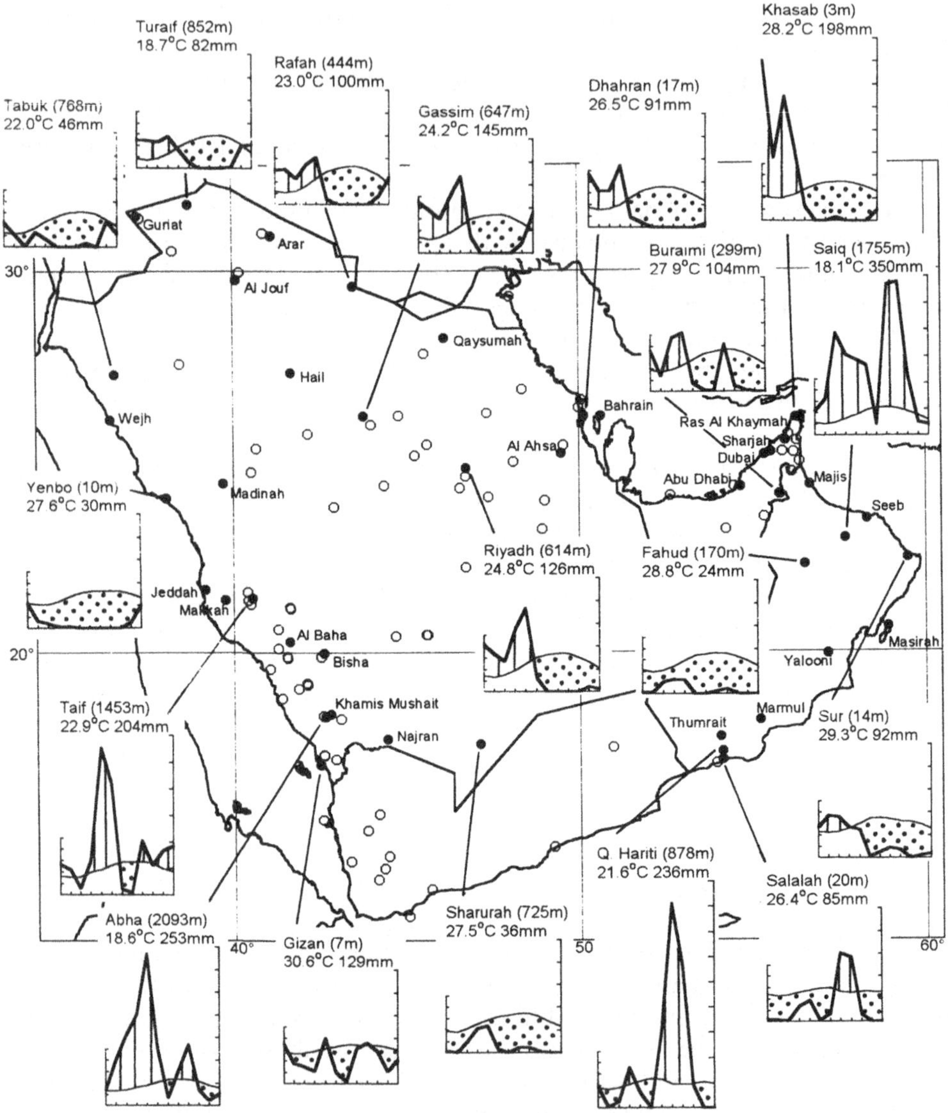

Figure 2.6 Locations of the 43 meteorological stations used for the statistical analysis of climate (see Table 2.2) for 1985-1995 (filled circles, with station names), and of 70 other meteorological stations (open circles) which were used as a further aid in the construction of the climate maps (Figures 2.7-2.12), and for which information was obtained from a variety of sources (Al-Jerash 1989, Böer 1997, König 1988), not necessarily for 1985-1995. Climagrams are given on the map for 18 stations which represent the main climate types of the Arabian Peninsula, and on the facing page for the remaining 25 stations.

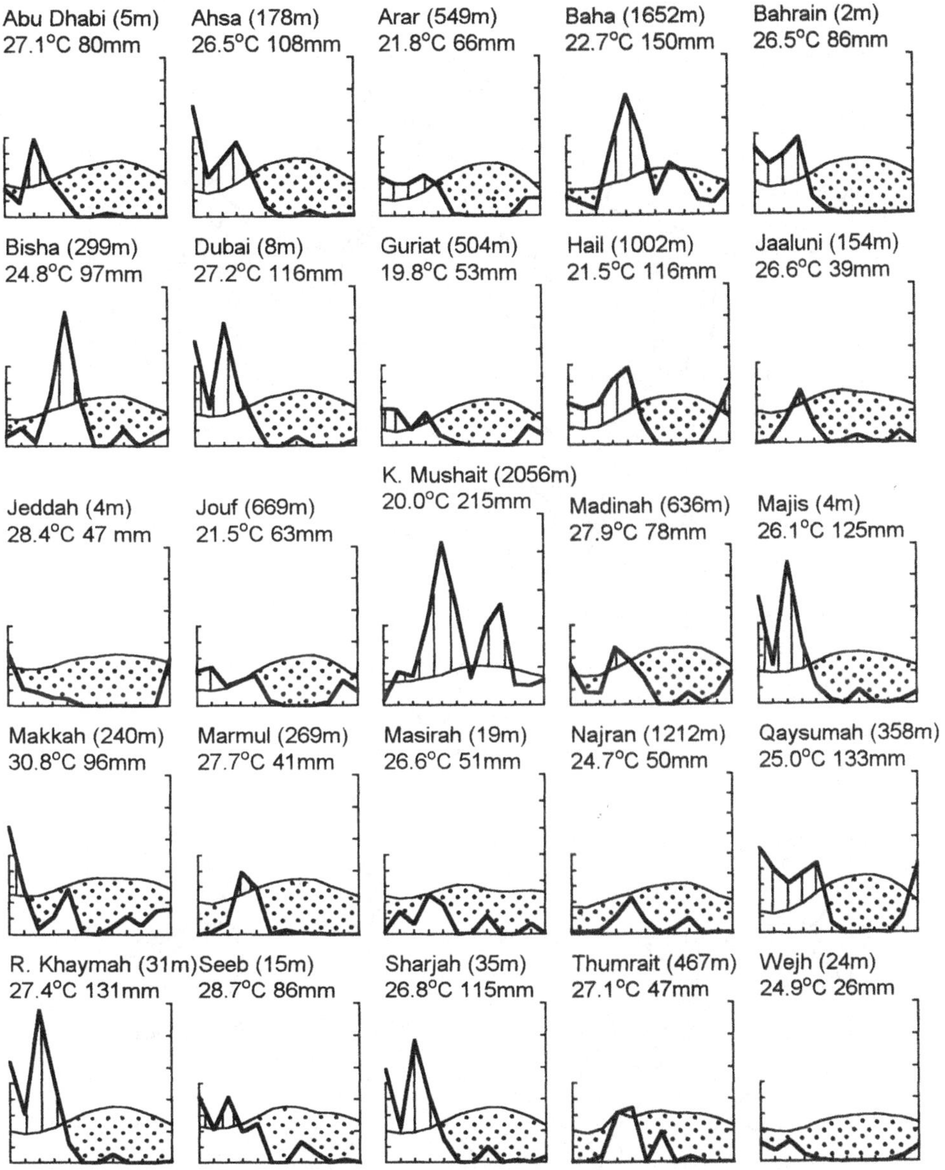

Figure 2.6 continued.

As would be expected, rainfall patterns (Figure 2.7) closely follow that of topography. Mean annual total rainfall is <50 mm in the hyperarid interior of the Peninsula and on the northwestern coast, 50-150 mm over a large area from the Rub' al Khali to the north and northeast, *c.* 100 mm along the southwestern, southern, southeastern and eastern coasts, *c.* 250 mm in the semiarid northern

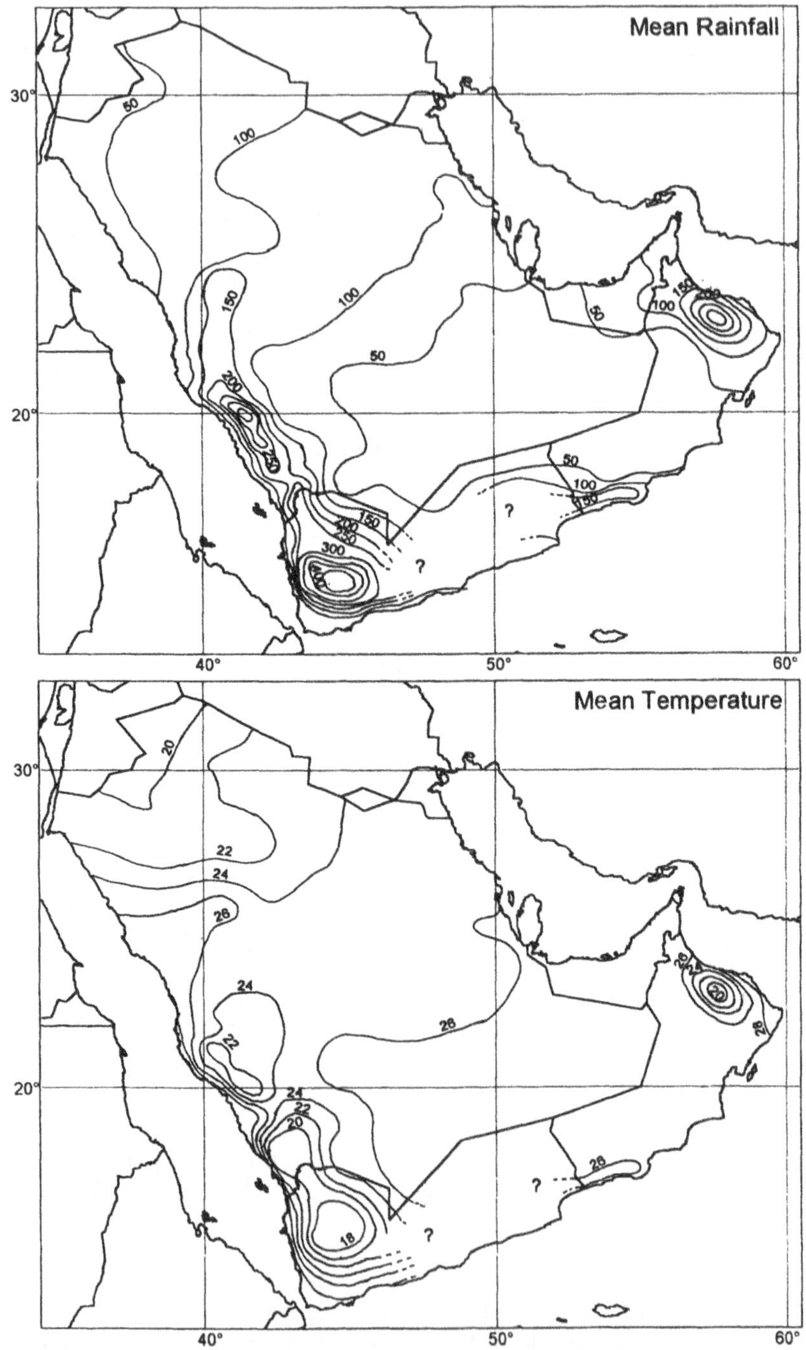

Figure 2.7 Contour maps of mean total annual rainfall and mean annual temperature over the Arabian Peninsula. Dashed lines and ? in this and subsequent contour maps indicate the absence of suitable data from large areas of Yemen.

mountains of Oman and the southern mountains spanning the Yemen-Oman border, and generally >400 mm in the southwestern highlands, attaining totals >1,000 mm in a small area of the extreme southwest. Rainfall figures for the Rub' al Khali are based largely on information from stations at its periphery.

The seasonal distribution of rainfall can be examined by contour maps of the proportion of the annual rainfall that falls in each of our four seasons (Figure 2.11). The influence of the European and Asian winter is seen from the north of the Peninsula across to the southeast, with 40-50% of the annual rainfall occurring during the winter in the north, rising to 40-70% across the eastern coast and coastal areas of northern Oman. Winter is also the main rainfall period for the hyperarid coastal plain bordering the Red Sea north of latitude 20°. The majority of this winter rain comes from the mobile upper troughs and low pressure 'systems' of the dominant westerly flow, and to a lesser extent, along the eastern coast, from occasional perturbations in the normally clear Polar Continental air (see section 2.2.2.1)

The north of the Peninsula receives most of the remainder of its rainfall of 50-150 mm during the spring, largely due to the continued influence of the weather associated with the westerly flow. Spring is also the main rainfall period for the southwestern mountains, for the areas landward of the mountainous regions from the Hadhramaut through to Dhofar and for the central desert regions of Oman bordering the Rub' al Khali, 50-80% of the annual rainfall falling at this time. Geographical extrapolation of the contours suggests that this is also the time when rain is most likely to fall in the Rub' al Khali. Rain falling over central Oman and possibly over the Rub' al Khali is due to the influence of the Oman Convergence Zone, whilst the rain over the southwestern highlands arises from both local convective activity and from the continuing influence of the westerly flow (see section 2.2.2.2).

Summer rainfall is scarce north of latitude 20°N, with <10% of the annual rain falling at this time, largely due to the effect of the central Arabian heat low (see section 2.2.2.3). Only the southern regions receive any summer rainfall, this being almost entirely due to the influence of the southwest monsoon, with 50-70% of the annual total falling across the escarpment mountains of the Yemen-Oman border regions and 20-50% of the annual rainfall total falling in the southwestern highlands.

The transitional months of October and November contribute very little to the total annual rainfall, with <10% falling in the area circumscribed by longitude 42°E and latitude 22°N. The low proportion of the annual rainfall that falls at this time in the north and northwest of the Peninsula is the first indication of the returning influence of the westerly flow.

In a water-limited environment the variability of rainfall is a major influence on vegetation, an aspect of climate in which climagrams are unfortunately misleading, since they involve the averaging of rainfall over a number of years. The 12-month autocorrelation coefficient (Figure 2.12) is uniformly low over the whole Peninsula, with only 11 of the 43 stations (Table 2.2) having values indicative of any seasonality (i.e. >0.183, see section 2.3.2 and Figure 2.4).

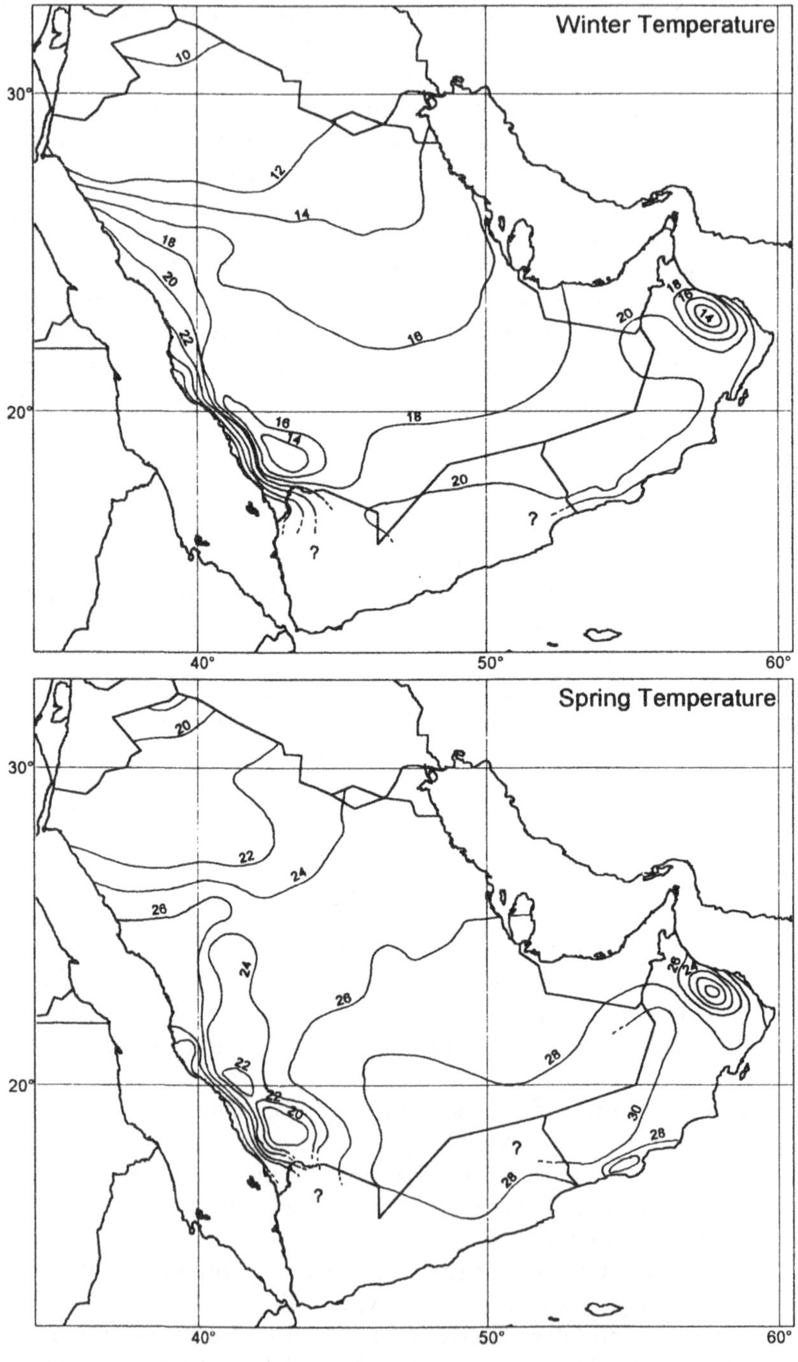

Figure 2.8 Contour maps of mean temperature for the four meteorological seasons over the Arabian Peninsula. This page: winter (December-February) and spring (March-May). Facing page: summer (June-September) and autumn (October-November).

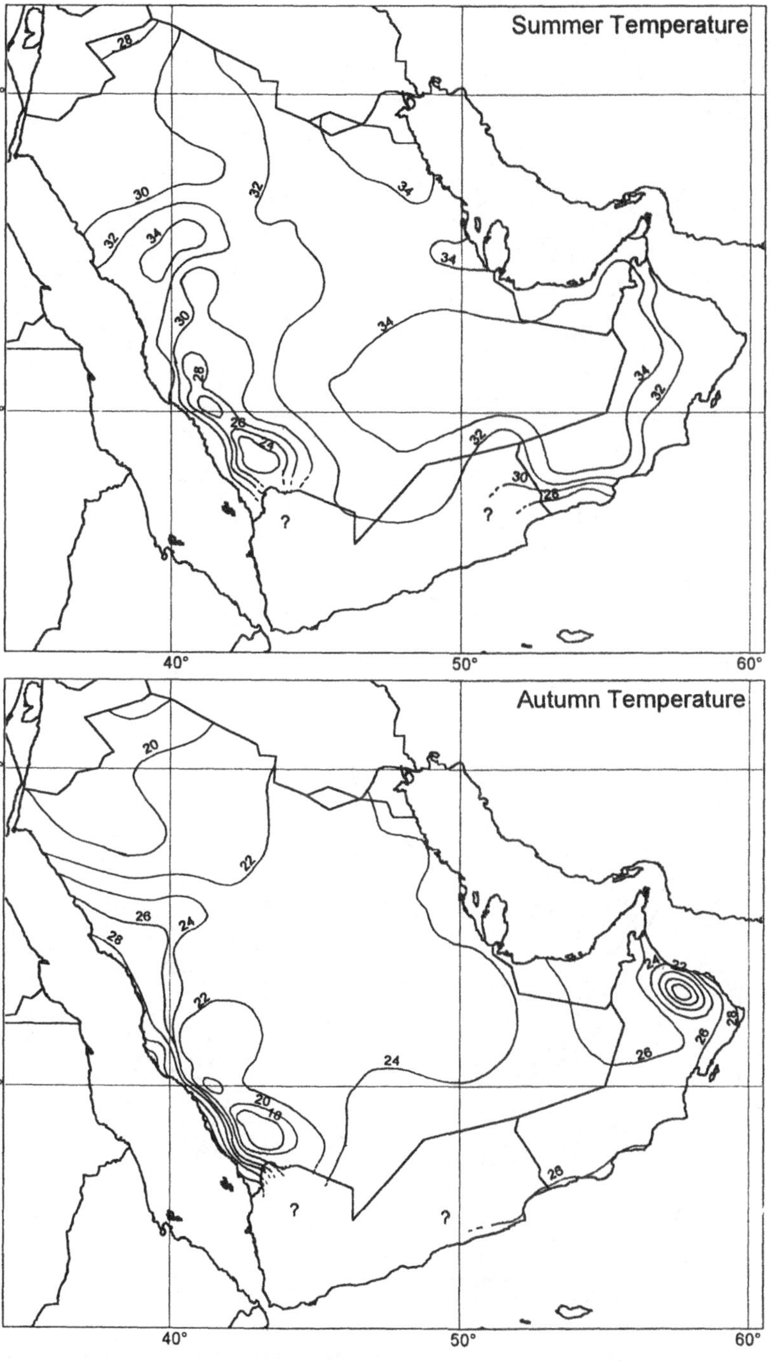

Figure 2.8 continued.

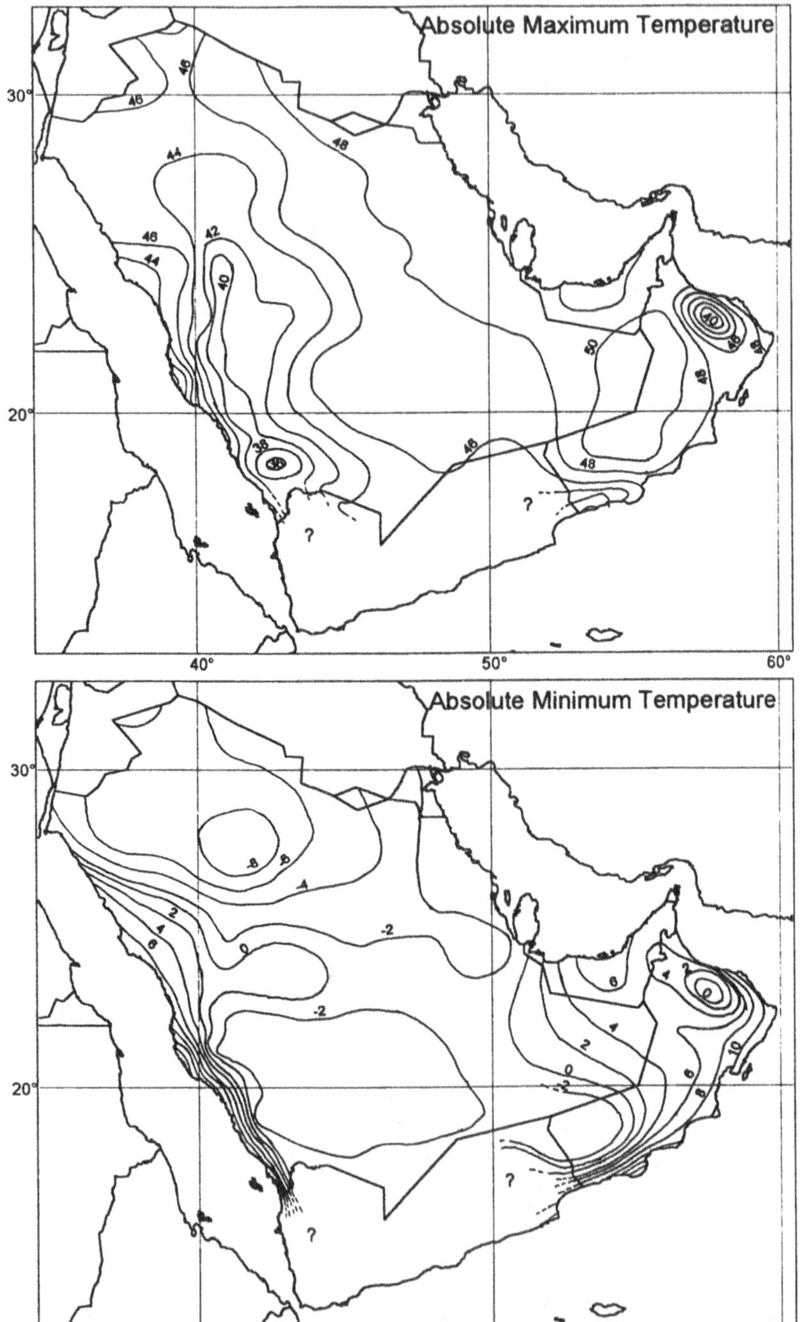

Figure 2.9 Contour maps of absolute maximum and minimum temperatures over the Arabian Peninsula.

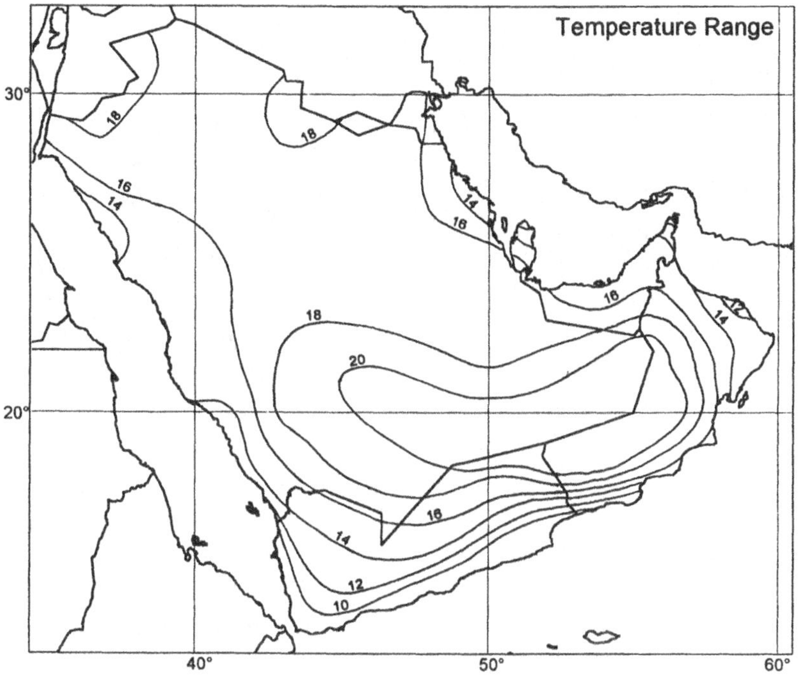

Figure 2.10 Contour map of maximum monthly temperature range over the Arabian Peninsula. Range is calculated as the mean of the greatest monthly range occurring in each year.

Seasonality is most marked along the relatively small area of the southern coast and escarpment mountains influenced by the southwest monsoon, and to a lesser extent in the southwestern highlands, though perceptible in many other areas.

Inter-annual variability of rainfall is greatest in central, eastern and southeastern regions, with coefficients of variation of total annual rainfall exceeding 100% in the central deserts of Oman bordering the Rub' al Khali, and by inference within the Rub' al Khali itself. Inter-annual variability is considerably less in the north, and at about the 25th parallel, identified as the area northwards of which annual plants become more important in the vegetation of the deep sand areas (see Chapter 8), the coefficient of variation is 50-70%, falling to <50% in the northeast.

An aspect of rainfall which has not been documented for the Peninsula, but which is fundamental to the timing of growth and flowering, is the 'spottiness' of desert rainfall. To the north of the Peninsula in the hyperarid (rainfall 30-35 mm yr^{-1}) southern area of the Jordan rift valley, between one half and two thirds of the total rainfall is of a highly localized type, coming mostly from small convective cells (Sharon 1972). The area covered by individual cells seems to be randomly distributed in space, since the long-term mean rainfall is uniform throughout the region. It is likely that there is a similar pattern of rainfall occurrence over those

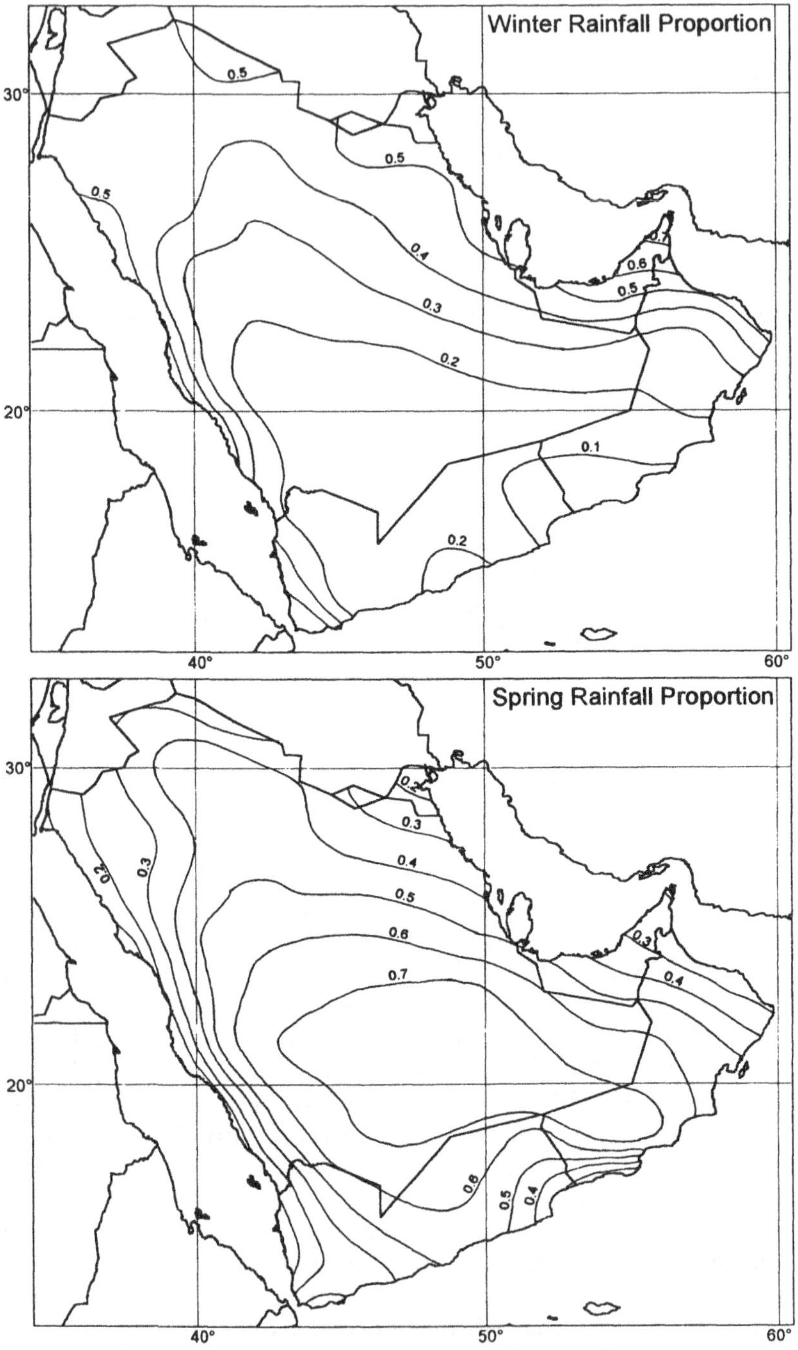

Figure 2.11 Contour maps of the seasonality of rainfall for the four meteorological seasons (see legend to Figure 2.8 for definition) over Arabia. Seasonality is expressed as the proportion of the mean total annual rainfall that occurs in a particular season. This page: winter and spring. Facing page: summer and autumn.

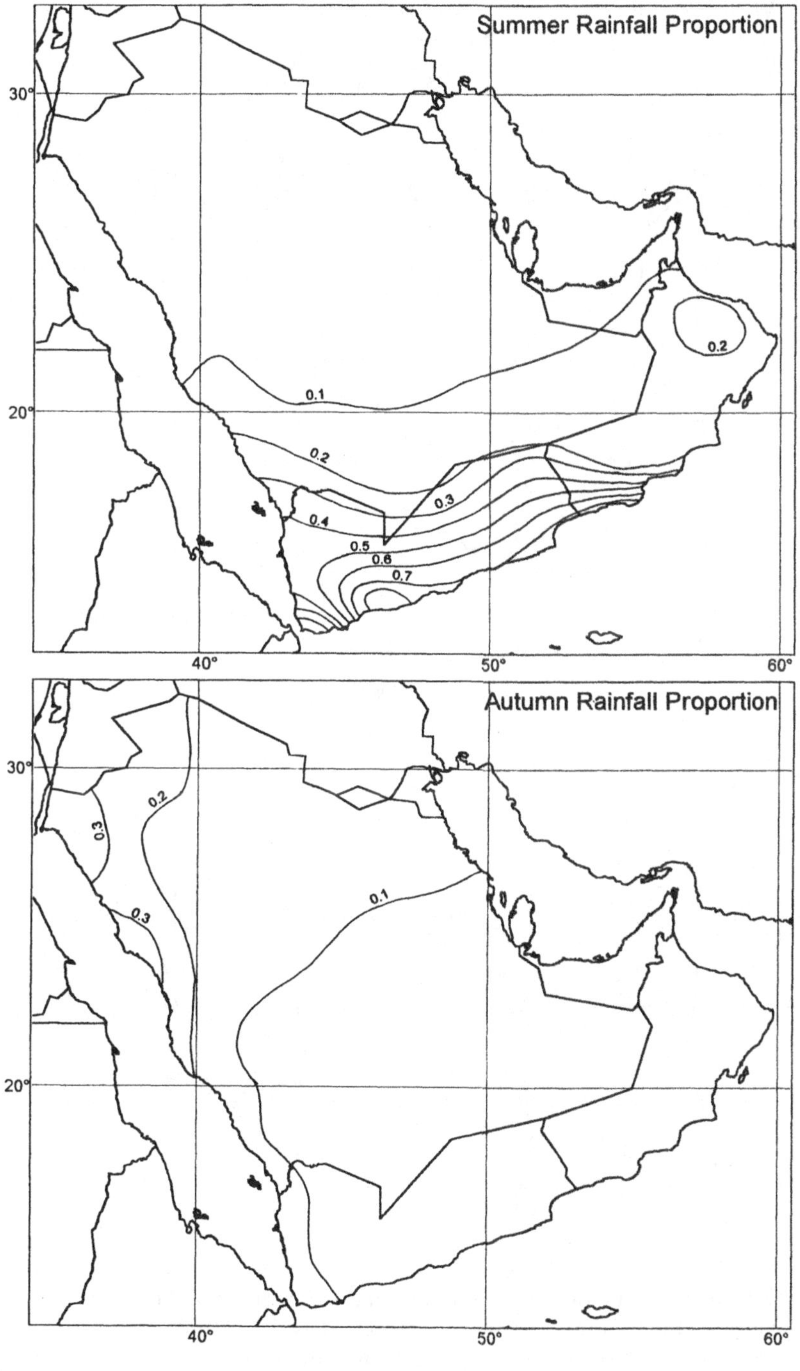

Figure 2.11 continued.

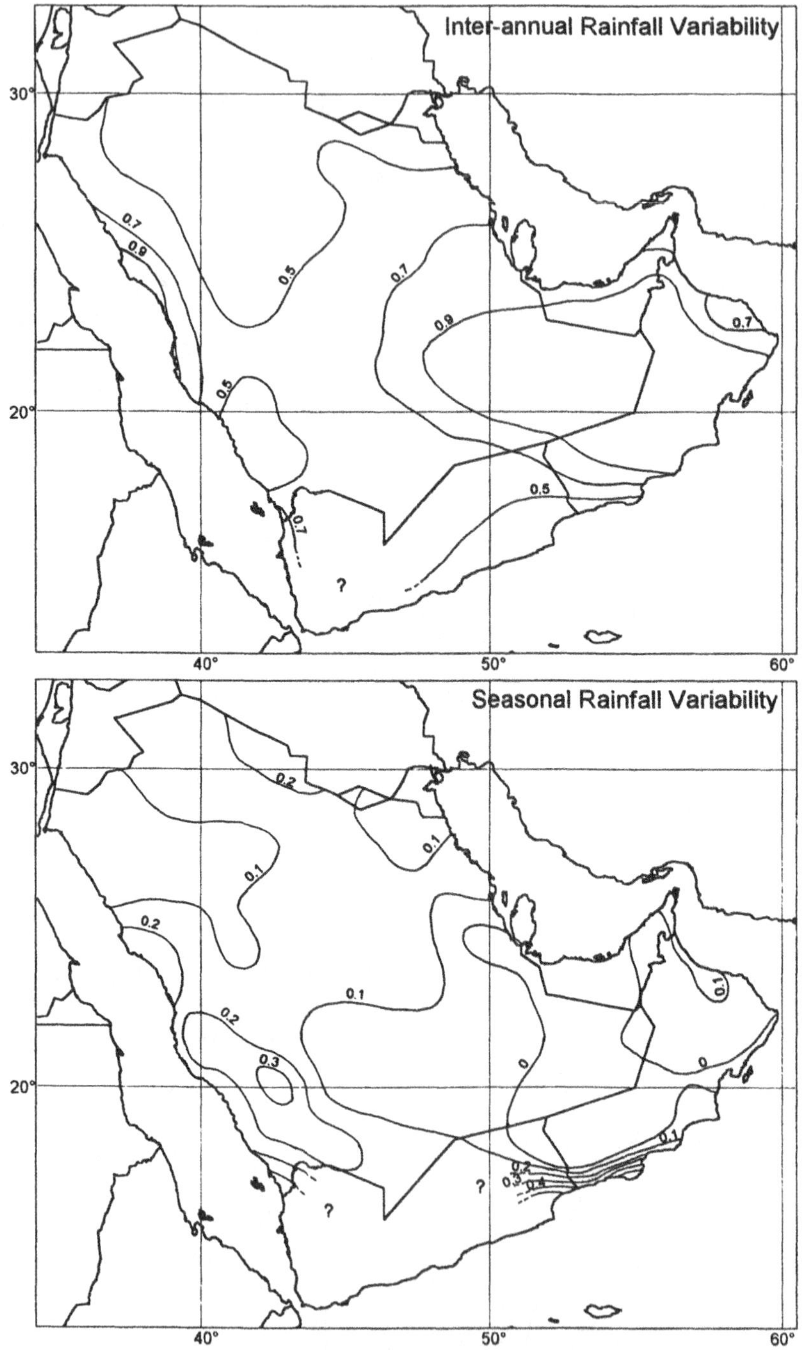

Figure 2.12 Contour maps of the inter-annual and seasonal variability of rainfall over the Arabian Peninsula. Inter-annual variability is expressed as the coefficient of variation of total annual rainfall and seasonal variability as the 12-month autocorrelation coefficient of total monthly rainfall (see section 2.3.2 for details).

areas of the Peninsula with high inter-annual rainfall variability (Figure 2.12). Under such circumstances there will be a consequent 'spottiness' of plant growth and flowering.

2.3.5 Fog

Fog is a rare meteorological phenomenon in Arabia, but at locations where it occurs regularly the contribution to available moisture and its consequent influence on plant growth can be substantial. Given the paucity of data on both the prevalence of fog and the amount of moisture which it contributes to the ecosystems of the Peninsula, the occurrence and density of lichens is currently the best available indicator of fog occurrence. Lichen growth is known to be closely associated with fog distribution in deserts (e.g. in the Namib, Schieferstein & Loris 1992). The dense growth of lichens on the bark of *Prosopis* trees in the woodlands at the eastern edge of Ramlat al Wahibah, the abundance of epilithic lichens in the south of Masirah Island, the widespread occurrence of lichens in the eastern areas of the central desert of Oman, and the extraordinary fruticose lichens in the juniper woodlands at the higher altitudes of the southwestern highlands (see Chapter 6) are all testament to the occurrence and influence of fog, mist and low cloud.

Unfortunately, although fog and mist are monitored at many, though not all, of the Peninsula's meteorological stations (Table 2.2), their contribution to occult precipitation has only been measured on a regular basis at Ja'aluni in the central desert of Oman. Figure 2.13 illustrates the occurrence of fog at this station for 1992-1993, during which fog occurred on an average of 54 days, more than twice the number of days recorded at any other station in the Peninsula (Table 2.2). Fog at Ja'aluni occurs on nights when relatively moist Tropical Maritime air (i.e. from the south) is drawn inland on southerly sea breezes and cooled below its dew-point at night by radiational cooling. It is relatively rare during the peak summer months when nights are too short for sufficient radiational cooling too occur. A further key ingredient for fog development is a wind speed of <8 knots and constancy of wind direction (Stanley-Price *et al.* 1988). This region is truly a 'fog desert', comparable to other fog deserts such as that of the Namib (Olivier 1995). The considerable amount of water (Figure 2.13) that condenses on tree foliage and drips to the ground augments the impoverished mean total annual rainfall of 39 mm, helping to maintain vegetation growth into dry periods (Stanley-Price *et al.* 1988).

The fog, or perhaps more correctly low cloud, which occurs in the Dhofar region during the period of the southwest monsoon contributes a large quantity of occult precipitation during June-August (see section 2.2.2.3 above and Stanley-Price *et al.* 1988). This seasonal phenomenon maintains highly productive grasslands and montane escarpment woodlands.

In other areas of the Peninsula, fog is known to occur in the southwestern highlands, though unfortunately the only data available comes from stations such

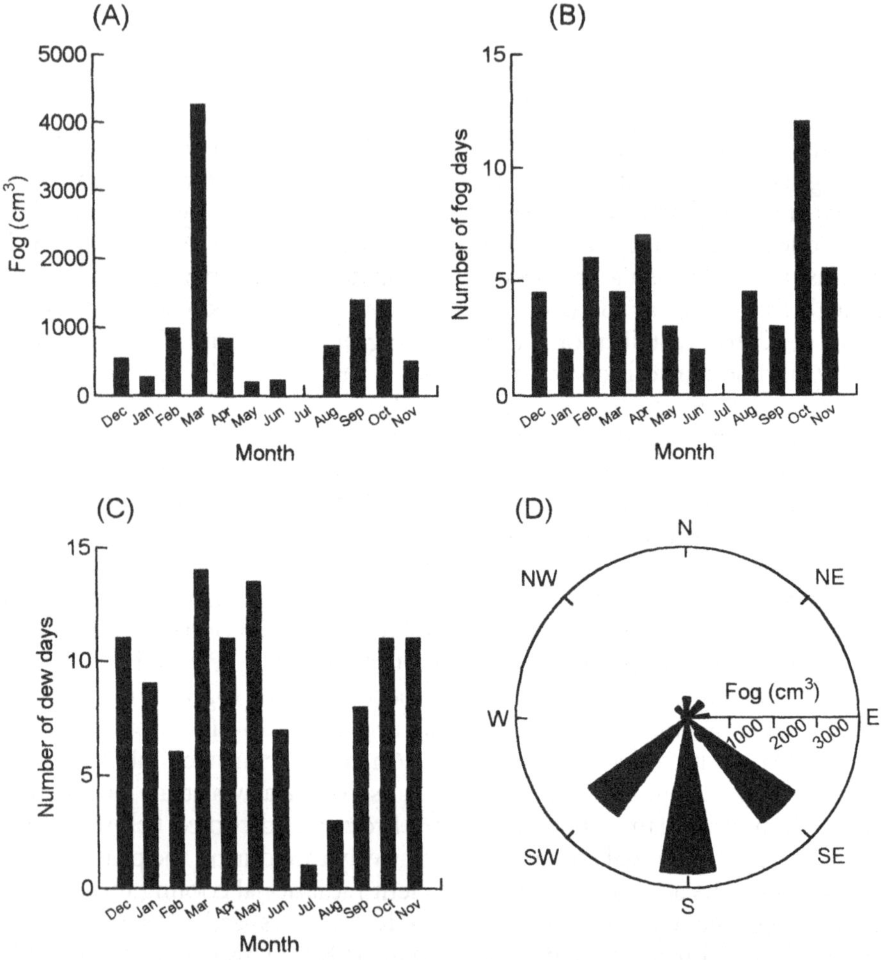

Figure 2.13 Fog and dew at Ja'aluni in the Jiddat al Harasees (see Figure 2.6 for location). Mean volume of fog moisture deposited on a multidirectional 1 m^2 mesh collector (A), mean number of days with fog (B) and mean number of days with dew in each month (C), and mean total volume of fog collected from the eight compass directions (D), for 1992-1993.

as Abha (at 2,093 m) which lie below the main cloud and fog zone, on both arid western and eastern coasts in the vicinity of Yenbo and Dhahran respectively, and in the far north at Qaysumah and Turaif. Fog occurs predominantly in the winter and autumn seasons at all locations except on the northwestern Red Sea coast in the vicinity of Yenbo and Wejh, where fog occurs predominantly in the summer.

2.4 Summary

Meteorology and climate over the Arabian Peninsula are dominated by the influence of Polar Continental and Tropical Continental air-masses, the former during the winter and the latter during the summer, with minor incursions of Polar Maritime and Tropical Maritime air. The timing of the influences of these air-masses allows us to divide the meteorological year over the Peninsula into four 'seasons': December-February, (winter), March-May (Spring), June-September (Summer) and October-November (Autumn). The march of temperature is seasonal and regular from year to year, with rainfall, occasionally brought by tropical storms and cyclones during May-June and October-November, being the meteorological event of main significance.

Mean total annual rainfall is <50 mm in the hyperarid interior of the Peninsula and on the northwestern coast, 50-150 mm over a large area from the Rub' al Khali to the north and northeast, *c*. 100 mm along the southwestern, southern, southeastern and eastern coasts, *c*. 250 mm in the semiarid northern mountains of Oman and the southern mountains spanning the Yemen-Oman border, and generally >400 mm in the southwestern highlands, attaining totals >1,000 mm in a small area of the extreme southwest. Inter-annual variability of rainfall is in general high, but particularly so in the central and southeastern regions, and seasonality of rainfall is generally low, except over the monsoon-affected southern regions, though perceptible in many areas.

Winter rainfall comes mainly from the combination of troughs in the upper atmospheric westerly flow in the north combined with incursions of low-level Tropical Maritime air from the Arabian Sea, bringing 40-50% of the annual rainfall in the north, 40-70% in the east and southeast and 40-60% along the narrow northwestern coastal strip. Spring is a transitional period, being affected by both the westerly flow and the changes over Asia and the Arabian Sea due to the northward advance of the thermal equator. This is an important season for rainfall over the southwestern mountains and central regions, the latter due to the influence of the Oman Convergence Zone, with 40-70% falling during this time. Winter and spring mean temperatures exhibit similar patterns, with a general latitudinal increase towards the Tropic from the northwest to the southeast, a 1-3°C altitudinal depression across the Nejd and pronounced decreases with altitude in the mountains.

Summer is dominated by Tropical Continental air and thermal low pressure centres, especially over central and southeastern Arabia. The summer temperature pattern is therefore markedly different from that of winter and spring, with uniformly high temperatures spanning 15 degrees of latitude. Less than 10% of the total annual rain falls in summer over most of the Peninsula, excepting the southern and southwestern areas influenced by the southwest monsoon, with some of these areas receiving 20-80% of their annual rainfall total during this period, and with 20% of the total annual rain falling at this time in the Hajar mountains of northern Oman. The summer thermal low over the Peninsula plays a fundamental role in the maintenance of the southwest monsoon. In the transitional autumn season the northwest to southeast temperature gradient is re-established; this is a quiet time for rainfall, with only the northern and northwestern areas receiving >10% of their rainfall at this time.

Fog occurs at a number of locations on the Peninsula, most notably in the southwestern highlands, in Dhofar and in the central desert of Oman. Fog in other locations, such as in the south of Masirah Island and in the woodlands on the eastern edge of Ramlat al Wahibah, is indicated by the occurrence there of high densities of lichens. In the 'fog desert' of central Oman there are on average 54 days of fog per year, with the resultant moisture having a marked effect on vegetation growth in what is otherwise a hyperarid area with <40 mm of rainfall per year.

Chapter 3

Geology and Geomorphology

Ingeborg Guba & Ken Glennie

3.1 Introduction

Most of the Arabian Peninsula is a desert, exceptions at a small scale being oases, and on a larger scale the highlands of Yemen and the southern flanks of the Hadhramaut that are influenced by the southwest monsoon (see section 2.2.2.3 for details of the monsoon). Although it is the limiting effects of an impoverished water supply that creates a desert, the Arabian Peninsula was not always arid; past climates have played a fundamental role in the formation of the landscape which we see today. The processes of weathering and erosion have shaped the geomorphology of the Peninsula and, along with the influence of the vegetation, these processes have produced the soils. The nature of these soils varies with rock type, and rock type depends on a variety of geological factors that include the environment of deposition, the effects of changes in climate through time and other geological processes. Since the underlying terrain has a major influence on soil formation and hence plant community composition, an understanding of the history and geographic distribution of rock and geomorphological types leads to a better understanding of the distribution of vegetation.

S. A. Ghazanfar and M. Fisher (eds.), Vegetation of the Arabian Peninsula, 39-62.

The terrains of the Arabian Peninsula can be divided into four main types: (1) the 'hard rocks' of the Precambrian Arabian Shield (see Table 3.1 for a geological timetable) in the southwestern region, which are composed of igneous (i.e. crystalline, formed from molten magma) and metamorphic (i.e. altered by heat, pressure or chemical action) rocks, (2) the *harrah,* which are much younger (Tertiary) rocks that cover large areas of the crystalline basement, (3) the 'soft rock' domain of the northeast, characterized by light-coloured sedimentary rocks, and (4) the *nafuds,* which are the vast sand seas. Some areas, especially the Oman mountains at the southeastern edge of the Peninsula, present a complex mixture of 'hard rock' and 'soft rock' terrain flanked by huge alluvial fans; the latter also occur adjacent to the southwestern mountains.

This chapter provides a brief description of the general geomorphology and geological evolution of the Peninsula, followed by regional descriptions of geology and geomorphology. The descriptions necessitate the use of various geological and geomorphological terms, and these are briefly defined upon first usage. The basic geographical, geomorphological and geological background is provided in four figures: Figure 3.1 indicates the locations of the major regions and structures mentioned in the text, and Figures 3.2-3.4 provide, respectively, relief, geomorphological and geological sketch maps of the Peninsula.

3.2. General Geomorphology

Located between northeast Africa and Southwest Asia, the Arabian Peninsula is separated from Africa by the Red Sea and from Asia by the Arabian Gulf, the latter extending from southern Mesopotamia (Syria, Jordan and Iraq) to the Strait of Hormuz and on to the Gulf of Oman. The southeastern rim of the Peninsula is limited by the Gulf of Aden and the Arabian Sea. The boundary in the northwest is the Gulf of Aqaba and the Jordan Rift, to the north the Taurus mountains of southern Turkey and to the northeast the Zagros mountains.

The long axis of this huge rectangular-shaped crustal block (i.e. a part of the earth's crust) is aligned NNW-SSE, the geomorphology of which is largely the result of uplift parallel to the Red Sea and Gulf of Aden, with a resulting gentle tilt to the ENE. The greatest amount of uplift occurred in the southwest, creating mountains up to 3,760 m high in Yemen. The uplift has given rise to steep escarpments inland of the Red Sea coast. These scarp mountains average *c.* 2,000 m south of 21°N near Mecca; further north they average *c.* 1,000 m with several peaks above 2,000 m. The scarp mountains overlook the Tihamah coastal plain. Towards the east and north the elevations of the Arabian Peninsula gradually decrease to form a linear depression comprising the lowlands of Mesopotamia, the coastal regions of the Arabian Gulf, and a depression southwest of the Oman mountains. This latter mountain range, which rises to 3,009 m along the southeast edge of the Peninsula, forms a separate geomorphological unit with a geological history that differs from that of the rest of Arabia.

Table 3.1 Stratigraphic table illustrating some of the major events that took place in Arabia's geological history. Note that the vertical time scale is not constant. HS is the depositional time span of the Hajar Supergroup of the Oman mountains and equivalent units along the northeast margin of Arabia. Note that carbonates are developed only in the tropics; siliciclastics dominate in latitudes greater than about 30°.

Age (Ma)	Era	Period	Major events in Arabia	Typical rock units
0-2	Cenozoic	Quaternary	Arabia mostly terrestrial.	Dune, wadi, lake and sabkha. ↑
25		Tertiary	Opening of Red Sea and Gulf of Aden.	Shallow marine carbonates. ↓
65	Mesozoic		Obduction along NE Arabia.	↑ Semail ophiolites/Hawasina.
		Cretaceous	Neo-Tethys starting to close.	↑
140		Jurassic	Stable tectonic conditions over most of Arabia.	H S Oceanic sedimentation NE of Arabia.
208		Triassic	↑ Neo-Tethys opening.	↓ ↓
251	Palaeozoic	Permian	Marine transgression. Calving of microcontinents. Mountain glaciation in Oman.	Widespread carbonates. Local diamictites in south. ↑
296		Carboniferous	Major unconformity (except central Saudi Arabia).	
360		Devonian		Local carbonates in north.
410		Silurian	Arabia south of Equator.	Mostly siliciclastics
440		Ordovician	Glaciation in central Arabia.	
500		Cambrian	↓	Local fluvio-aeolian Hormuz and Ara salts. ↓
590		Precambrian	Glaciation in Oman. Island arcs.	Carbonates and siliclastics. Mistal/Abu Mahara diamictites. Crystalline basement.
850				

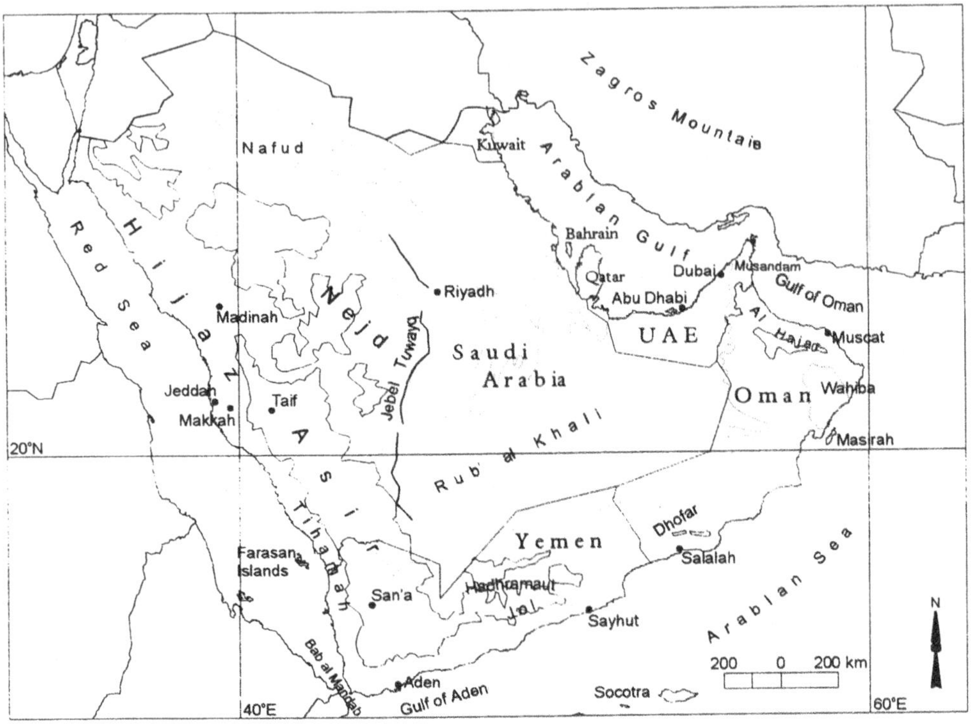

Figure 3.1 The Arabian Peninsula, with the location of the major regions, structures and cities. The highlands are indicated by the 1,000 m contour (see Figure 3.2 for further details) and the faint line indicates the major sand areas (see Figures 3.6 and 8.1 for further details). The cuesta region of east central Arabia is indicated by the solid lines to the west of Riyadh.

3.3. Geological Evolution

3.3.1 Outline of Arabian Plate History

The geographical term 'Arabian Peninsula' is not quite synonymous with the geological term 'Arabian Plate', which is a large area of the lithosphere, the earth's crust, that behaves as a single relatively rigid unit moving over the underlying slightly plastic mantle. Geologically, the plate boundaries of Arabia are defined (in clockwise order) by the spreading axes of the Arabian Sea, Gulf of Aden and Red Sea, the Gulf of Aqaba-Jordan Valley transform fault system and the system of thrusts where the Taurus mountains override the northern margin of Mesopotamia. Although geographically the Zagros mountains do not form part of Arabia,

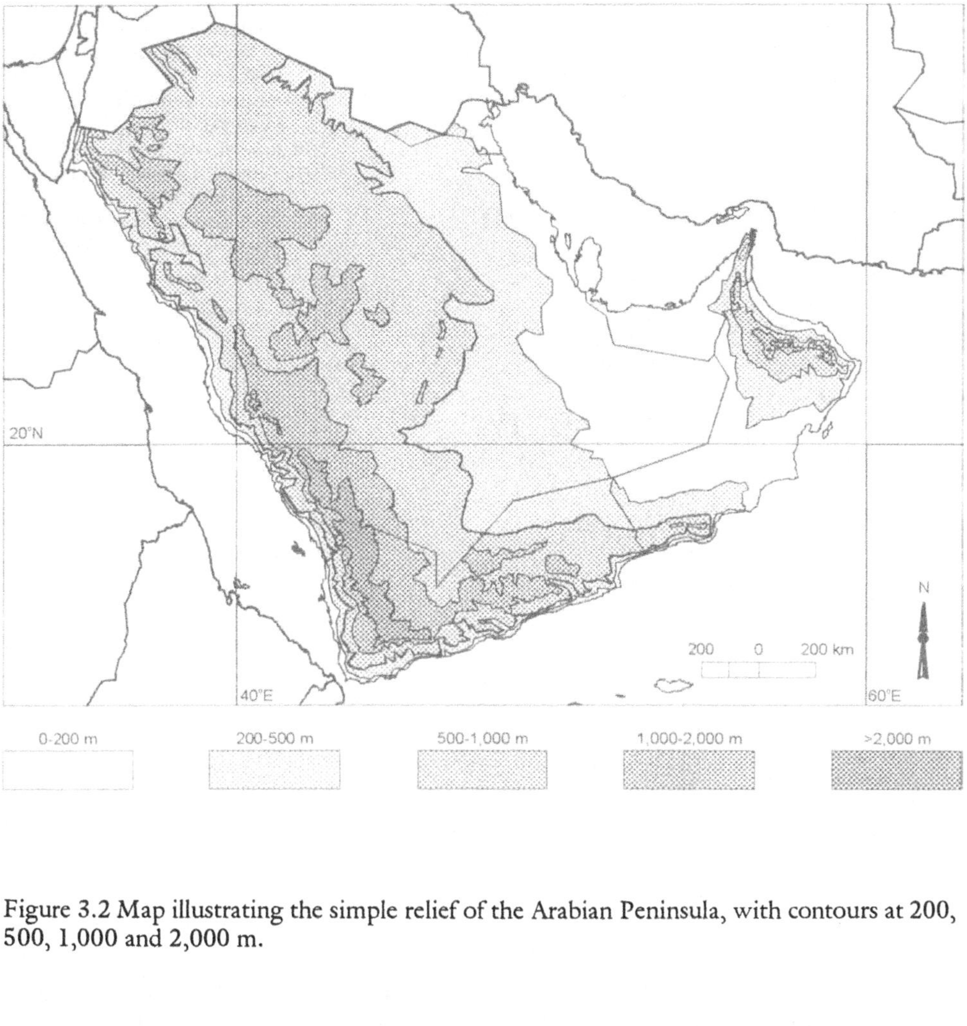

Figure 3.2 Map illustrating the simple relief of the Arabian Peninsula, with contours at 200, 500, 1,000 and 2,000 m.

they do so geologically, for they are underlain by the Arabian craton (i.e. the ancient continental crust) as far east as the Sanandaj-Sirjan zone of Iran (Glennie 1995, Glennie *et al.* 1990). In the Oman sector the Arabian Plate extends out under the western Gulf of Oman, where its northern limit in the Strait of Hormuz is now about to be subducted (i.e. underthrust) beneath Cenozoic sedimentary sequences offshore of Makran in southern Iran. The northern limit of the peninsular portion of Arabia can be considered to lie between the northern end of the Gulf of Aqaba and the mouth of the Tigris-Euphrates river.

For much of its earlier history, Arabia formed part of the megacontinent Gondwana, which included South America, Africa, India, Australasia and Antarctica. Under the influence of sea-floor spreading, Gondwana was virtually 'upside down' within the southern hemisphere relative to the present poles for

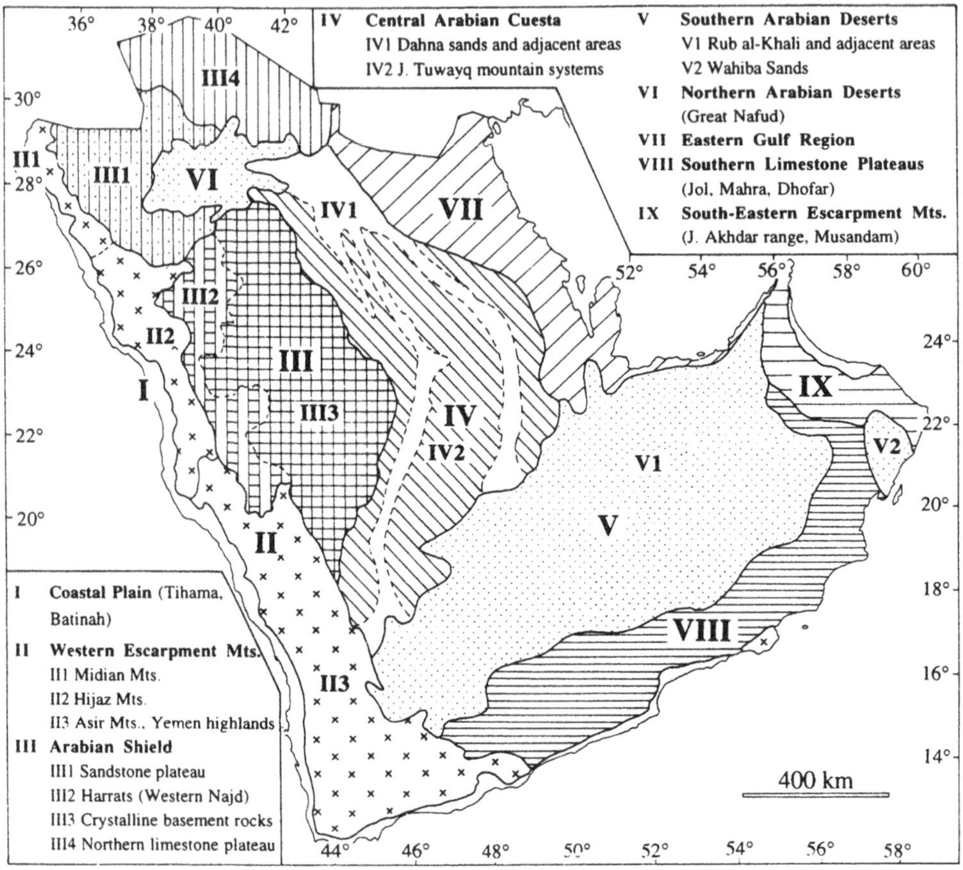

Figure 3.3 The geomorphological regions of the Arabian Peninsula (modified after Barth 1976).

much of the earlier Palaeozoic before crossing the South Pole (Scotese & Barret 1990, Scotese & McKerrow 1990). A major ocean, the Tethys, lay off the northeast margin of Arabia.

Gondwana began to break up into its current component plates in the Late Triassic, although in the Late Permian smaller microcontinents (Anatolia, central Iran, southern Afghanistan and perhaps Tibet, Glennie 1995, Glennie *et al.* 1990) were already calved off its northeastern margin with the creation of a new belt of oceanic crust, the Neo-Tethys, between. Since then, Neo-Tethys first widened, with Gondwana moving westward away from the axis of ocean-floor spreading, and then narrowed again as the southern Atlantic Ocean began to open and, with ocean-ocean subduction occurring in Neo-Tethys, Arabia moved relatively eastward. This latter sense of movement led to a major phase of obduction, when ocean-floor sediments (the Coloured Melange of Iran and the Hawasina of

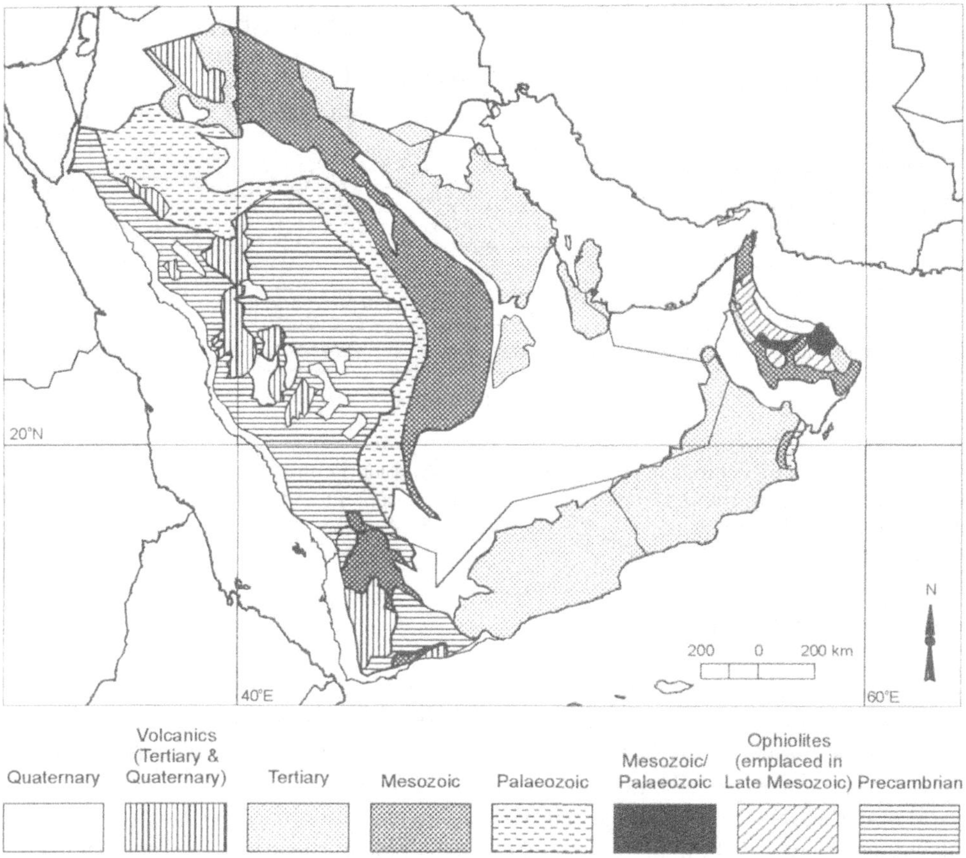

Figure 3.4 Geological sketch map of the Arabian Peninsula.

Oman) and overlying slabs of oceanic crust (the Semail and smaller ophiolites; i.e. igneous and metamorphic rocks often associated with deep sea and shelf sediments) were emplaced over the continental margin of Arabia during the Late Cretaceous. Obduction did not immediately lead to uplift of the Oman and Zagros mountain ranges, the flanks of their present sites being mostly below shallow seas until the Mid Tertiary, when other plate movements, including the creation of the Red Sea and Gulf of Aden, led to the separation of the Arabian Plate from Africa, with associated uplift of the western and southwestern mountains of Arabia, and folding and uplift of the Oman and Zagros mountain ranges (see Table 3.1 for a summary).

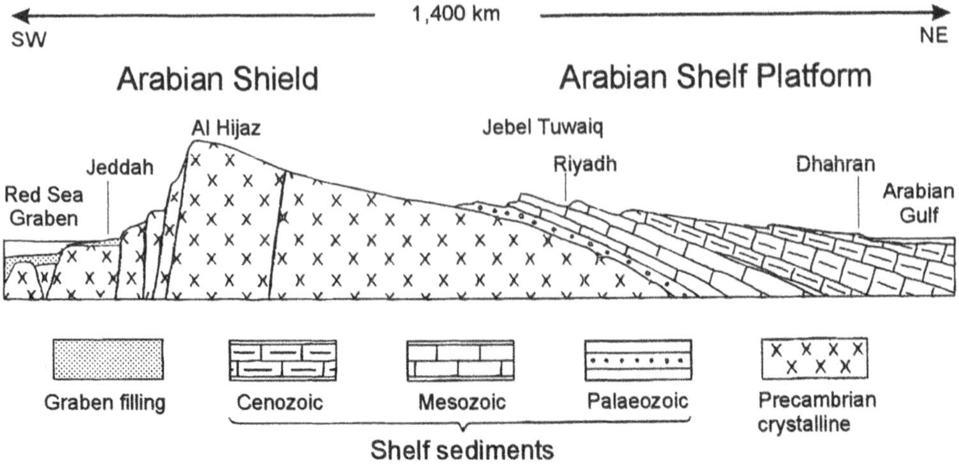

Figure 3.5 Schematic geological cross-section of the Arabian Peninsula from the Red Sea to the Arabian Gulf showing the outcrops of the crystalline basement in the southwest (the Arabian Shield) and the sedimentary rocks in the northeast (the Arabian Shelf Platform). Adapted from Jado and Zötl (1984).

3.3.2 Precambrian Beginnings

As the eastern half of the former Nubio-Arabian Shield, the original crust of the western Arabian Shield was constructed from a series of N-S trending fault-bounded rock units formed behind oceanic trenches comprising metamorphosed andesitic volcanics (i.e. silicates of alumina, lime and soda of volcanic origin), and a suite of diorites of plutonic origin (i.e. greenstones formed by fusion and subsequent slow crystallization at great depths below the surface), the oldest of which formed around 950 Ma (Beydoun 1991), but contain older Proterozoic rocks dated at about 1,700-2,300 Ma (Gass *et al.* 1990) . Crystalline rock sequences dated about 850 and 770 Ma occur in Jebel Ja'alan, south of the eastern Oman mountains, with similar rocks near Mirbat and on al Hallaniyah (formerly the Kuria Muria Islands) offshore of southwestern Oman.

The deposition of sedimentary rocks seems to have begun during the Late Precambrian around 650 Ma. The older rock units are alternations of shallow marine limestones and dolomites and siliciclastic (i.e. with grains of silica as opposed to carbonate) sandstones and shales. Some of these carbonates are stromatolitic (i.e. rich in algal laminae). At or just before the beginning of the Cambrian era, thick (Hormuz) salts (mainly halite) were deposited over much of the eastern half of Arabia, including what is now the southeastern Zagros mountains, and a NE-SW trending trough extending NNE from north of Mirbat to near the Oman mountains (Ara salt). Diapiric flow of these salts (a diapir forms when a column or wall of salt rises and deforms or even pierces overlying rock sequences) was responsible for many folds in overlying strata, and for the presence of salt domes at the surface in Oman, the Emirates, Zagros mountains and especially in the southern Arabian Gulf and western Strait of Hormuz, after which the extensive salts of the Gulf area are named.

3.3.3 Older Palaeozoic Sedimentary Sequences

The Palaeozoic succession of Arabia is both marine and terrestrial (fluvial, aeolian and lacustrine, especially in the south and southeast) in origin, and mainly siliciclastic in nature. Tillites (sedimentary rock composed of glacial till compacted into hard rock) in southern Arabia and exposures of a striated pavement in central Saudi Arabia are evidence of a Late Ordovician to earliest Silurian period of glaciation (Al-Laboun 1993). Another period of glaciation occurred in southern Arabia during the Late Carboniferous to Early Permian (Levell *et al.* 1988). Deep erosion resulted in the removal of some Palaeozoic sequences down to the Late Precambrian in some areas; the erosion was associated with a combination of uplift along the northeastern and southeastern margins of Arabia, southeastern Oman being the site of glaciation of probable mountain origin (Al Khlata Formation, Al-Belushi *et al.* 1996) and the repeated glacially-induced lowering of global sea-level as Gondwana moved across the South Pole during the Carboniferous and Early Permian (Crowell 1995, Schandelmeier & Reynolds 1997). Some of this erosion may have occurred during uplift of the plate margins preceding the crustal separation from northeast Arabia of the microcontinents Anatolia, central Iran and central Afghanistan, and the creation of an intervening area of new ocean (Neo-Tethys). Neo-Tethys widened, causing western Gondwana (Afro-Arabia and South America) to essentially drift westwards.

3.3.4 Mid-Late Permian, Mesozoic and Tertiary Deposition

Arabia was also drifting northward during the Permian, eventually crossing the equator in the Cretaceous. The glacial conditions of the Early Permian were replaced in the Later Permian by warm shallow seas that covered much of the area, with carbonate-rich sediments dominating deposition over most of Arabia in the Mid Tertiary (see maps in Schandelmeier & Reynolds 1997). Following the creation of Neo-Tethys, the climatic change is emphasised by deposition of the Late Permian Khuff dolomites over much of internal Arabia and limestones of the Saiq Formation in the Oman mountains. An evaporite sequence (the Gothnia and Hith formations) developed across eastern Arabia at the end of the Jurassic, and an erosional event occurred in the Mid Cretaceous. From the Late Permian until the Mid Cretaceous, carbonate turbidites (sediments deposited by a turbidity current) of the Hawasina, derived from unconsolidated sediments of the continental margin, and thin deep-water radiolarian cherts characterised the sedimentary sequences deposited over the ocean floor of Neo-Tethys (Glennie 1995, Glennie *et al.* 1974).

Following obduction of the allochthonous Hawasina and Semail nappes (i.e. folded rocks which were not deposited where they are found) in Oman and equivalent formations in Iran, shallow-water carbonates were deposited over much of Arabia during the Palaeogene, with deeper water limestones over the Zagros area. Under the influence of early folding of the Zagros range (a reaction to the opening of the Red Sea) the axis of deep-water deposition migrated westward in the later Palaeogene until about the Mid Miocene. Shallowing with time, deposition first became more evaporitic (i.e. characteristic of deposits of salts resulting from the evaporation of a body of water) and then completely terrestrial by the end of the Miocene (see the series of palaeogeographic maps in Jones & Racey 1994).

Mid Tertiary uplift of the western edge of Arabia flanking the newly forming Red Sea resulted in erosion and the deposition of continental clastics (i.e. sediments consisting of broken pieces of older rocks) to both the west and east of the rising highlands. Adjacent to the newly forming Gulf of Aden and Red Sea, volcanic rocks were erupted throughout the Cenozoic. As the Oman and Zagros mountain ranges began to rise during the Late Neogene they shed much of their products of erosion westward into the newly forming NW-SE trending axial basin of Arabia. By the Late Miocene, most of Arabia was terrestrial.

3.3.5 Quaternary

Highland areas are subject to erosion and lowlands to deposition. To judge from the extent of the alluvial fans spreading from Arabia's mountain ranges, there must have been a much higher rainfall during the earlier Quaternary than in more recent times. Much of the Rub' al Khali basin is underlain by Pliocene alluvium (i.e. deposits of transported matter left by water flowing over land) that was reworked into dunes in the Late Pleistocene; many dunes were separated by interdune lakes (McClure 1978). Finer-grained sediments were deflated from the surfaces of alluvial fans and from the beaches of the Arabian Gulf and transported mostly to the south and southwest under the influence of the *shamal* (the North Wind) to be deposited as sand dunes (Figure 3.6).

The foregoing suggests that over much of Arabia, the Quaternary climate had periods of both high rainfall (as evidenced by the occurrence of alluvial fans and interdune lakes) and aridity (as evidenced by the occurrence of extensive sand seas). How could this happen, and why is much of Arabia now desertic, when it has not always been so?

The earth's surface can be considered as a heat engine in which the sun drives a giant convection cycle, hot air rises over the equator, flows at high altitude to the poles, where it gets cold and dense, and descends to begin its return journey to the equator at ground level. The earth's circumference is much shorter in temperate latitudes than at the equator. Thus poleward-moving air becomes compressed and heavier, and part is forced to descend at about 30°N and S of the equator. This air warms as it descends, and thus is capable of absorbing more

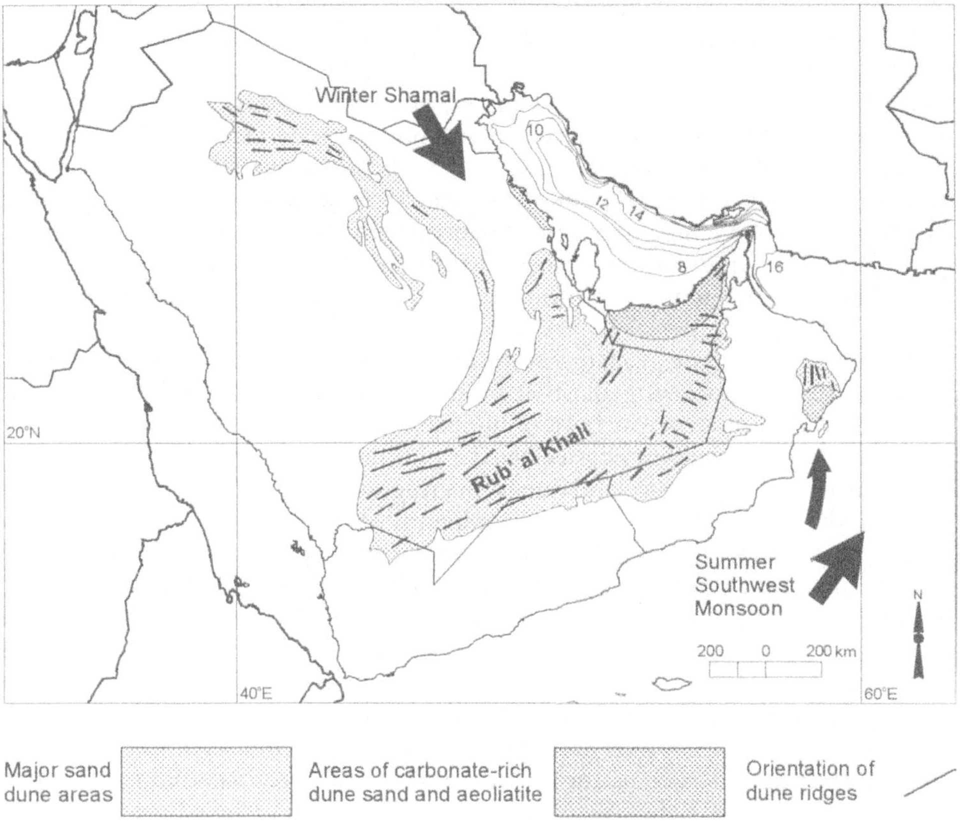

Figure 3.6 The present distribution of the major sand dune areas of the Arabian Peninsula (for a more detailed map of all sand areas, see Figure 8.1), the occurrence of carbonate-rich dune sand and aeolianite, and contours showing the approximate dates of flooding of different parts of the Arabian Gulf between 16 and 8 ka; the present sea level was reached at about 6 ka. Arrows indicate the general directions of the winter *shamal* and summer southwest monsoon. Modified after Glennie (1996) and Holm (1953).

moisture and, under cloud-free skies, of desiccating the land over which it flows as it completes its convection cycle back to the equator. The earth rotates, however, creating a velocity difference between the rotating land surface and the wind flowing across it towards the equator, resulting in northeast trade winds in the northern desert belt and southeast 'trades' over the Kalahari in southern Africa (Glennie 1987). The surface of Arabia is crossed by the northeast trade winds (the *shamal*, which, as indicated by dune morphology, veers to the southwest over the Rub' al Khali; Figure 3.6).

Monsoons (winds that reverse their direction seasonally) are driven by temperature and associated air-pressure differences between large land masses (which heat and cool relatively rapidly in summer and winter), and adjacent large oceans, whose waters change temperature only slowly. The plains of northern India heat rapidly in summer and 'suck' air across the Arabian Sea as the southwest monsoon, bringing heavy rain to the Western Ghats of India; the thermal low over the Rub' al Khali prevents the southwest monsoon from penetrating far into Arabia, where today it brings moderate rainfall only to the coastal areas of Dhofar (see section 2.2.2.3 for further details). During glacial periods, however, the main path of the monsoon lay south of Arabia and was effective over the Wahibah area of Oman only in the waning phases of recent glaciations. The winds of the so-called winter 'northeast monsoon' are relatively weak and dry. They tend to reinforce the *shamal*, carrying Asian dust to Oman and the UAE.

During high-latitude glaciations, the area of the north polar high atmospheric pressure expands, squeezing all other pressure belts, both high and low, towards the equator. Adjacent narrow belts of high and low pressure induce strong winds, which in trade-wind deserts desiccates the ground more intensely, deflates loose sediment and sweeps it into sand dunes that migrate or extend down-wind; under these conditions, rainfall is almost permanently absent.

The water that forms snow over growing ice caps is not returned to the oceans. Global sea-level fell by as much as 120 m during the last glacial maximum (*c.* 25-16 ka BP), and the Arabian Gulf dried out, exposing a vast source of unconsolidated sediment to the deflationary effects of strong dry winds; this sediment was deposited as dune sand in the northeastern Rub' al Khali. In contrast, during interglacial periods sea-level was close to that of today, winds were much weaker and precipitation much more prevalent except in the interior of the largest deserts such as the Sahara. Such conditions occurred over Arabia from about 10-6 ka BP, the so-called 'Climatic Optimum', when dune sands were stabilised by vegetation and game was plentiful. Since then, the climate of Arabia has steadily become more arid.

Over the last 550 ka, there have been six major near-polar glaciations (Glennie 1996). Luminescence dating (Juyal *et al.* in press) coupled with palaeowind studies (Figure 3.7) indicates that during at least the last two glaciations, the *shamal* was actively deflating sand from the exposed surface of the southern Arabian Gulf and building sand dunes in the Emirates. The warm, clear shallow waters of the modern Arabian Gulf form an ideal site for the organic 'manufacture' of calcareous microfauna and flora (Evans 1995). The exposure of such shallow sea floors at times of glacial low sea-levels resulted in dunes that are rich in carbonate grains for up to 80 km inland from the present coastline of both the Emirates and the Wahibah. The older of these sands were cemented into rock (aeolianite) during humid interglacial periods. Aeolianites are also exposed in northern Qatar and along the Arabian Gulf coast of Saudi Arabia (Figure 3.6).

3.4 Regional Geology and Geomorphology

As is clear from the foregoing and can be inferred from the geological map (Figure 3.4) and Table 3.1, not all parts of Arabia have had the same geological history and thus do not expose the same range of rocks. The geology and geomorphology and surface-shaping aspects of the various regions are described in further detail below.

3.4.1 The Red Sea Coastal Plain

Bounded inland by the scarp mountains of the Arabian Shield, the Tihamah coastal plain is narrowest in the north near the Gulf of Aqaba and widens in the south to reach its maximum width of 40 km near Jizan. The lowermost level is a coralline plain, about 3 m above sea-level, and inland there is a pediment (a gently sloping surface of eroded bedrock) eroded from crystalline basement, and Tertiary rocks covered with alluvial sand and gravel. 170 km northwest of Jizan, basaltic lava flows cover small areas of the coastal plain.

The relief in the Tihamah area is related to the tectonic movements that resulted from the opening of the Red Sea. A zone of steeply inclined tensional faults at the foot of the escarpments trends parallel to the shoreline and is concealed by coastal sediments. In this zone, blocks of the basement of the Arabian Shield have been faulted down, step-by-step, towards the Red Sea (Figure 3.5).

3.4.2 Mountains of Western Arabia

The scarp mountains (the Hijaz and Asir ranges), trending NW-SE parallel to the Red Sea, form the western edge of the Arabian Shield. With a width of 40-140 km, they include the western edge of the Nejd Pediplain and the Hijaz plateau. The mountains rise with steep steps above the Tihamah plain to their highest elevation at 3,760 m in Yemen. Elevations decrease from Yemen northwest to Mecca, and northwest of Medina the escarpment features are interrupted by block-faulted mountains that extend across to the Red Sea.

Like the coastal plain, the geomorphology of the western Arabian mountains is the result of the tectonic activity that led to the parting of the Arabian-Nubian Shield and the creation of the Red Sea. Along the NW-SE faults, blocks were down-faulted towards the Red Sea as a system of horsts (elongate high blocks bounded by faults) and grabens (valleys formed by downward displacement of fault-bounded blocks).

The great variety of hard rocks in this region can be divided into plutonic (granite and related types), metamorphic (gneiss) and volcanic rocks (Torrent & Sauveplan 1977). Included within the basement complex are broad belts of sedimentary rocks (Powers *et al.* 1966), which are now contorted, intruded and metamorphosed and occur mainly as schists (i.e. crystalline rocks whose component minerals are arranged in a more or less parallel manner). These Precambrian rocks are largely overlain by young volcanic rocks; the areas of occurrence of which are known as ***harrah***. The granites are more rapidly eroded by physical weathering, thereby facilitating the development of higher peneplains (plains formed by erosion), whereas the metamorphic rocks form sharper ridges because of their foliation (layering by segregation of different minerals into streaks and lenses of differing hardness).

3.4.3 Mountains of Southern Arabia

The mountains of southern Arabia include the Yemeni highlands, the Hadhramaut plateau and its extension into Dhofar in southern Oman. The Yemeni highlands extend for 450 km along the Gulf of Aden eastward from Bab al Mandab, rising to average elevations of about 2,100 m. The Hadhramaut plateau, to the east of the Yemeni highlands, is a huge elevated block that dips gently northwestwards to the Rub' al Khali basin and to Dhofar in the northeast. Elevations decrease to the east. The plateau is abruptly terminated to the south because of faulting associated with the opening of the Gulf of Aden.

The east-west trending Hadhramaut valley cuts through the Hadhramaut plateau for a distance of 800 km and joins the Arabian Sea at Sayhut. The valley is the main catchment area for rains carried by the southwest monsoon that precipitate over the highlands and plateau. To the north of the plateau lies al Ahkaf, a flat plain transitional to the Rub' al Khali.

Dhofar comprises two large structural domains (Powers *et al.* 1966): a broad plateau tilted to the north with large grabens in the south, and a complex coastal belt cut by inclined faults related to episodes of extension during rifting of the Gulf of Aden.

The mountains of Dhofar form a raised plateau reaching an altitude of 2,000 m in the south and dipping to the north beneath the sands of the Rub' al Khali. Drainage is to the north and northeast, with alluvial fans, up to 10 km wide, merging with the Rub' al Khali. The southern boundary of the plateau is an escarpment, which in places drops dramatically 400-800 m into the Arabian Sea. The plateau is deeply incised by a dense wadi system. To the south of the Dhofar plateau lies a narrow coastal plain *c.* 60 km long, varying in width up to a maximum of about 15 km.

As in other areas of the Arabian Shield, the Yemeni highlands are underlain by the Precambrian Basement. These basement rocks show evidence of strong deformation, intrusion and metamorphism followed by peneplanation long before the first sediments were deposited. This area seems to have been less stable than the western and southern parts of the Arabian Shield, operating apparently as a hinge for differential movements between the African and Arabian Plates (Powers *et al.* 1966). The plateau is covered in north Yemen by flat-lying Jurassic, Cretaceous and Tertiary sedimentary sequences (Powers *et al.* 1966). Extrusive volcanic rocks of Late Cretaceous to Recent age characterise central and southern Yemen.

The mountains of the Hadhramaut are underlain by Precambrian igneous and metamorphic rocks. Outcrops of these crystalline rocks are limited to the southern coastal margin, for example near Mirbat in Dhofar and in the nearby Hallaniyah Islands. The peneplaned surface of the Precambrian Shield is covered with thick sequences of sedimentary rocks comprising mainly limestones, marls (soil of clay mixed with carbonate of lime), sandstones, bauxitic clay (hydrous oxide of alumina and iron), evaporites, shales, turbidites and conglomerates (Roger *et al.* 1994). The superficial Quaternary deposits show a heterogeneous assemblage of sediments of different origins: slope talus, alluvium and aeolian deposits, travertine (concretionary limestone deposited from water holding lime in solution) and littoral marine sediments.

3.4.4 Oman Mountains and Adjacent Areas

The Oman mountains define a narrow 'S' shape in southeast Arabia adjacent to the Gulf of Oman. Known as al Hajar, this mountain range extends about 700 km from Musandam in the north to Sur in the southeast, reaching an elevation of 3,009 m at Jebel Shams in the central part of the range. Northwest of Muscat, the range narrows to a width of 30-40 km, and is accompanied on its northeastern flank by the Batinah coastal plain. Southeast of Muscat the mountains descend steeply into the Gulf of Oman. On their southwestern flank they descend with equal steepness towards the apices of a bajada (a series of alluvial fans). Three major structures involving sedimentary rocks can be distinguished: Ru'us al Jibal (Musandam) in the far north, a flat-topped steep-flanked anticline (i.e. a fold from which strata slope in opposite directions) that plunges to the NNE; the NW-plunging dome of Jebel al Akhdhar, which includes Jebel Shams, and its northwesterly continuation the Hawasina Window (where Hawasina rocks are exposed between the flanking Semail ophiolites), and Sayh Hatat, a dome-like structure on the northern flank of which lies Muscat (see structural cross sections in Glennie *et al.* 1974).

Late Precambrian sedimentary rocks are exposed within the cores of both Jebel al Akhdhar and Sayh Hatat, the latter additionally containing strata of Ordovician age. Above an intervening erosional surface, an autochthonous (i.e. deposited where found) sequence of carbonates belong to the Hajar Supergroup, which range in age from Mid or Late Permian to Mid Cretaceous (Table 3.1). Sequences of similar age occur beneath the desert surface of interior Oman, and extend across much of eastern Arabia and the Arabian Gulf; they contain the oil and gas accumulations of the area.

In the Late Cretaceous, the Hajar Supergroup was tectonically overlain by two allochthonous rock units, the Hawasina and the Semail. The Hawasina comprises mostly an imbricate sequence of carbonate turbidites whose time span of deposition was also Mid or Late Permian to Mid Cretaceous. At the top of the pile of individual Hawasina nappes are mountain-size blocks of white, shallow-marine limestone of Late Permian and Late Triassic ages that are generally referred to as 'exotics'; exotic because they look so different from the darker deep-marine Hawasina sediments and rocks of the Semail that surround them.

The Hawasina is overlain by the Semail Nappe, an ophiolite comprising former oceanic crust. Sediment-transport directions indicate that the Hawasina turbidites were deposited on the floor of Neo-Tethys (Hawasina Ocean of Glennie *et al.* 1974). The sedimentary sequences became imbricated as they were scraped off the underlying older oceanic crust during its Late Cretaceous subduction down an oceanic trench. At the same time, new oceanic crust (the future Semail ophiolite) was forming behind the plane of subduction. These two allochthonous units were obducted over the Hajar Supergroup during the attempted subduction of Arabia's continental margin down the Semail oceanic trench. However, the continental crust was too thick and buoyant to be subducted, and subduction ceased within this trench only to start again further to the northeast within Neo-Tethys in what is now the Makran area of Iran. When the Makran subduction trench became active, compressive stresses were released within the Semail trench, thus allowing the continental margin to rise isostatically (like releasing a piece of wood below water), bringing its overburden of Hawasina and Semail rocks with it to about sea-level (for further explanation see Glennie 1995).

In the current flank areas of the mountains shallow-marine carbonates dominated deposition during the Late Cretaceous and Palaeogene and now overlie sequences of both the Hawasina and the Semail; the Hawasina extends considerably farther south than the Semail. These young carbonates are well developed in the eastern Hajar northwest of Sur, where they overlie both autochthonous and allochthonous sequences, and as isolated mountain-size sequences and Cuesta to the north and south respectively of Ibri, where they are underlain by the Hawasina. Such cuestas are the result of Neogene uplift of the Oman mountains into a relatively simple anticline.

The Late Cretaceous and Tertiary sedimentary cover is the most extensive lithostratigraphic unit cropping out in Oman, covering about two thirds of the country (Le Métour *et al.* 1995), and, apart from the Oman mountains and Huqf area, constituting virtually all the surface geology (Figure 3.4). This unit comprises marine limestone, marl, chalk, calci-turbidites, lacustrine limestone and terrigenous clastic sediments.

Rocks lithologically similar to the Hawasina and Semail of the Oman mountains are exposed on Masirah Island off the Arabian Sea coast of Oman. The ophiolite however seems to represent oceanic crust newly formed during a phase of separation of what is now southern Afghanistan from Arabia during the Mid Jurassic, emplaced onto the downfaulted edge of the Arabian margin during a phase of compression early in the Cretaceous, and then having received a tectonically emplaced slide of Hawasina-like sediments (for further discussion, see Glennie 1995).

The surface of virtually the whole of Oman became terrestrial by the Late Miocene. Since then, floodwaters have transported enormous amounts of eroded material down both sides of the rising Oman mountains: northeast to the Batinah Coast and the Gulf of Oman, and radially, south towards the Gulf of Masirah, southwest to the salt-covered Umm as Samim and west beneath the northeastern dunes of the Rub' al Khali and the southeastern Arabian Gulf. The largest drainage system of the Oman interior is Wadi Halfayn, which occasionally flows into the Gulf of Masirah east of the northern Huqf, an area in which Late Precambrian, Palaeozoic and Mesozoic strata are exposed at the surface (Figure 3.4).

3.4.5 Central Plateau

The large area of central-western Arabia is referred to as the Central Plateau. It includes the Hijaz plateau, the Hisma plateau and the Nejd Pediplain. It is bounded by the western mountains, to the south by the Yemen highlands and Hadhramaut plateau, and to the east by the Cuesta Region. The topography of this huge area is characterised by the peneplaned substrate of the Arabian Shield and the gentle slope of the surface towards the northeast (Figure 3.5).

The Hijaz plateau extends from Taif southwards to Abha, where it is about 180 km wide, at elevations of 1,500 to 3,100 m. The Hisma plateau in northwest Arabia is, like the Hijaz plateau, elevated most at its southwestern corner. The Nejd Pediplain is nearly 600 km wide and includes the large Harrat Rahat volcanic lava field, which extends all the way from Madinah to Mecca, covering 24,000 km^2 (Chapman 1978a, 1978b).

The main drainage system is northeastward following the gentle slope of the plateau. Wadi ar Rimah is the largest of the draining wadis, carrying water all the way from Madinah to the Arabian Gulf basin near Basrah (Chapman 1978a, 1978b). The other major wadis, Ranyah, Bishah and Tathlith, flow as Wadi ad Dawasir eastwards into the Rub' al Khali. The largest lava flow, Harrat Rahat, covers an old valley system in which groundwater flows beneath the basalt (hard black volcanic lava) mainly to the north (Durozoy 1972).

Vast areas of the Central Plateau expose the basement rocks of the Arabian Shield, comprising mainly Precambrian plutonic rocks such as granites and granodiorites, and strongly deformed metamorphic rocks. Basaltic lava flows cover large areas. There have been eruptions during recorded history, such as the basalt flow near Madinah in AD 1256 (Hötzl & Zötl 1984).

3.4.6 Cuesta Region

The term 'cuesta' describes a long narrow hill with a steep slope on one side and a gentle slope on the other. In the cuesta region of central Arabia there are several escarpments oriented north-south in the vicinity of Riyadh (Figure 3.5). To the east, this area is bounded by ad Dahna, and to the west by the Central Plateau region. The cuestas are composed of sedimentary rocks that curve around the crystalline shield of the Central Plateau. The longest escarpment is that of Jebel Tuwayq, west of Riyadh. It is 800 km long, with a maximum elevation of about 1,500 m and an average of 840 m. Northeast of Riyadh is the 250 km long al 'Aramah escarpment, with an elevation of about 540 m.

The cuestas are erosional escarpments formed of platform sediments overlying the basement of the Arabian Shield (Figure 3.5). The Tuwayq escarpment represents the boundary between the outcrops of Precambrian basement to the west and the sedimentary cover to the east. The cuestas are capped by weathering-resistant Upper Jurassic and Upper Cretaceous limestones overlying less resistant rocks, mainly sandstones. Large sinkholes and caves at the base of the scarps mark the presence of soluble sediments such as carbonates, gypsum and anhydrite. Quaternary terraces in wadis expose gravel beds, aeolian sands and sabkha sediments (Hötzl, Kramer & Maurin 1978, Schyfsma 1978).

Several major wadis transect the cuestas from west to east: Wadi Nisah, Wadi Birk, Wadi ad Dawasir and Wadi al Hinuw. At the same time, desert streams, which appear to be dry at the surface, flow in the sub-surface. There is therefore a considerable amount of groundwater and an abundance of springs. With the rich volcanic soil washed in from the west, this area enjoys remarkable fertility.

3.4.7 Summan Plateau

The Summan Plateau is located in the coastal region of the Arabian Gulf in the northeastern part of the Peninsula. It is a long hard-rock plateau, 80-250 km wide, extending north-south. The plateau slopes from an elevation of about 400 m in the west to 250 m in the east, where it is terminated by an escarpment. On the eastern side of the plateau, the drainage system of Wadi al Widyan carries water into the plains of Iraq and to the Euphrates River. In the west, drainage is into Wadi as Sirhan.

The even surface of the Summan Plateau is composed of Late Miocene or Pliocene flat-lying sandy limestones with the typical karst topography of sink holes, solution cavities and caves. In the northwest, the plateau is overlain by an extensive gravel desert underlain by sedimentary rocks and Pliocene basalt.

3.4.8 Arabian Gulf Coast

The Gulf Coast along the northeastern edge of the Arabian Peninsula is 50-100 km wide. The foreshore is very gently inclined so that changes in the limited tide and in wind direction cause the waterfront to shift back and forth by several kilometres. The region gradually rises from the sea to an elevation of about 300 m along the edge of the Summan Plateau. The waters of the Arabian Gulf cover the northeastward dipping sedimentary sequences, which reach a thickness of some 10,000 m overlying the crystalline basement complex (Figure 3.5 and Beydoun 1991). Rainwater infiltrates the upturned edges of aquifers in the southwest and flows by gravity to the northeast, where it is known to discharge as freshwater artesian springs on the floor of the Gulf.

The rocks exposed along the coast of the Arabian Gulf comprise Tertiary sedimentary sequences, including those uplifted by domes of Hormuz salt at Jebel Dhanna in southwestern Abu Dhabi and Jebel Ali in Dubai. Late Miocene sabkha and aeolian sediments are overlain by fluvial sequences in Jebel Barakah that contain a rich variety of vertebrate fossils (Whybrow & Hill in press). The cover of unconsolidated sediments is of four main types:

(1) Quaternary beach gravel and sand, found southeast of Jubayl, where coastal erosion has exposed low terraces covered additionally by coquinas (rocks composed of marine shells) and individual shells. As is explained by Evans (1995), much of the carbonate sand found around the coasts of the Arabian Gulf was formed organically on the floor of the Gulf.
(2) Quaternary gravel of limestone and quartz derived from the older rocks of the Shield area, filling several shallow basins.
(3) Coastal sabkhas or salt flats, common along the coast from Kuwait to Musandam, the most extensive occurring along the coast of Abu Dhabi (see also section 3.4.9.2 below).
(4) Aeolian sands which cover the area inland from the sabkhas in the Emirates, where they form the northeastern part of the Rub' al Khali. Details of the dunes of this area are discussed in the next section.

3.4.9 Dune Sand Areas (Unconsolidated Quaternary Sediments)

3.4.9.1 Sand dunes

Approximately one third of the Arabian Peninsula is covered with unconsolidated Quaternary deposits consisting of large sand seas and widespread sheets of gravel. The main sand seas are the Great Nafud, al Jafurah, ad Dahna and the Rub' al Khali, literally the 'Empty Quarter' (Figures 3.3 and 3.6, see also Figure 8.1 for a more detailed map of all sand areas). The Great Nafud occupies the northeastern part of the Arabian Peninsula with an area of about 57,000 km^2. Ad Dahna is a narrow belt of dunes and shifting sand about 1,300 km long, forming an arc from the Great Nafud east of Riyadh to the Rub' al Khali in the south. Al Jafurah lies within the coastal region of the Arabian Gulf. The Rub' al Khali covers about 600,000 km^2 of southern Arabia. It is bounded by the bajada flanking the Oman mountains in the east, the Hadhramaut plateau in the south and the Hijaz plateau to the west. All these sand seas were deposited by winds associated with the *shamal*. South of the eastern Oman mountains is a relatively small sand sea, the Wahibah Sands, with an area of about 9,400 km^2, that was formed by winds of the southwest monsoon (here blowing almost due north) rather than the *shamal*.

The resultant of wind direction and velocity, together with the exposed rock topography, control the size and type of aeolian sand dunes, which can be classified into two basic types:

Transverse: These dunes have their long axes transverse to the prevailing wind. They are common in al Jafurah and in the northeastern Rub' al Khali. With a shortage of sand, the transverse dune breaks up into the crescent shaped barchan dunes, the horns of which extend downwind as linear ridges. Transitions between transverse and barchan forms are known as barchanoid dunes. Giant barchanoid dunes up to 150 m high are characteristic of al Liwa in the western Emirates.

Longitudinal: Since transverse dunes are unstable when exposed to strong winds, like the horns of barchans, 'blow-outs' extend down wind and develop into longitudinal dunes or *irqs*. Such dunes, 200-300 km long and with a crestal spacing of about 2 km, are typical of the southeastern Rub' al Khali (Glennie 1987). Shorter dunes of this type, up to 100 m high and with crestal spacing of nearer 1 km, characterise the northern Wahibah Sands (see Dutton 1988b, and papers therein, for a detailed report on the area).

On a mega scale, the lateral extent and aligned nature of the longitudinal dunes of the southeastern Rub' al Khali and Wahibah Sands indicate that they were formed by winds of relatively uniform direction parallel to the dunes' axes. Dunes of this size are not being built today but were formed by strong winds during the last glaciation. Since then, regional wind patterns have altered and become more changeable, resulting in more complex dune surfaces. In areas where both the *shamal* and southwest monsoon have a seasonal influence, such as in the eastern Rub' al Khali, the former longitudinal dunes have been modified by variable winds into series of star dunes. In other areas large barchanoid dunes may now have smaller dunes, of both barchan and linear types, climbing their windward slopes.

The dune patterns of the Emirates (Figure 3.7) help to unravel some of the Late Quaternary history of this part of eastern Arabia. Except in and around the giant barchanoid dunes of al Liwa, most of the other dunes seem to have originated as longitudinal types. Small, active, N-S trending dunes occur east of Sabkhat Matti. To the northeast, giant dunes trend NW-SE over most of the area but deviate to the north as the northern Oman mountains are approached. These dunes are thought to have been formed by strong *shamal* winds at the height of the last glaciation, and have since suffered strong deflation and reworking. In the northeast, the crests of these dunes have been eroded and redeposited in the interdune areas as small E-W trending linear dunes. Further south, many of these dunes have now developed avalanche slopes or more linear extensions on their southern sides under the influence of the *shamal*. These complications indicate changes in the effective wind pattern since the last glaciation. Similar changes in

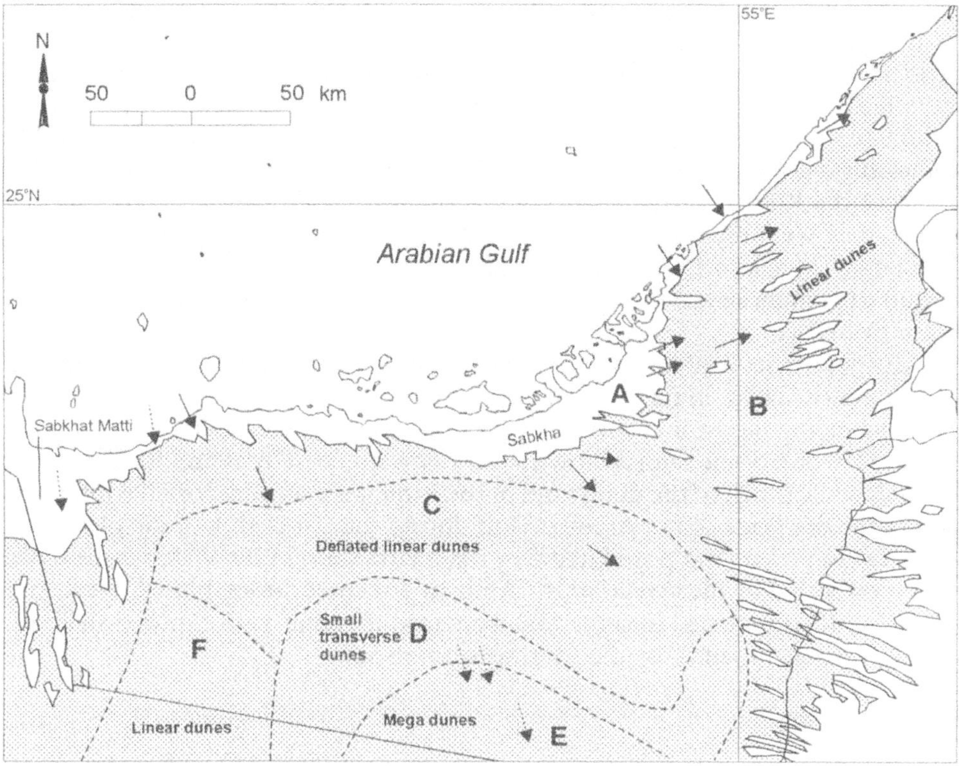

Figure 3.7 Dune sands and sabkhas of the Emirates can be subdivided into several discrete areas: A, coastal sabkhas grading landward (including Sabkha Matti) into inland sabkhas; B, deflated linear megadunes and interdune areas; C, dunes of type B partly swamped by small transverse dunes whose axes trend approximately E-W; D, a close-knit sand sea of small transverse dunes; E, mega barchanoid dunes and interdune sabkhas of al Liwa; F, N-S trending small linear dunes east of Sabkhat Matti. Solid Arrows: Glacial-age wind directions deduced from bedding attitudes of carbonate dunes. Dashed arrows: wind directions deduced from uniform dips in dune sand beneath some sabkha surfaces.

wind direction can be deduced from the shape of the dunes in the 'Uruq al Mutaridah in the eastern Rub' al Khali.

Carbonate dune sands, now largely cemented, underlie the unconsolidated dune sands of the Emirates for up to 80 km from the coast. Dating, and their content of microfossils, indicates that they were derived from the floor of an exposed Arabian Gulf during at least the last two glaciations, and became cemented during interglacial times of higher rainfall or higher water table. In the Wahibah area, similar carbonate sands, exposed on the Arabian Sea coast as cliffs of hard rock, were carried landward by the southwest monsoon during the waning phases of glaciations.

When dune sands are buried below the water table, the sand grains acquire a coating of reddish iron oxide. By reworking such sands, whole dunes can become reddened. The Great Nafud is a vast area of reddish sand and moving dunes, with no streams or oases and sparse vegetation. The quartz sands of ad Dahna are bright red-orange while those of al Jafurah are buff to tan in colour, possibly indicating a less reddened source in sands beneath the present Arabian Gulf. The sands of the Rub' al Khali are also of red-orange colour and of medium to fine grain; they were reworked from the underlying Pliocene fluvial and lacustrine sediments (McClure 1978). Some sands of al Liwa dunes and in Sabkhat Matti are up to granule size, indicating a probable source in fluvial sands derived from western Arabia.

Most of the aeolian sand is derived by deflation of young fluvial, lacustrine and coastal sands. These in turn are the products of weathering and erosion of older rocks including sandstones that originated by erosion of the basement (Chapman 1978a, 1978b). Some sand is blown southeastward from Syria and Jordan by the *shamal*, while much of the dune sand in the Emirates was transported southward across the dry floor of the Arabian Gulf during the last high-latitude glaciation (Glennie in press). It is postulated that the quartz sand in the Wahibah area was first transported by the *shamal* to the Arabian Sea coast during glacial times, and then secondarily carried into the Wahibah area, together with exposed shallow-marine carbonate sands, by the southwest monsoon.

3.4.9.2 Sabkhas

The coastal sabkhas of the Arabian Gulf, so well exposed along the coast of Abu Dhabi, are characterised by a thin crust of salt (halite) and a rubbery mat of almost black algae underlain by sand, silt or clay, with a locally cemented hard layer of gypsum (hydrous calcium sulphate) crystals about 50 cm below the surface. Marine shells, typically the gastropod *Cerithium*, are common, and at ground temperatures above about 50°C, the gypsum alters to anhydrite. The coastal sabkha is flooded with marine water during *shamal*-induced storm tides or spring tides, when the algae are rejuvenated and start active growth. The sabkha surface becomes desiccated during intervening periods; pools of water evaporate to form salt crusts, algal mats crack and curl and aeolian sand and dust may be trapped beneath them. The coastal sabkhas along the Abu Dhabi coast are thought to have formed because the postglacial flooding of the Arabian Gulf cut off the supply of sand to dunes further south, deflation removed dune sand down to the water table, and increasing aridity over the past 5 ka caused strong evaporation.

Inland, away from any direct marine influence, the coastal sabkha transforms subtly into an inland sabkha, which lacks marine shells unless wind-blown (but only over limited distances inland). The largest inland sabkha, Sabkhat Matti in southwestern Abu Dhabi, is about 50 km across and tapers southward into Saudi

Arabia for almost 150 km from the coast. The Umm as Said sabkha south of Doha in Qatar has dunes migrating across it from NW to SE (Shinn 1973).

In the 'Uruq al Mutaridah in the eastern Rub' al Khali is a large group of NW-SE trending inter-dune sabkhas. As is also the case further north, the associated dunes show evidence of modification by winds blowing from the north. To their east is the giant sabkha of Umm as Samim, which has a length of over 100 km and a width of *c.* 50 km, a large part of which is covered with polygonal halite. These inland sabkhas are thought to have originated as deflation hollows during the last glacial period, and became flooded (perhaps even forming temporary lakes) because of higher post-glacial rainfall and an associated rise in the water table some 5 to 10 ka BP; hyperarid conditions during the past 5 ka ensured their conversion to sabkhas. It is interesting that, apart from occasional fluvial water from Wadi Umayri, an additional supply is derived from an underlying aquifer, the Paleocene Umm ar Radhuma Formation, which is overlain by evaporites of the Eocene Rus Formation, from which some of the salt is probably derived (Heathcote & King in press).

3.5 Summary

Geologically, the Arabian Peninsula forms the greater part of a crustal plate bordered, clockwise, by the Red Sea, the Gulf of Aqaba-Jordan Fault, the Taurus mountains, the crush zone along the northeast margin of the Zagros mountains, the Gulf of Oman, the Arabian Sea and the Gulf of Aden. The Red Sea, Gulf of Aden, Arabian Sea and Gulf of Oman coasts are relics of the progressive crustal break-up of the former megacontinent of Gondwana.

Geographically, Arabia is considered to extend eastward to the west coast of the Arabian Gulf and the first folds of the Zagros mountains flanking the Tigris-Euphrates river plain. The basement rocks of the Arabian Plate extend beneath the Zagros mountains, however, so that geologically, this range is really an integral part of Arabia.

Much of the varied geology and geomorphology of Arabia is the outcome of two major sequential events:

1) Following the creation of Neo-Tethys, the northeastern continental margin of Arabia supplied carbonate turbidites and radiolarian cherts to the floor of the ocean. During the latter half of the Cretaceous, much of that sediment (the future Hawasina of the Oman mountains and Coloured Series of Iran) was scraped as tectonic slices from the underlying oceanic crust, which was subducted towards the northeast down an oceanic trench. Near the end of the Cretaceous, nappes of Hawasina/Coloured Series within the upper trench were emplaced over the down-going continental margin of Arabia, which, however, was too thick and buoyant to be completely subducted. These nappes, in turn, were overlain by oceanic crust newly formed behind the subduction trench. The creation of another system of subduction trenches farther to the east reduced the compressive forces within the

original subduction trench and permitted uplift of the partly contained continental margin and, in Oman, its cover of Hawasina and Semail nappes, to about sea-level.

2) The Mid Cenozoic separation of Africa and Arabia resulted in the creation of the Red Sea and Gulf of Aden, and was associated with a gentle tilt of Arabia to the northeast. As a consequence, uplifted Precambrian rocks are now exposed in western Arabia, where basaltic lavas also flowed to the surface along major fractures. Because tilting caused erosion in the west and deposition in the east, the surface rocks become younger to the northeast. As the Red Sea widened, the northeast margin of Arabia was compressed against adjacent continental and oceanic crust. The rigid basement sequences of Arabia underthrust the continental blocks of central Iran, with ensuing folding of the thick rock sequences of the Zagros mountains overlying a slide plain of Late Precambrian Hormuz salt. In contrast, the continental margin of Oman was adjacent to relatively weak oceanic crust in the Gulf of Oman. Movement of the new Arabian Plate, driven by the opening of the Red Sea and Gulf of Aden, resulted in subduction of Gulf of Oman oceanic crust beneath the Makran coast of Iran, and at the same time initiated apparently simple uplift of the Oman mountains by folding.

The Mid Cenozoic uplift of western Arabia, coupled with uplift in the Oman mountains and the initiation of folding of the Zagros mountains, rejuvenated erosion, much of the products of which were deposited in or on the flanks of the NW-SE trending axial depression of interior Oman, the Arabian Gulf and Tigris-Euphrates valley. Fluvial sedimentation dominated the latest Tertiary and early Quaternary, to be succeeded, especially during near-polar glaciations, by dune activity, with interglacial periods coinciding with weaker winds and higher rainfall. Increasing aridity and deflation over the past 5 ka has brought about the conversion of former lakes to inland sabkhas and the creation of sub-horizontal coastal sabkhas parallel to the water table.

About a third of the Arabian Peninsula is covered with unconsolidated Quaternary deposits consisting of large aeolian sand seas, widespread sheets of fluvial gravel and sabkhas, both coastal and inland; the remainder is mostly barren rock. The dunes of the main sand seas, the Great Nafud, al Jafurah, ad Dahna and the Rub' al Khali, all of which occur mostly in Saudi Arabia, migrated southward under the influence of the *shamal*. The much smaller Wahibah Sands in eastern Oman was created by the winds of the southwest monsoon.

CHAPTER 4

Biogeography and Introduction to Vegetation

Harald Kürschner

"*We are especially ignorant of the vegetation in the Arabian Peninsula*"
(Zohary 1973)

4.1 Introduction

Plant migrations during past geological periods, long-term geological and climatic changes, recent climate and hydrology, edaphic, geological and topographical structures and the effects of man and his livestock have all played a part in the establishment and development of the flora and vegetation of Arabia. The varied and often dramatic terrain of the 2.7 million km^2 of the Peninsula (Figures 3.1-3.4) is subject to a climate that ranges from the hyperarid to the humid subtropical (see Chapter 2), and the consequent complexity of the landscape is reflected in the variety of its plant life.

The natural vegetation of Arabia includes evergreen and deciduous woodlands, drought-deciduous open thorn woodlands, sclerophyllous and succulent shrublands, dwarf shrublands and open xeromorphic grasslands (Figure 4.1). However, the introduction of agriculture and the transformation of vast areas into rangelands over long periods of time has had a dramatic influence on the development of the vegetation landscape that we see today. The natural vegetation in many areas has been obliterated and can only be reconstructed from remnants of apparently natural vegetation and from the distribution of relict species.

S. A. Ghazanfar and M. Fisher (eds.), Vegetation of the Arabian Peninsula, 63-98.

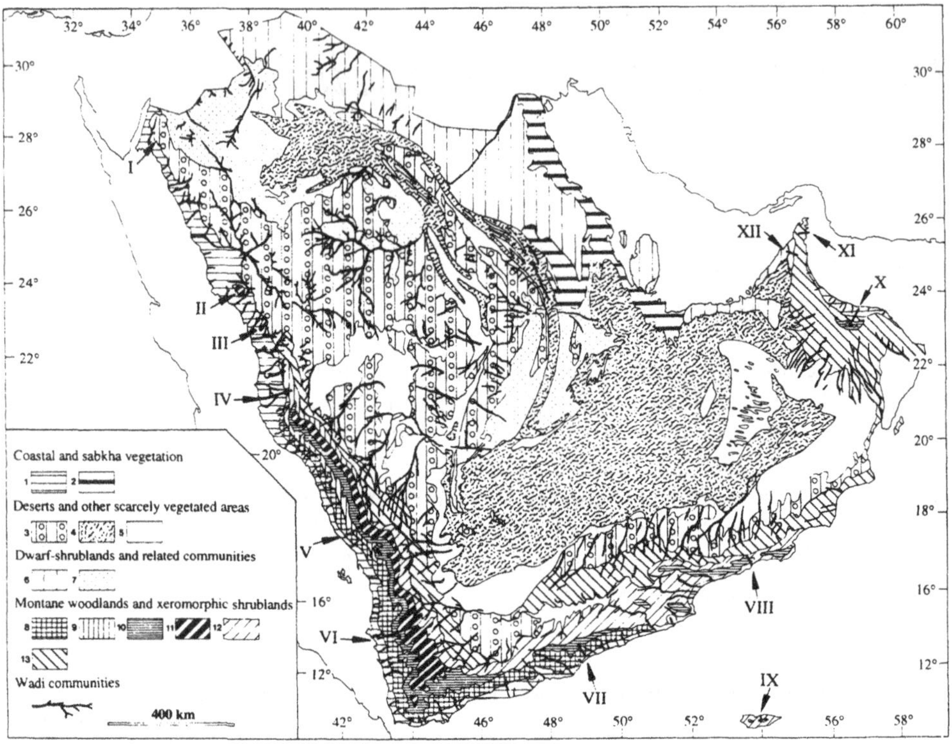

Figure 4.1 Main vegetation types of the Arabian Peninsula (modified after Frey & Kürschner 1989). 1, Coastal and sabkha vegetation and very open drought-deciduous thorn woodlands of the coastal plains and lower foothills. 2, Open xeromorphic grasslands. 3, Very open xeromorphic dwarf-shrublands, rock and gravel deserts. 4, Sand deserts. 5, Scarcely vegetated dunes and sandy gravel plains. 6, Very open xeromorphic dwarf-shrublands. 7, Open xeromorphic dwarf-shrublands. 8, Drought-deciduous thorn woodlands and shrublands (*Acacia-Commiphora* woodlands). 9, Drought-deciduous thorn woodlands (*Acacia* woodlands). 10, Semi-evergreen sclerophyllous woodlands and sclerophyllous scrub. 11, Evergreen needle-leaved woodlands, mixed with thorn woodlands (completely destroyed in the Yemeni highlands). 12, Mixed formations of drought-deciduous shrublands, xeromorphic succulent shrublands and open xeromorphic grasslands. 13, Very open xeromorphic semi-desert shrublands. The locations of transects I-XII of Figure 4.7 are indicated.

For a long time, the main phytogeographical subdivision of Southwest Asia followed Zohary (1966, 1973), who divided Arabia into two floral regions: the Saharo-Arabian region (east Saharo-Arabian subregion) covering the semi-arid and arid interior of Arabia, and the Sudanian region, further divided into a Nubo-Sindian province (along a small belt of the Arabian coast) and an Eritreo-Arabian province (covering mountainous southwestern Arabia and Socotra, Figure 4.2). Unfortunately this treatment did not correspond particularly well with the system

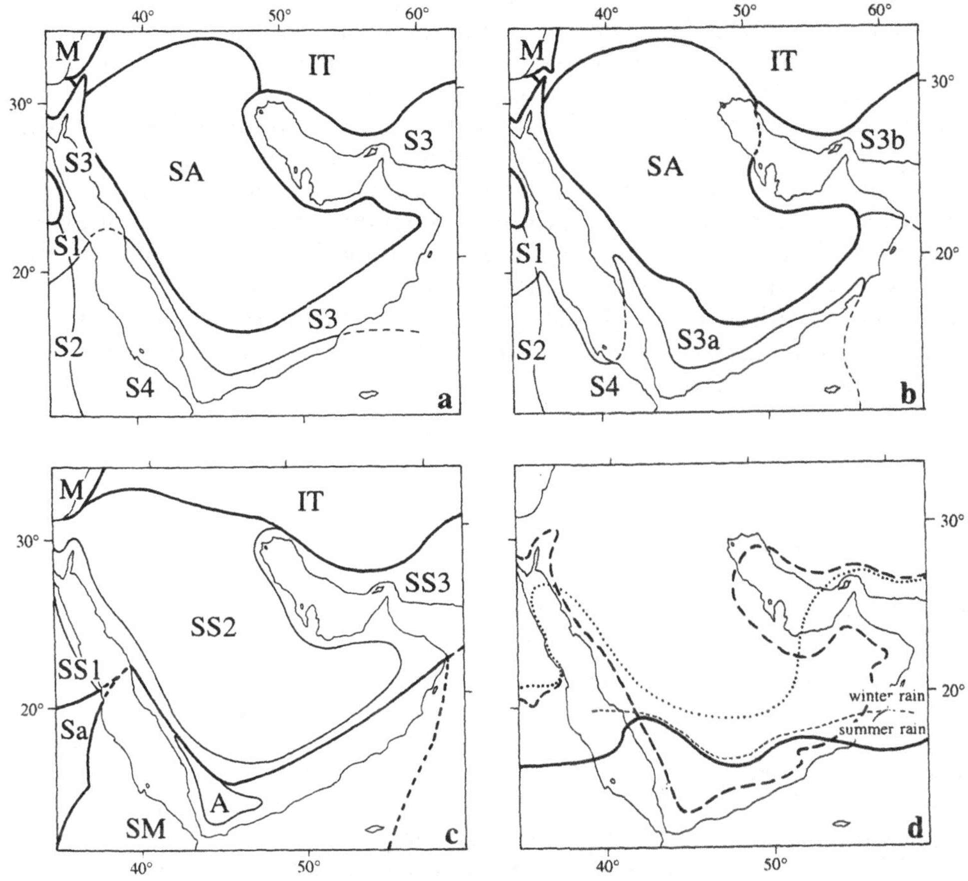

Figure 4.2 Principal phytochoria and floristic borders on the Arabian Peninsula as outlined by (a) Zohary (1966, 1973), (b) Kürschner (1986a), (c) Léonard (1989), and (d) borderlines between subtropical and tropical climatic zones after Troll and Pfaffen (1965, dotted line), Wissmann (1948, long-dashed line), Creutzburg and Habbe (1964, continuous line), and Deil (1996, short-dashed summer rain/winter rain line). A, Afromontane archipelago-like regional centre of endemism; IT, Irano-Turanian region/Irano-Turanian regional centre of endemism; M, Mediterranean region/Mediterranean regional centre of endemism; Sa Sahel regional transition zone; SA, Saharo-Arabian region; SM, Somalia-Masai regional centre of endemism; S, Sudanian region (S1, Saharo-Sudanian province; S2, Eu-Sudanian province; S3, Nubo-Sindian province, a, western Nubo-Sindian sub-province, b, eastern Nubo-Sindian subprovince/Omano-Makranian subprovince; S4, Eritreo-Arabian province); SS, Saharo-Sindian regional zone (SS1, Saharan regional subzone, SS2 Arabian regional subzone, SS3, Nubo-Sindian local centre of endemism).

proposed for Africa by White (1983), which was based on regional centres of endemism separated by transition zones and regional mosaics (a regional centre of endemism is a phytochorion that has more than 50 % of its species confined to it and a total of more than 1,000 endemic species; see also section 12.2).

Léonard (1989) showed that White's phytogeographical system could be extended to Southwest Asia, subdividing the Arabian Peninsula into three main phytogeographical units: the Saharo-Sindian regional zone (SS) including the Arabian regional subzone (SS2) and the Nubo-Sindian local centre of endemism (SS3), the Somalia-Masai regional centre of endemism (SM) and the Afromontane archipelago-like centre of endemism (A) (Figure 4.2).

In this chapter the taxa characteristic of these three main phytochoria and their divisions are first described, along with the geographical and historical affinities of each zone. The role of Mediterranean and Irano-Turanian intruders in the vegetation of Arabia is then considered, and the vegetation history and the occurrence of disjunctions and relict species are described. In this historical and biogeographical light, the main vegetation types and indicator plant species of each area are then summarised.

4.2. Biogeography

4.2.1 PHYTOCHORIA

4.2.1.1 Saharo-Sindian Regional Zone

The largest phytochorion of the Arabian Peninsula is the Saharo-Sindian (SS) regional zone, divided into an Arabian regional subzone (Figure 4.2: SS2) that corresponds to the Saharo-Arabian region *sensu* Zohary (1966, 1973) and the Saharo-Sindian *sensu* Eig (1938), and a Nubo-Sindian local centre of endemism (Figure 4.2: SS3), comparable with the Nubo-Sindian province of the Sudanian region *sensu* Zohary (1966, 1973) and the Sudano-Deccanian *sensu* Eig (1938). The Nubo-Sindian province was later split into a western and eastern subprovince (the Omano-Makranian subprovince of the Nubo-Sindian province, Kürschner 1986a).

The Arabian regional subzone is a more or less well defined entity, typical of most of extra-tropical Arabia. It is influenced by a bi-seasonal, Mediterranean type of climate with a rain maximum during winter and spring (see Figure 2.11). The flora is poor in species, with an autonomous stock of Saharo-Arabian taxa derived from an ancient Mesogean floral stock (old Mediterranean, formerly distributed along the northern and southern coasts of the Tethys sea). Typical derivates from this stock are *Agathophora alopecuroides, Anastatica hierochuntica, Anabasis articulata, Astragalus spinosus, Asteriscus graveolens, A. pygmeus, Calligonum* spp. (probably of Irano-Turanian origin), *Cornulaca* spp., *Diplotaxis harra, Fagonia* spp., *Gymnocarpos decandrus* (Figure 4.3), *Gymnarrhenia micrantha, Haloxylon salicornicum, Helianthemum lippii, Lasiopogon* spp., *Moltkiopsis ciliata, Moricandia sinaica, Morettia canescens, Neurada procumbens, Notoceras bicorne, Oligomeris linifolia, Paronychia arabica, Polycarpaea repens, Pteranthus dichotomus, Reaumuria* spp., *Reboudia* spp., *Retama raetam, Rhanterium eppaposum, Rhazya stricta, Salsola tetrandra, Salvia lanigera, Savignya parviflora* (Figure 4.3), *Sclerocephalus arabicus,*

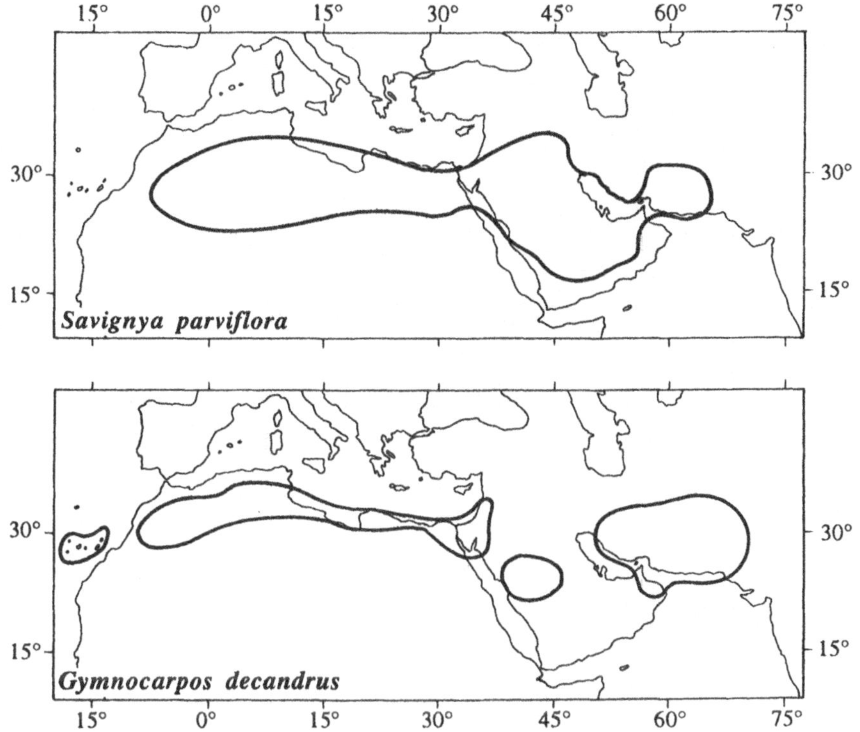

Figure 4.3. Saharo-Sindian species. Distribution of *Savignya parviflora* (after Hedge & King 1983) and *Gymnocarpos decandrus* (after Petrusson & Thulin 1966).

Scrophularia deserti, Stipagrostis spp., *Tamarix* spp. (probably of Irano-Turanian origin), *Traganum nudatum* and *Zilla spinosa*.

The Arabian regional subzone is bordered by a narrow but continuous area along the coastal belt of the Peninsula that belongs to the Nubo-Sindian local centre of endemism, which is often difficult to delimit from the adjacent Arabian regional subzone (see for example the delimitation of the phytochoria for central Oman, Ghazanfar 1992b). Typical of this subzone is a xero-tropical arborescent vegetation (old "*Acacia* flora" of palaeotropical origin) of desert habitats, in which temperatures are high enough to support this type of xero-tropical vegetation, but rainfall too low to maintain a typical tropical vegetation (the *Acacia-Commiphora* flora of the Somalia-Masai regional zone). These formations of scattered trees associated with a ground vegetation of dwarf shrubs and grasses ('Pseudo-savannas' *sensu* Zohary 1973) are largely dependent on a "...certain hydrothermic regime, in which the thermal element is the decisive one, while the rain factor can be compensated for by underground moisture" (Zohary 1973). Typical taxa restricted to this belt are *Aerva persica, Acacia raddiana, A. tortilis, Calotropis procera, Capparis* spp., *Cassia* spp., *Caylusea hexagyna, Chrozophora* spp., *Cleome* spp., *Cocculus pendulus, Cymbopogon* spp., *Eremopogon foveolatus, Forsskaolea*

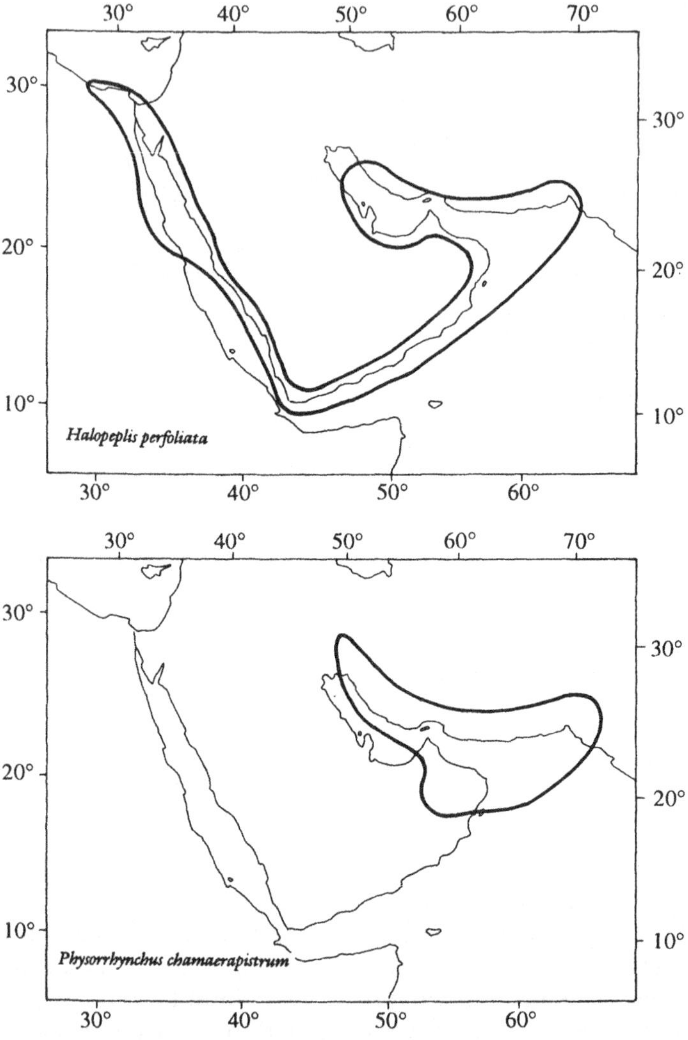

Figure 4.4 Distribution of *Halopeplis perfoliata*, a Saharo-Sindian species with a Nubo-Sindian distribution pattern, and of *Physorrhynchus chamaerapistrum*, a Saharo-Sindian species with an Omano-Makranian distribution pattern (after Kürschner 1986a).

tenacissima, Halopeplis perfoliata (Figure 4.4), *Indigofera* spp., *Lasiurus scindicus, Launaea* spp., *Leptadenia pyrotechnica, Monsonia* spp., *Panicum turgidum, Pennisetum divisum, Pergularia tomentosa, Periploca aphylla, Polygala erioptera, Salvadora persica, Salvia aegyptiaca, Schouwia purpurea, Tetrapogon villosus* and *Tephrosia* spp.

Various taxa of eastern and southeastern Arabia show floristic links to Makran in southern Iran and to Pakistan, and are not uniformly distributed within the whole Nubo-Sindian local centre of endemism. These Omano-Makranian elements (Kürschner 1986a) indicate the existence of a probable migration route for xerophytic species during the Pleistocene, when a low sea-level brought a drying up of most parts of the Arabian Gulf. Typical taxa of this type are *Atriplex griffithii* subsp. *stocksii, Agriophyllum minus, Centaurea wendelboi, Convolvulus acanthocladus, C. cephalopodes, C. spinosus, Cornulaca aucheri, C. amblyacanthus, Ebenus stellatus, Enneapogon persicus, Euphorbia larica, Farsetia heliophila, Grantia aucheri, Helichrysum makranicum, Jaubertia aucheri, Launaea bornmuellerianum, Leptorhabdos parviflora, Limonium stocksii, Lonicera aucheri, Nerium oleander, Ochradenus aucheri, Otostegia aucheri, Physorrhynchus chamaerapistrum* (Figure 4.4), *Prosopis cineraria, P. koeltziana, Pseudogaillonia hymenostephana, Pseudolotus makranicus, Pteropyrum scoporium, Pycnocycla* spp., *Salsola drummondii, Salvia macilenta, S. mirzayanii, Schweinfurthia papillionacea, Sphaerocoma aucheri, Taverniera cuneifolia, Tephrosia persica, Teucrium orientale* subsp. *taylorii* and *T. stocksianum.*

4.2.1.2 Somalia-Masai Regional Centre of Endemism

The southwestern and southern parts of the Arabian Peninsula are strongly influenced by a subtropical to tropical climate with spring and summer rains (see Figure 2.11), and are thus dominated by a xero-mesic tropical flora of palaeotropical origin that extends into Oman. These areas belong to the Somalia-Masai regional centre of endemism (including the Eritreo-Arabian province of the Sudanian region, *sensu* Zohary 1966, 1973), that in fact represents the impoverished northern part of an African flora (Figure 4.2: SM). Examples of species with this conspicuous distribution pattern, linking southwest Arabia with the other side of the Red Sea, are *Acacia* spp. (eg. *A. asak, A. etbaica, A. hamulosa, A. mellifera* and *A. oerfota*), *Adenia venenata, Aloe* spp., *Barbeya oleoides, Berberis holstii, Boswellia sacra, Breonadia salicina, Cadaba beccarinii, C. glandulosa, C. longifolia, C. mirabilis, C. rotundifolia, Caesalpinia eriantha, Carissa edulis, Cassythia filiformis, Celosia* spp., *Celtis africana, C. toka, Commiphora* spp. (eg. *C. foliacea, C. gileadensis, C. habessinica, C. kataf* and *C. myrrha*), *Diospyros mespiliformis, Dobera glabra, Dodonaea viscosa, Euclea racemosa* subsp. *schimperi* (Figure 4.5), succulent *Euphorbia* spp., *Ficus cordata* subsp. *salicifolia, F. glumosa, F. ingens, F. palmata, F. sur, F. sycomorus, F. vasta, Forsskaolea viridis, Grewia* spp., *Jatropha pelargonifolia, Kalanchoe citrina, K. glaucescens, Maytenus parviflorus, Ochna inermis, Olea europaea* subsp. *africana, Osyris compressa, Pilea tetraphylla, Pouzolzia mixta, P. parasitica, Premna resinosa, Psiadia punctulata, Pterolobium stellatum, Sansevieria* spp., *Sarcostemma viminale, Sphaerocoma hookeri, Stephania abyssinica, Tarchonanthus camphoratus, Tinospora bakis* and *Trichilia emetica,* most of them important associates of the *Acacia-Commiphora* and semi-evergreen sclerophyllous woodlands of the montane vegetation between 500-1,800 (-2,000) m (see sections 6.3. & 6.4). Some of these, such as *Balanites aegyptiaca, Cadaba glandulosa, Maerua crassifolia, Maytenus senegalensis* and *Moringa peregrina,* have a wider distribution towards the north, partly extending into the adjacent extra-tropical regions.

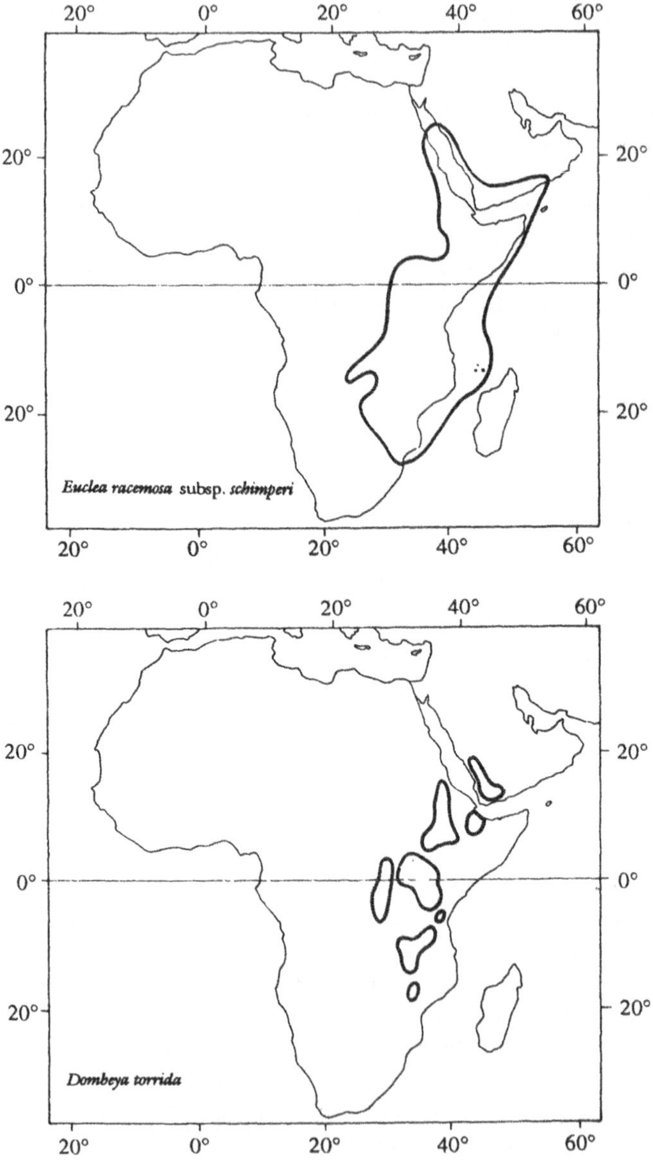

Figure 4.5 Distribution of *Euclea racemosa* subsp. *schimperi*, a Somalia-Masai species, and of *Dombeya torrida*, an Afromontane species. Modified after White & Léonard (1991).

4.2.1.3 Afromontane Archipelago-like Regional Centre of Endemism

The highest parts of the escarpment mountains of Asir and the Yemen highlands are islands of Afromontane vegetation, floristically and ecologically comparable to those of the high mountains of northeast and east Africa (Figure 4.2: A). These areas are characterized by a high rainfall (generally >300 mm per year, see Figure 2.7), leading to a continuous vegetation cover rich in trees, shrubs and herbs. Most prominent are evergreen needle-leaved woodlands, dominated by *Juniperus procera*. Further elements of the Afromontane distribution typical of southwestern Arabia are *Acacia origena, Aeonium leucoblepharum* (Figure 6.21: D), *Alchemilla cryptantha, Bersama abyssinica, Buddleja polystachya, Cadia purpurea, Campanula edulis, Catha edulis, Clutia richardiana, Crassula* spp., *Cussoni holstii, Delosperma abyssinicum, Dombeya torrida* (Figure 4.5), *Erica arborea, Felicia dentata, Helichrysum abyssinicum, Hypericum quartinianum, H. revolutum, Jasminum abyssinicum, Maesa lanceolata, Myrica humilis, Myrsine africana, Pavetta abyssinica, Pittosporum viridiflorum, Rhamnus staddo, Rosa abyssinica, Salvia merjamie, Senecio hadiensis, S. harazianum, Solanum schimperanum, Teclea nobilis, Thesium radicans* and *T. stuhlmannii*. Besides the arborescent and herbaceous Afromontane flora, the strong connections with Africa are also indicated by various bryophytes which often grow as epiphytes in the *Juniperus* woodlands (Kürschner 1984b, 1996b, see Chapter 5).

4.2.2 Mediterranean and Irano-Turanian Intruders

Towards the north, northeast and southeast, the Saharo-Sindian regional zone of Arabia comes into close contact with the Mediterranean regional centre of endemism (the East Mediterranean province of the Mediterranean region *sensu* Zohary 1966, 1973) and, separated by the Arabian Gulf, with the Irano-Turanian regional centre of endemism (the Mesopotamian and Irano-Anatolian provinces of the Irano-Turanian region *sensu* Zohary 1966, 1973; Figure 4.2). Floristic elements from these neighbouring phytochoria, even if rare, also extend into Arabia. Among the woody species only a few taxa occur as marginal intruders into the Arabian Peninsula, indicating former migration routes and possibly a formerly wider distribution of Mediterranean and Irano-Turanian vegetation. No typical communities of these phytochoria exist in Arabia today.

Enclaves of woody Mediterranean species can be found in the mountains of northwestern and western Arabia, where scattered and isolated stands of *Juniperus phoenicea* in the Midian, Hijaz and northern Asir mountains form the southernmost outposts of the arborescent Mediterranean element. Further Mediterranean taxa in Arabia are *Anagyris foetida, Hypericum hircinum, Lonicera etrusca, Malcolmia chia, Myrtus communis, Petrorrhagia cretica, Rhus tripartita, Umbilicus horizontalis* and *Velezia rigida*.

Irano-Turanian taxa occur in northeast and northwest Arabia, as well as in the mountains of northern Oman and Musandam. Examples of this chorotype are *Acanthophyllum* aff. *bracteatum* (northern Saudi Arabia), *Achillea fragrantissima, A. santolinoides, Artemisia sieberi, Prunus arabica* (*Amygdalus arabica*) (Musandam), *P. korshinskyi* (*A. korshinskyi*) (Jebel al Lawz), *Astracantha echina* subsp. *arabica, Biennertia cycloptera, Crataegus sinaica* (Midian mountains), *Halocharis sulphurea,*

Juniperus excelsa subsp. *polycarpos* (Jebel al Akhdhar), *Noaea mucronata, Pistacia atlantica* and *P. khinjuk* (Jebel al Lawz). The floristic link between southeast Arabia and the mountains of southwest Iran is also indicated by the distribution of *Dionysia mira* and *Viola cinerea* in both northern Oman and the Zagros mountains and Kuh-e Ghenou, and by the close relationship in *Hypericum* section *Campylosporus* of the Socotran endemics *H. balfourii* and *H. socotranum* with *H. dogonbadanicum* of the southwestern Zagros mountains (Robson 1987).

4.2.3 Vegetation History, Disjunctions and Relict Species

Probably the oldest section of the Arabian flora is the semidesert and desert flora of the Arabian regional subzone that was derived from a Cretaceous Mesogean floral stock located along the coasts of the Tethys sea (Zohary 1973). At that period most parts of central Arabia (the Arabian Platform) were covered by the Tethys sea, and taxa of this Mesogean element could not therefore colonize central Arabia before the Alpine orogeny and the withdrawal of the Tethys sea in the Late Cretaceous-early Eocene (see Table 3.1). Since the African-Arabian Shield was not yet divided by the Red Sea, a palaeo-African vegetation from the southwest extended eastwards into the future Arabian Peninsula throughout the middle-late Eocene and Oligocene. This palaeotropical vegetation persisted in the western and southwestern highlands as a precursor of today's Sudanian (i.e. Nubo-Sindian) vegetation (Mandaville 1990).

It is now widely accepted that Arabia experienced its main floro-historical events during the Tertiary. This was the main period for plant invasion, especially for the extra-tropical Saharo-Sindian flora that developed under the increasing aridity of the Mid to Late Miocene. Today, Arabia is a transition zone from the Holarctic to the Palaeotropical Kingdom, and from an extra-tropical to a tropical climate. The border between the southern, subtropical and tropical flora of Palaeotropical origin and the northern, extra-tropical flora of temperate, Holarctic origin is difficult to define. According to Zohary (1973), the northern border of the Palaeotropical flora is indicated by open thorn woodlands, dominated by various *Acacia* spp. and with a strong decline in the abundance and diversity of annual plants so characteristic of the Saharo-Arabian deserts. Limiting factors are temperature and rainfall, with the latter often compensated for by underground water in wadis. Most of the large wadis and depressions of the hot arid deserts of central and eastern Arabia are therefore today covered by *Acacia*-dominated plant associations restricted to alluvial soils that have freshwater available in the root zone. This makes the boundary between the Arabian and the Nubo-Sindian flora very problematic, because most of the thorn woodlands consist of a phytogeographical mixture of a Sudanian arborescent flora and a Saharo-Arabian ground layer (Kürschner 1986a).

Fossil pollen of Miocene horizons indicates that a proto-Sudanian flora existed in Arabia during the Mid Miocene (Mandaville 1984). Various families of tropical origin and distribution (Arecaceae, Combretaceae, Meliaceae, Myrtaceae and

Sapotaceae) occurred that are today rare or absent. In the eastern Rub' al Khali (near az Zumul), a hyperarid area today, pollen of the former genus *Psilatricolporites,* of the families Arecaceae, Myrtaceae, and of the water fern *Ceratopteris* have been found 150-200 m below the present surface, indicating freshwater marshlands of a humid, tropical to subtropical climate. Fossil mangroves of Early to Mid Miocene age (near Dawmat al Awdh) and Upper to Late Miocene (on the Gulf coast at Jebel Barakah) indicate a coastal environment of broad alluvial flood plains with swamps, open savanna grasslands, streams and mangroves in the intertidal zone (Whybrow & McClure 1981). Further evidence for such a Late Tertiary tropical vegetation in eastern Arabia is provided by fossils of turtle, fish, crocodile, rodents, hyrax, rhinoceros, pig, giraffe and bovids (Hamilton *et al.* 1978). The vegetation was replaced during the Late Miocene by a more drought-adapted Saharo-Arabian flora.

During Late Pliocene or Early Pleistocene times, the former Sudanian (i.e. Nubo-Sindian) flora partly recovered its former area by migration via the large trans-Tuwayq wadi systems (Wadi al Batin, Wadi as Sabha and Wadi ad Dawasir) from refuge areas in the mountains of western Arabia, and this flora is still present within this Arabian regional subzone. During the more humid period of 3.5-1.2 million years ago (Hötzl *et al.* 1978), these great wadi systems provided routes for a limited and selective reintroduction of an arborescent Sudanian flora from the west to the east, as indicated by wadi communities of *Acacia gerrardii, A. raddiana, A. tortilis, Capparis* spp. and *Cleome* spp. in the main channels of the trans-Tuwayq wadis, and the unique isolated stands of *Suaeda monoica* at Jawb al Asul in the mouth of Wadi ad Dawasir (Mandaville 1990).

After this pluvial episode central and eastern Arabia became more arid, only interrupted by increased precipitation during the Pleistocene (see Chapter 3). These hotter and dryer conditions favoured many Chenopodiaceae and led to the development of the present pre-adapted, Saharo-Arabian desert flora derived from the Mesogean stock.

Due to the relatively late separation of the Arabian Peninsula from the African continent in the Early Miocene, some 25 million years ago, the southwestern and southern mountainous part of Arabia exhibit close relationships with the Somalia-Masai regional zone and the Afromontane archipelago-like regional centre of endemism. The border of these phytochoria with the Saharo-Sindian phytochorion is frequently marked by the distribution of the palaeo-African *Acacia-Commiphora* woodlands. These strong floristic affinities are indicated by numerous mesic-African relicts and genera of probable Palaeo-African origin that today often have a disjunct distribution pattern, but which in earlier times had a much wider range. Even if the historical causes underlying disjunctions are complex, and long-distance dispersal may be responsible for some disjunctions, most of these taxa reflect Tertiary migration routes dating from before the formation of the Red Sea. Cribb (1979) has shown that 16 of the 18 known orchid species of southern Arabia are of African origin, whereas Asian links are represented only by *Epipactis veratrifolia*. Further mesic-African relicts are *Commicarpus grandiflorus, C. squarrosus, C. stenocarpus, Eulophia guineensis* var.

purpurata, Geranium mascatense, Habenaria cultrata, Vernonia arabica, V. areysiana, V. spathulata, and various grasses such as *Aristida migiurtina, A. stenophylla, Arthraxon hispidus, A. lancifolius, Eragrostis mahrana, Fingerhuthia africana, Hackelochloa granularis, Loudetia flavida, Rottboellia cochinchinensis, Sporobolus airiformis, Tripogon purpurascens* and *T. subtilissimus.*

Examples of taxa of Palaeo-African origin that have their closest relatives in Africa and a speciation centre in southern Arabia (local centres of endemism are Ibb, Hajayrah, Jebel 'Ureys, Jol, Dhofar and central and northern Oman, *cf.* Miller & Nyberg 1991 and section 12.2), and that often extend eastwards across the Arabian Gulf, are genera such as *Anogeissus* (*A. leiocarpa* in Africa and *A. benthii* and *A. dhofarica* in Arabia), *Campylanthus* (*C. salsoloides* in Macaronesia, *C. incanus* and *C. spinosus* in Africa, *C. junceus* in Africa and Arabia, *C. chascaniflorus*, *C. pungens*, *C. sedoides* and *C. yemensis* in Arabia and *C. ramosissimus* in Southwest Asia, Miller 1980), *Dorstenia* (*D. barnimina* and *D. foetida* in Africa and Arabia and *D. gigas* and *D. socotrana* in Socotra), *Dracaena* (*D. draco* in Macaronesia, *D. ombet* in Africa and Arabia and *D. cinnabaria* and *D. serrulata* in Arabia and Socotra), *Farsetia* (14 taxa in Africa, *F. longisiliqua* in Africa and Arabia, *F. burtonae*, *F. dhofarica*, *F. latifolia*, *F. linearis* and *F. socotrana* in Arabia and Socotra, *F. heliophila* in Arabia and Southwest Asia, *F. jacquemontii* and *F. macranthera* in Southwest Asia and *F. aegyptiaca* and *F. stylosa* in Africa, Arabia and Southwest Asia, Jonsell 1986), *Gymnocarpos* (*G. decandrus* in Macaronesia, Africa, Arabia and Southwest Asia, *G. parvibractus* in Africa, *G. argenteus*, *G. mahranus*, *G. dhofarensis* and *G. rotundifolius* in Arabia and *G. bracteatus* and *G. kuriensis* in the Socotra archipelago, Petrusson & Thulin 1966), *Maytenus* (13 taxa in Africa, *M. arbutifolia*, *M. senegalensis*, *M. undatus* in Africa and Arabia and *M. dhofarensis*, *M. forsskaolina* and *M. parviflora* in Arabia, Sebsebe 1985), *Ochradenus* (*O. randonioides* in Africa, *O. arabicus*, *O. harsusiticus* and *O. spartioides* in Arabia, *O. aucheri* in Arabia and Southwest Asia and *O. baccatus* in Africa, Arabia and Southwest Asia, Miller 1984b), and *Schweinfurthia* (*S. pedicellata* in Africa and Arabia, *S. imbricata*, *S. latifolia* and *S. spinosa* in Arabia, *S. papillionacea* in Arabia and Southwest Asia and *S. pterosperma* in Africa, Arabia and Southwest Asia, Miller, Short & Sutton 1982).

There are also a few taxa which show a reverse migration, i.e. from Arabia to Africa. The best known example is the Afromontane *Juniperus procera,* that originated from the northern Irano-Turanian *J. excelsa.* During the Miocene-Pliocene a pre-*J. excelsa* expanded southwards through the western mountains of Arabia to Ethiopia, and further along the East-African rift valley mountains as far south as Zimbabwe (the southernmost native population of any known juniper, Kerfoot 1961). This migration and the identity of these African junipers have recently been demonstrated by RAPD analysis (Adams *et al.* 1993).

Besides these distribution patterns, there are also some remarkable disjunctions between families, genera, species or subspecies in Arabia and those in the arid regions of southern Africa (De Winter 1971, Monod 1971, Werger 1978). Examples of such disjunctions that occur in Arabia and in the extreme deserts and semi-desert areas of Namibia and the Karoo (the Karoo-Namib region) belong to families such as the Neuradaceae, Salvadoraceae, Wellstediaceae (*Wellstedia socotrana* in East Africa and Socotra, *W. dinteri* in Southwest Africa), genera such as *Aizoon, Citrullus, Echidnopsis, Erodium, Fagonia, Forskaolea, Kissenia, Lotononis, Matthiola, Moringa, Oropetium, Stipagrostis, Tamarix, Tetrapogon, Thamnosma* and *Trigonella*, species such as *Commicarpus squarrosus, Corbichonia decumbens, Enneapogon desvauxii, Fingerhuthia africana, Hermannia modesta, Kissenia spathulata, Mesembryanthemum nodiflorum, Oligomeris linifolia, Polycarpon tetraphyllum, Schismus barbatus, Sterculia africana, Stipa capensis, Stipagrostis obtusa, S. uniplumis, Suaeda fruticosa*, the closely related species *Tricholaena teneriffae* (North Africa and Arabia) and *T. capensis* (South Africa) or the vicariant subspecies or varieties of *Cleome angustifolia* (subsp. *angustifolia* in Arabia, subsp. *petersiana* and subsp. *diandra* in southern Africa), *Stipagrostis ciliata* (var. *ciliata* in Arabia, var. *capensis* in southern Africa), and *S. hirtigluma* (var. *hirtigluma* in Arabia, var. *pearsonii*, var. *patula* in southern Africa). The present distributions of these taxa are interpreted as relicts of formerly continuous distributions. The two arid regions, Arabia and southern Africa, were probably connected in the past (either in the Tertiary or more recently) by an arid corridor (the arid tract of Balinsky 1962) stretching across Africa from Somalia via Kenya, Tanzania and Zambia to Botswana, southwest Africa and South Africa (Werger 1978 and references therein). Long-distance dispersal can be excluded as an explanation for these disjunctions since most of the taxa show well-developed achoric dispersal mechanisms.

Beside the strong African affinities expressed in various disjunct distribution patterns and in the presence of mesic-African relicts, relicts of a mesic-tropical Asian flora of the Tethys sea coasts from the Upper Cretaceous and the Palaeogene also occur in southern Arabia. Taxa and derivates of this tropical Asian relict flora are *Anogeissus dhofarica, Dyerophytum indicum, Epipactis veratrifolia, Leptorhabdos parviflora, Malvastrum coromandelianum, Monotheca buxifolia, Plectranthus rugosus, Tecomella undulata,* and grasses such as *Apluda mutica, Capillipedium parviflorum, Cymbopogon jwarancusa* subsp. *olivieri, Dimeria ornithopoda* and *Oplismenus burmannii.*

Most conspicuous amongst the taxa which show a reverse migration is *Monotheca buxifolia,* the main centre of which is the mountains of Afghanistan and northern Pakistan. After a gap of several hundred kilometres it occurs again in the Jebel al Akhdhar range of northern Oman, where it is the dominant species of the semi-evergreen sclerophyllous woodlands, and in southwest Saudi Arabia (near Zahran al Janub). Across the Red Sea it has scattered stands in Somalia (Mount Surud, al Madu range and the Gebis range), Djibouti (Mount Goudah) and Ethiopia.

These taxa of Palaeo-African, mesic-African and mesic-tropical Asian origin indicate the great importance of southern Arabia as a bridge on the west-east and reverse migration track and as a refuge area for xerotropical elements. During the Tertiary period, Asia, Arabia and Africa were connected by land bridges that allowed the migration of African and Asian taxa along the coasts of the Tethys sea. With the drier climate that developed during the Oligocene, a southward migration of these tropical elements occurred and only a few mesic and xeric derivatives of the Palaeo-African savanna element (*Acacia-Commiphora* flora, Zohary 1973) and the tropical Asian stock (Laurus element, Zohary 1973) could survive in special habitats. A later migration of the tropical Asian taxa during the Pleistocene via the dry Arabian Gulf can be excluded, since the present semi-arid to arid climate had already become established by this time. The present subtropical climatic conditions of southern Arabia has allowed the survival of these xeric and mesic-tropical relicts, and lead at the same time to the evolution of various endemics.

4.3 Main Vegetation Types and Indicator Plant Species

Based on geomorphological diversity, phytogeography and climate, the vegetated landscape of the Arabian Peninsula can be classified into five main types. These are: the coastal and sabkha vegetation, the relatively sparse vegetation of the deserts, dwarf shrublands and related communities, montane woodlands and xeromorphic shrublands, and wadi communities (Figure 4.1). This section provides an overview of the dominant plant formations and principal habitats of these vegetation types, each of which is dealt with in detail in subsequent chapters. This summary is based both on published works and personal knowledge from several excursions to Arabia. The following studies have been particularly useful: Al-Hubaishi and Müller-Hohenstein (1984), Baierle *et al.* (1985), Batanouny (1981), Chaudhary (1983), Deil (1986a), Deil and Müller-Hohenstein (1996), Frey and Kürschner (1986, 1989), Gabali and Al-Gifri (1991), Ghazanfar (1991a, 1992a), Hepper (1977), König (1987), Kürschner (1986a & b, 1997), Mandaville (1977, 1984, 1985, 1986, 1990), Mies and Zimmer (1993), Miller and Morris (1988), Miller and Cope (1996), Munton (1988), Popov (1957), Radcliffe-Smith (1980), Scholte *et al.* (1991), Vesey-Fitzgerald (1955, 1957a & b), Villwock (1991), White and Léonard (1991) and Zohary (1973).

4.3.1 Coastal Plain and Sabkha Vegetation

The vegetation of the inlets, islands, capes, bays, sabkhas, coastal plains and alluvial fans (Figure 4.1: 1) is one of the best studied vegetation types of the Arabian Peninsula (Aleem 1978, Babikir & Kürschner 1992, Batanouny 1981, Frey & Kürschner 1986, Frey *et al.* 1984, Frey *et al.* 1985, Ghazanfar in press-a, Halwagy & Halwagy 1977, Kürschner 1986c, Mahmoud *et al.* 1982, Younes *et al.* 1983, Zahran *et al.* 1983). These areas are dominated by mosaic-like communities of mangroves, halophytes, drought-deciduous thorn woodlands and open xeromorphic shrublands and grasslands.

One of the most conspicuous formations typical of the subtidal and intertidal zones of the coast are stands of mangrove, usually of *Avicennia marina* (Figure 4.6: 1), with *Rhizophora mucronata* occurring only sporadically in the southwest.

Mangroves extend to 25 °N on the Red Sea coast (in the Hanak area), with an isolated outpost in the Gulf of Aqaba on the Sinai coast, and to the coasts of Bahrain in protected bays and estuaries along the Arabian Gulf. On the coastal side of the mangroves there is often a series of distinctive and sharply zoned halophytic formations, consisting mainly of succulent chenopods (Figure 4.6: 2-3). The nature of the zonation and species composition of these formations depends on soil salinity, water-holding capacity and available water in the soil, height above sea-level and inundation by high tide, distance from the shore and texture of the deposits. Typical species, sometimes forming large and monospecific stands, are *Aeluropus lagopoides, Arthrocnemum macrostachyum* (Figure 4.6: 2), *Biennertia cycloptera, Cressa cretica, Halocnemum strobilaceum, Halopeplis perfoliata* (Figure 4.6: 2-3), *Juncus rigidus, Salicornia europaea, Salsola* spp., *Sporobolus arabicus, Suaeda* spp. and *Zygophyllum* spp.

The aeolian sands, alluvial wadi fans and foothills of the mountains are dominated by open, drought-deciduous thorn woodlands (Figure 4.6: 4) and shrublands, often intermixed with xeromorphic grasslands (the *Panicum turgidum* community). There are three types of thorn woodlands and shrublands in the coastal areas:

Firstly, a Mediterranean/Saharo-Sindian community type, restricted to the northwestern part of the Peninsula. Here, the sands and gravelly areas are dominated by *Acacia raddiana, A. tortilis* and *Retama raetam* (cf. Jebel al Lawz, Figure 4.7: 1a).

Secondly, a Sudanian community type, typical of the western Tihamah coastal plain and the southern coast, characterized by *Acacia* spp., *Balanites aegyptiaca, Maerua crassifolia*, and *Ziziphus spina-christi*. Examples are:

Jebel Radwa (up to 300 m) with *Acacia asak, A. ehrenbergiana, A. tortilis, Balanites aegyptiaca* and *Maerua crassifolia* (Figure 4.7: II).

Jebel Subh (up to 500 m) with *Acacia asak, A. ehrenbergiana, A. tortilis, Balanites aegyptiaca, Leptadenia pyrotechnica, Maerua crassifolia* and *Ziziphus spina-christi* (Figure 4.7: III).

Jiddah-Taif transect (up to 500 m) with *Acacia asak, A. ehrenbergiana, A. hamulosa, A. tortilis, Balanites aegyptiaca* and *Capparis* spp (Figure 4.7: IV).

Yemen escarpment mountains (al Luhayya to Ma'rib, up to 300 m) with *Acacia ehrenbergiana, A. tortilis, Balanites aegyptiaca* and *Dobera glabra* (Figure 4.7: VI).

Hadhramaut (al Mukalla to Jol plateau, up to 300 m) with *Acacia tortilis, Balanites aegyptiaca, Merua crassifolia* and *Ziziphus spina-christi* (Figure 4.7: VII)

Thirdly, an eastern community type, with Sudanian/Omano-Makranian taxa such as *Prosopis cineraria*. Examples are:

Jebel al Akhdhar (up to 450 m) with *Acacia ehrenbergiana, A. tortilis, Prosopis cineraria* and *Ziziphus spina-christi* (Figure 4.7: X).

Musandam (up to 250 m) with *Acacia ehrenbergiana, A. tortilis, Euphorbia larica, Grewia tenax* subsp. *makranica, Moringa peregrina, Prosopis cineraria* and *Ziziphus spina-christi* (Figure 4.7: XI)

Dubai-Musandam mountains (up to 250 m) with *Acacia ehrenbergiana, A. tortilis* and *Prosopis cineraria* (Figure 4.7: XII) (formerly open drought-deciduous thorn woodlands, but today mostly destroyed, Deil & Müller-Hohenstein 1996).

Open xeromorphic grasslands, dominated by the perennial tussock grass *Panicum turgidum,* are widely distributed in the eastern coastal lowlands east of the Summan limestone plateau from Kuwait to the Jafurah sands (Figure 4.6: 5, coastal white sand association, Vesey-Fitzgerald 1957a; *thumam* grass- shrubland, Mandaville 1990). These *Panicum* communities are typical of well-drained sands and are important grazing resources. They include woody associates such as *Calligonum comosum* (locally very abundant and dominating the sands), *Leptadenia pyrotechnica,* and *Lycium shawii*. Other associates are *Aristida mutabilis, Centropodia forsskalii, Cutandia memphitica, Cyperus conglomerates, Dipterygium glaucum, Fagonia indica, Monsonia nivea, Moltkiopsis ciliata, Polycarpaea repens* and the annual psammophytic *Plantago* spp.

4.3.2 Deserts and other Scarcely Vegetated Areas

The sand deserts (*ergs*) and the rock and gravel deserts (*harrahs* and *hamadas*) cover a greater proportion of the land surface of the Arabian Peninsula than any other landform type (Figure 4.1: 3-5). These areas are generally hyperarid, receiving <100 mm rainfall per year, and have a mosaic of sparse vegetation (*mode contracté*, Monod 1954) which is difficult to treat in a single physiognomic classification. Plant life is mostly confined to the depressions, wadis, runnels and rocky pavements that receive water from large catchment areas. In general, these areas are treeless and the communities are often composed of very few species.

4.3.2.1 Rock and Gravel Deserts

Western and central Arabia consists of crystalline basement rocks of the Arabian Shield covered by extensive lava flows (the *harrahs* of the Western Nejd region, Figure 4.6: 6), isolated inselbergs, Cambrian sandstones (the Hisma Range in the northwest), and the central cuesta region dominated by a series of low sandstone and limestone escarpments (Jebel Tuwayq, see Figures 3.3, 4.7: 7). In most of these regions all fine weathering products have been removed by wind, leaving a desertic region of diverse topography covered with rocks, gravel deposits and thin sands. Prominent amongst these is the sparsely vegetated western Nejd region (Harrat Nawasif, Harrat Rahat, Harrat Khaybar and Harrat Uwayrid).

The typical vegetation of these rock and gravel deserts (Figure 4.1: 3) are open, xeromorphic dwarf shrublands intermixed, after rain, with perennials and annuals. Towards the north and northwest in the Widyan al Hamad and Midian areas, these dwarf shrublands are influenced by the Irano-Turanian phytochorion, as indicated by the dominance of *Artemisia sieberi* and *Achillea fragrantissima.* Central Arabia, on the other hand, is dominated by various *Haloxylon salicornicum* communities of Saharo-Sindian origin. *Haloxylon salicornicum* is the most abundant species, associated with taxa such as *Astragalus spinosus, Fagonia bruguieri, Farsetia aegyptiaca, Gymnocarpos decandrus, Halothamnus bottae, Salsola cyclophylla* and *Stipagrostis* spp.

The rocky pavements of the cuesta regions are typified by a sparse vegetation of *Anastatica hierochuntica, Anvillea garcinii* and *Blepharis ciliaris,* and small grasses such as *Oropetium africanum, O. capense* var. *arabicum* and *Tripogon multiflorus.*

4.3.2.2 Sand Deserts

Extensive sand deserts of the Saharo-Sindian belt are typical of three main areas of the Arabian Peninsula (Figures 3.3, 3.6, 8.1, 4.1: 4-5): the Great Nafud (Figure 4.6: 8-9), the Rub' al Khali, and connecting both, the crescent-shaped Dahna sands.

The Rub' al Khali lies in one of the driest regions in Arabia, with an annual average total rainfall of *c.* 35 mm. Despite these hyperarid conditions, three plant communities can be found, forming very open desert shrublands (Mandaville 1986). The communities consist of *c.* 37 species and lack annuals. The *Calligonum crinitum* subsp. *arabicum* community (*abal* shrubland) is widespread in the western, southern and southeastern parts and replaces the more northerly distributed shrubs of *C. comosum.* Typical of the central and northeastern part is a *Cornulaca arabica* community (*hadh* saltbush shrubland), that is endemic to the Rub' al Khali and one of the most important vegetation types, covering thousands of square kilometres (Mandaville 1990). Associated with this, particularly on better drained sands, are *Calligonum crinitum* subsp. *arabicum, Cyperus conglomeratus, Dipterygium glaucum, Tribulus arabicus,* and *Limeum arabicum.* Towards the north and within the transition zone to the Dahna sands, this endemic formation is replaced by an *Haloxylon persicum* community (*ghada* shrublands), often forming islands of small patches or higher hummocks. This community is also typical of the eastern coastal lowlands (Figure 4.1: 2), characterizing deeper and less stable sand habitats.

The dominant plant community of the Dahna sands and parts of the Great Nafud is a *Calligonum comosum-Artemisia monosperma* community (*abal-adhir* sand shrublands). It consists of fairly wide-spaced shrubs of *Calligonum comosum* and *Artemisia monosperma,* intermixed with tussocks of perennial grasses such as *Centropodia forsskalii, Cyperus conglomeratus* and *Stipagrostis drarii.* After sufficient rainfall the northern parts of the Dahna sands supports a good cover of annuals such as *Anthemis scrobicularis, Cutandia memphitica, Eremobium aegyptiacum, Horwoodia dicksoniae, Monsonia nivea* and *Plantago* spp. *Scrophularia hypericifolia* is typical here, connecting the communities of the Dahna sands with those of the Great Nafud.

In contrast to the Rub' al Khali, the Great Nafud is much richer in species (*c.* 150 taxa, Chaudhary 1983) with a high proportion of annuals. After rain, the sands are often covered by a dense layer of psammophytic annuals, representing 40-75 % of the flora. Two plant communities can be distinguished, a *Haloxylon*

Figure 4.6 (1) Mangroves *(Avicennia marina)* near ad Dhakhira, Qatar. (2) Halophytic formations (sabkha vegetation) dominated by *Arthrocnemum macrostachyum* and *Halopeplis perfoliata* near Muscat (Qurm Nature Reserve, Oman). (3) *Halopeplis perfoliata,* a succulent halophytic chenopod typical of coastal sabkhas of the Arabian Peninsula.

Figure 4.6 cont'd. (4) Very open, drought-deciduous thorn woodlands dominated by *Acacia tortilis* in the coastal area of Qatar. (5) Open xeromorphic grasslands (the ***Panicum turgidum*** community) in the eastern lowlands of Saudi Arabia (the coastal white sand association).

Figure 4.6 cont'd. (6) Rock desert (harrah) in the western Nejd region, near Baha, Saudi Arabia. (7) Central cuesta region (limestone escarpments) of the Jebel Tuwayq mountain systems near Qariyat al Faw, Saudi Arabia.

Figure 4.6 cont'd. (8) Sand desert (Great Nafud) west of Hail, Saudi Arabia. (9) Sand desert (Irq Banban, Dahna sands) northeast of Riyadh, Saudi Arabia. (10) Sandy gravel plains with scattered *Acacia* shrublands northeast of Najran, Saudi Arabia.

Figure 4.6 cont'd. (11) Drought-deciduous *Acacia-Commiphora* thorn woodlands in the Asir mountains near al Farsha, Saudi Arabia. (12) Drought-deciduous *Acacia* thorn woodlands woodlands dominated by *Acacia origena* in the Asir mountains (Jebel Ashap) near Abha, Saudi Arabia.

Figure 4.6 cont'd. (13) Semi-evergreen sclerophyllous woodlands dominated by *Barbeya oleoides* in the Asir mountains of Saudi Arabia. (14) Monsoon-influenced evergreen needle-leaved *Juniperus procera* woodlands in the Asir mountains (Jebel Sawdah) near Abha, Saudi Arabia.

Figure 4.6 cont'd. (15) Xeromorphic succulent shrublands dominated by ***Euphorbia balsamifera*** subsp. ***adenensis*** on the east-facing slopes of the Asir mountains near Zahran al Janub, Saudi Arabia. (16) ***Dracaena ombet*** near al Farsha, Saudi Arabia). (17) ***Euphorbia ammak*** near Zahran al Janub, Saudi Arabia.

Figure 4.6 cont'd. (18) *Acacia tortilis* wadi in the Midian mountains (Jebel Shar) of Saudi Arabia. (19) *Hyphaene thebaica* wadi with surrounding *Acacia-Commiphora* woodlands in the Tihamah.

persicum-Artemisia monosperma-Stipagrostis drarii community which is typical of deeper sands and dunes, and a *Calligonum comosum-Artemisia monosperma-Scrophularia hypericifolia* community typical of sandy plains.

The sand deserts are surrounded by scarcely vegetated dunes, sandy flats, and sandy gravel plains (Figure 4.1: 5), which in general carry more plant life (Figure 4.6: 10). These areas form a transition zone between the open xeromorphic dwarf shrublands and rock deserts. Occasionally, in depressions or small wadis and runnels on the margins of the sands, isolated trees and shrubs, such as *Acacia* spp., *Calligonum* spp., *Prosopis cineraria* and *Ziziphus spina-christi,* occur, forming open woodlands in southeastern Oman (Munton 1988a, 1988b).

4.3.3 Dwarf Shrublands and Related Communities

Open xeromorphic dwarf shrubland and related communities are restricted to the northern (the Hisma Range and northern plains), northeastern (the Summan plateau and the coastal lowlands) and central parts (the cuesta region of the Jebel Tuwayq mountain systems) of the Peninsula (Figure 4.1: 6-7). Towards the north, they exhibit a more or less random distribution over the land surface (*mode diffus*, Monod 1954). Within these dwarf shrublands, two types can be distinguished, the open xeromorphic dwarf shrublands (Figure 4.1: 7) and the very open xeromorphic dwarf shrublands ('pseudosteppes', Figure 4.1: 6).

The first type occurs on sandy weathering products of the Cambrian sandstones, sandflats and gravel surfaces overlain with windblown sand, often forming a transition zone (Figure 4.1: 4-5). Indicator species here are *Calligonum comosum* and *Haloxylon salicornicum.*

The second type is a *Haloxylon salicornicum* and *Rhanterium eppaposum* community which occurs in the Summan plateau and the coastal lowlands. *Haloxylon salicornicum* (*rimth* saltbush shrublands) is restricted to more sandy areas, where the ground water is not far below the surface and there is an annual rainfall of 100-140 mm. This community is present from Kuwait to the edge of the Rub' al Khali, and on the Arabian Shield close to the plant formations of the *harrahs* (Figure 4.1: 3), where *Haloxylon salicornicum* dominates the moderately saline, sandy or clayey-sandy, depressions and plains. Associates here are *Acacia* spp., *Lycium shawii* and *Ziziphus nummularia.* The latter species often forms well-developed island-like stands in the marly-clayey basins of the flat Summan limestone plateau (*Ziziphus*-basins, Mandaville 1990), scattered through the *Haloxylon*-pseudosteppe.

The *Rhanterium eppaposum* community (*arfaj* shrublands) covers huge areas of the Arabian Platform, but does not extend far onto the Arabian Shield. This community is best developed in the northern plain region and Kuwait, but may still be found along the eastern edge of the Dahna sands, the coastal lowlands, southern Qatar and the hinterland of Dubai. In contrast to the *Haloxylon*-pseudosteppe it favours better drained soils on higher ground where bedrock is not far below the surface (Mandaville 1990) and rainfall is 75-100 mm per year.

The *Rhanterium* dwarf shrublands are of great importance as a grazing resource, and therefore it cannot be ascertained whether this community is a remnant of the natural climax or not.

Vast areas of northern Arabia, such as the al Qar'ah area of Kuwait, are without woody plants. The vegetation is largely restricted to ephemerals and annuals which grow after the winter rains and are dominated by *Stipa capensis* (*Stipa* steppe, Vesey-Fitzgerald 1957a). As with the *Rhanterium* pseudosteppe, the natural status of these formations is uncertain. Perhaps they represent the last stage in the degradation of an unknown vegetation type (Guest 1966).

4.3.4 Montane Woodlands and Xeromorphic Shrublands

Remnants of forests, dense woodlands and related plant formations are restricted to the mountainous regions of the Peninsula (the Midian, Hijaz and Asir mountains, the Yemen highlands, Jol plateau, Dhofar mountains, the Hajar range and Musandam) and to the Haggier mountains of Socotra (Figures 4.1: 8-13). The montane vegetation exhibits a distinct altitudinal zonation that is illustrated here by 12 (I-XII) transects (Figure 4.7, see Figure 4.1 for locations). The numbers used in the units of these transects correspond to those used for the main vegetation types in Figure 4.1, except for number 14 (Cold-deciduous woodlands of Irano-Turanian origin) which is not shown in Figure 4.1.

4.3.4.1 Drought-deciduous Thorn Woodlands and Shrublands

Drought-deciduous thorn woodlands and shrublands are typical of the foothills and lower to middle mountain slopes (Figure 4.1: 8). These constitute the *Acacia-Commiphora* woodlands that indicate the xero-tropical African influence in the southwestern and southern mountains of Arabia (Figure 4.6: 11-12, 18-19). Physiognomically and floristically they are clearly related to the communities of the Boscio-Commiphoretea of Sudan and Ethiopia (Deil & Müller-Hohenstein 1985, Knapp 1968). Today near habitation they are often destroyed and replaced by ruderal succulent communities. North of Makkah and in northern Oman the *Commiphora* spp. gradually disappear and are replaced by woodlands dominated by *Acacia* spp. However, isolated stands, even if floristically impoverished, still exist in the Hijaz mountains (Jebel Radwa and Jebel Subh, Figure 4.7: II-III). At higher altitudes *Acacia* woodlands and semi-evergreen, sclerophyllous woodlands replace the *Acacia-Commiphora* woodlands. Zonation and species composition is dependent both on altitude and geographical position.

Acacia-Commiphora *Woodlands of Lower Mountain Slopes.* Examples of the *Acacia-Commiphora* woodlands of lower mountain slopes (Figure 4.7: 8a) are:

Jebel Radwa (300-900 m): *Acacia ehrenbergiana*, *A. etbaica*, *A. hamulosa*, *Commiphora kataf*, *C. myrrha*, *Delonix elata* and *Moringa peregrina*.

Jebel Subh (500-1,000 m): *Acacia abyssinica, A. ehrenbergiana, A. etbaica, A. hamulosa, A. tortilis, Commiphora kataf* and *C. myrrha.*

Jiddah-Taif transect (500-1,200 m): *Acacia asak, A. etbaica, A. hamulosa, A. tortilis, Commiphora kataf, C. myrrha, Euphorbia cuneata, Grewia tenax, Hibiscus micranthus, Maytenus senegalensis* and *Melhania denhami.*

Jebel Sawdah (up to 400 m): *Acacia ehrenbergiana, A. hamulosa, A. mellifera, A. tortilis, Boscia arabica, Cadaba farinosa, C. longifolia, Commiphora gileadensis, C. kataf, C. myrrha, Dobera glabra, Euphorbia cuneata, E. triaculeata* and *Maerua crassifolia.*

Yemen escarpment mountains (al Luhayya to Ma'rib, 300-1,000 m): *Acacia abyssinica, A. asak, A. mellifera, A. tortilis, Adenium obesum, Aloe* spp., *Cadaba* spp., *Combretum molle, Commiphora gileadensis, C. kataf, C. myrrha, Delonix elata, Euphorbia* spp., *Maerua crassifolia* and *Phoenix reclinata* (lower slopes of Jebel Bura).

Hadhramaut (al Mukalla to Jol plateau, 300-1,000 (-1,800?) m): *Acacia ehrenbergiana, A. mellifera, Adenium* sp., *Barleria* spp., *Boswellia carteri, Commiphora habessinica, Maerua crassifolia* and *Sterculia africana.*

Dhofar mountains (Salalah to Thumrait, up to 500 m): *Acacia tortilis, Boscia arabica, Cadaba farinosa, Caesalpinia erianthera, Commiphora foliacea, C. gileadensis* and *C. habessinica.*

Socotra (400-800 m): *Boswellia armeero, Commiphora* spp., *Croton socotranus, Dendrosicyos socotranus, Dracaena cinnabari, Jatropha unicostata* and *Sterculia africana* var. *socotrana.*

Acacia-Commiphora *Woodlands of Upper Mountain Slopes.* Examples of the *Acacia-Commiphora* woodlands of upper mountain slopes (Figure 4.7: 8b) are:

Jebel Radwa (900-1,500 m): *Acacia etbaica, Moringa peregrina* and *Rhus tripartita.*

Jebel Sawdah (400-1,100 m): *Acacia asak, A. etbaica, A. tortilis, Commiphora gileadensis, C. myrrha, Dobera glabra, Euphorbia cuneata, Grewia villosa* and *Moringa peregrina.*

Yemen escarpment mountains (al Luhayya to Ma'rib, 1,000-1,600 m): *Acacia asak, A. mellifera, Cadia purpurea, Carissa edulis, Commiphora habessinica, C. kataf, C. myrrha, Cordia abyssinica, Grewia* spp. and *Terminalia brownii.*

In wadis: *Breonadia salicina, Cordia abyssinica, Ficus vasta* and *Mimusops schimperi.*

4.3.4.2 Drought-deciduous Thorn Woodlands

Pure stands of montane *Acacia* woodlands that lack *Commiphora* species are typical of the west-facing upper slopes of the Asir mountains and the east-facing transition zone that gradually slopes towards the interior dwarf shrublands and rock deserts (Figures 4.1, 4.7: 9). They have been studied in detail only in the Asir mountains (König 1987) but often form mixed stands which are difficult to separate (Figure 4.6: 12). The typical species on Jebel Sawdah (Figure 4.7: V) between 1,100-1,600 m are *Acacia asak, A. etbaica, Adenium obesum, Anisotes trisulcus, Barleria bispinosa, B. trispinosa, Cadia purpurea, Grewia mollis, G. tembensis, G. villosus, Maytenus* spp. and *Pyrostria* sp. In the Yemen highlands (Figure 4.7: VI) this *Acacia* belt is widely destroyed and replaced by terraced agriculture or mixed with the semi-evergreen sclerophyllous woodlands. The rain-shadow sites of the higher escarpments, plateaus and east-facing transition zone are dominated by *Acacia* spp. such as *A. asak, A. gerrardii* and *A. origena* that often form widely distributed thorn woodlands.

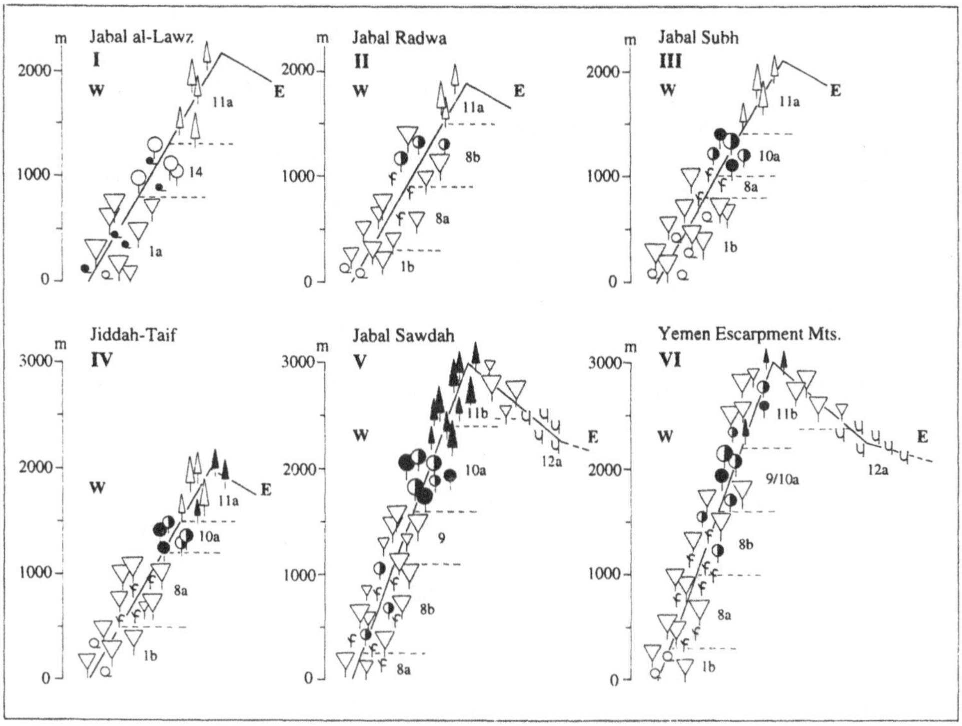

Figure 4.7. Montane woodlands and shrublands (transects I-VI) of the northern, western and southwestern escarpment mountains (sources: I-III original; IV after König 1987; V after König 1987 and Kürschner 1984b; VI after Al-Hubaishi & Müller-Hohenstein 1984 and Scholte *et al.* 1991). Numbers correspond to those used for vegetation types in Figure 4.2, except for 14 which is not shown in Figure 4.2. 1, Very open, drought-deciduous thorn woodlands of the coastal plains and lower foothills (a, Mediterranean/Saharo-Sindian type; b, Sudanian type; c, Sudanian/Omano-Makranian type). 8a, *Acacia-Commiphora* woodlands of lower mountain slopes. 8b, *Acacia-Commiphora* woodlands of upper mountain slopes. 10a, Semi-evergreen (sclerophyllous) woodlands, dominated by *Barbeya oleoides* and *Olea europaea* subsp. *africana*. 10b, Semi-evergreen (sclerophyllous) woodlands, dominated by the endemic *Anogeissus dhofarica*. 10c, Semi-evergreen (sclerophyllous) woodlands restricted to Socotra. 10d, Semi-evergreen (sclerophyllous) woodlands dominated by *Monotheca buxifolia*. 11a, Evergreen needle-leaved woodlands, dominated by the circum-Mediterranean *Juniperus phoenicea*. 11b, Evergreen needle-leaved woodlands dominated by the Afromontane *Juniperus procera*. 11c, Evergreen needle-leaved woodlands dominated by the Irano-Turanian *Juniperus excelsa* subsp. *polycarpos*. 12a, Mixed formations dominated by *Boswellia* spp., *Dracaena* spp., *Euphorbia ammak* and *E. balsamifera* subsp. *adenensis*. 12b, Mixed formations dominated by the endemic *Anogeissus dhofarica*. 12c, Mixed formations dominated by the endemic *Dendrosicyos socotranus*. 12d, Mixed formations dominated by the Omano-Makranian *Euphorbia larica*. 14, Cold-deciduous woodlands of Irano-Turanian origin.

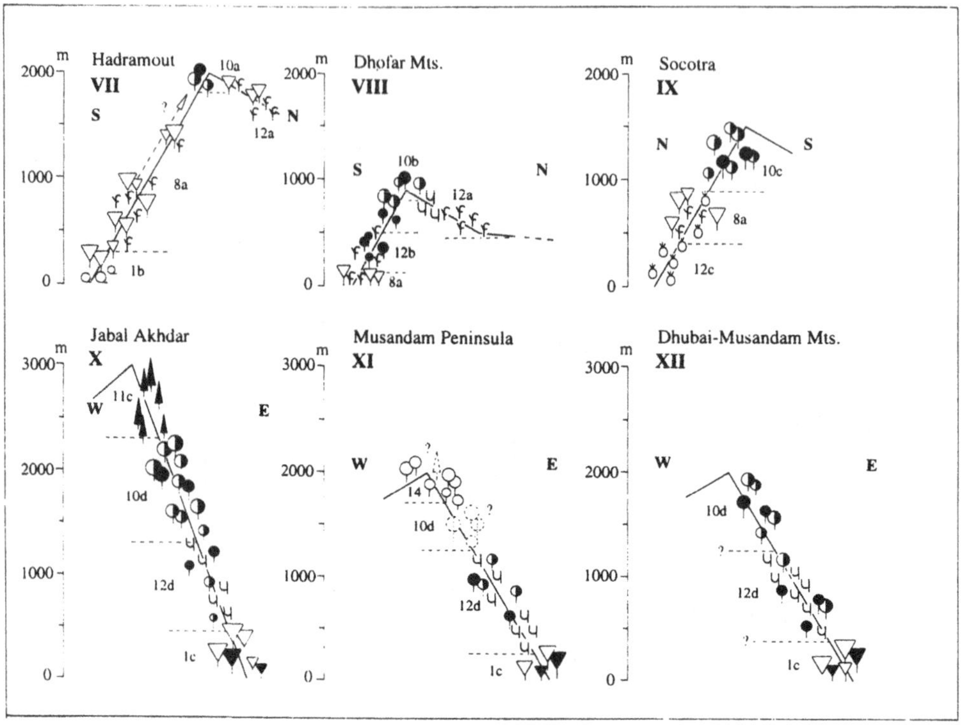

Figure 4.7 cont'd. Montane woodlands and shrublands (transects VII-XII) of the southern and southeastern escarpment mountains and Socotra (sources: VII after Gabali & Al-Gifri 1991 and Villwock 1991; VIII after Miller & Morris 1988 and Radcliffe-Smith 1980; IX after Mies & Zimmer 1993 and Popov 1957; X after Ghazanfar 1991b, Kürschner 1997 and Mandaville 1977; XI after Mandaville 1985; XII after Deil & Müller-Hohenstein 1996).

4.3.4.3 Semi-evergreen Sclerophyllous Woodlands and Sclerophyllous Scrub

The strong Afromontane affinities of southwestern and southern Arabia are further demonstrated by the semi-evergreen sclerophyllous woodlands and sclerophyllous scrub which form a narrow belt between the lower *Acacia-Commiphora* woodlands and the upper *Juniperus* woodlands (Figures 4.1: 10 & 4.7: 10). Characteristic species in western and southern Arabia are *Barbeya oleoides* (Figure 4.6: 13) and *Olea europaea* (*Olea chrysophylla* and *Olea europaea* subsp. *africana*), whereas the southeastern mountains of Jebel al Akhdhar are characterized by *Monotheca buxifolia*, a mesic tropical Asian relict that has strong affinities with the mountains of eastern Afghanistan (Kürschner 1986a). Towards the northwest, remnants of this formation can be found in the Hijaz mountains (Jebel Subh) southwest of Medinah.

Semi-evergreen Woodlands with Barbeya oleoides *and* Olea europaea. Examples of the semi-evergreen sclerophyllous woodlands, dominated by *Barbeya oleoides* and *Olea europaea* (Figure 4.7: 10a) are:

Jebel Subh (1,000-1,400 m): The northernmost remnants, with *Dodonaea viscosa, Olea europaea* subsp. *africana* and *Psiadia punctulata.*

Jiddah-Taif transect (1,200-1,500 m): *Acacia etbaica, Commiphora myrrha, Dodonaea viscosa, Grewia tenax, Olea europaea* subsp. *africana* and *Psiada punctulata.*

Jebel Sawdah (600-2,400 m): *Acokanthera schimperi, Barbeya oleoides, Buddleja polystachya, Carissa edulis, Celtis africana, Debregeasia saenab, Dodonaea viscosa, Dombeya schimperiana, Euclea racemosa* subsp. *schimperi, Jasminum floribundum, Maesa lanceolata, Nuxia congesta, Olea europaea, Pistacia falcata, Rhus abyssinica, R. retinorrhoea, Tarchonanthus camphoratus* and *Teclea nobilis.*

Yemen escarpment mountains (al Luhayya to Ma'rib, 1,600-2,200 m, mixed with drought-deciduous thorn woodlands): *Acacia abyssinica, A. gerrardii, A. origena, Barbeya oleoides, Buddleja polystachya, Dodonaea viscosa, Ehretia abyssinica, Euclea racemosa* subsp. *schimperi,* arborescent *Euphorbia* spp., *Olea europaea* and *Rhus abyssinica.*

Hadhramaut (al Mukalla to Jol plateau, 1,800-2,100 m): *Anogeissus benthii, Dodonaea viscosa, Euphorbia ammak, E. balsamifera* subsp. *adenensis, Grewia* spp., *Olea europaea* subsp. *africana* and *Sterculia africana.*

Semi-evergreen Woodlands with Anogeissus dhofarica. Semi-evergreen sclerophyllous woodlands dominated by the endemic *Anogeissus dhofarica* (Figure 4.7: 10b) are restricted to the seaward-facing escarpments of the Dhofar mountains (Salalah to Thumrait transect) at 500-900 m. Associated species are *Carissa edulis, Dodonaea viscosa, Euclea racemosa* subsp. *schimperi, Olea europaea* and *Rhus somalensis.*

Semi-evergreen Woodlands Restricted to Socotra. These semi-evergreen sclerophyllous woodlands are typical of the upper slopes of the Haggier mountains of Socotra (>800-1,500 m) (Figure 4.7: 10c). Dominant species are *Boswellia armeero, Buxus hildebrandtii, Carphalea obovatus, Cephalonoton socotranus, Cocculus balfourii, Croton* spp., *Euryops socotrana, Gnidia socotrana, Hypericum mysorense, H. scopulorum* and *Sideroxylon fimbriatum.*

Semi-evergreen Woodlands Dominated by Monotheca buxifolia. Examples of the semi-evergreen sclerophyllous woodlands dominated by *Monotheca buxifolia* (Figure 4.7: 10d) are:

Jebel al Akhdhar (1,350-2,350 m): *Dodonaea viscosa, Ebenus stellatus, Grewia villosa, Juniperus excelsa* subsp. *polycarpos, Monotheca buxifolia, Myrtus communis, Olea europaea* subsp. *africana* and *Sageretia spiciflora.*

Musandam: Today only *Dodonaea viscosa* is typical of this belt, but perhaps remnants of the formation may still occur.

Dubai-Musandam mountains: Relicts of this formation occur in the Hajar mountains, consisting of *Dodonaea viscosa, Monotheca buxifolia* and *Moringa peregrina.*

4.3.4.4 Evergreen Needle-leaved Woodlands

Well-developed, evergreen needle-leaved woodlands exist only in the Asir mountains of Saudi Arabia (*Juniperus procera* woodlands) and the Jebel al Akhdhar range of northern Oman (*Juniperus excelsa* subsp. *polycarpos* woodlands) (Figures 4.1: 11, 4.7: 11). In the Yemen escarpment mountains these woodlands have been completely destroyed and replaced by terraced agriculture and isolated stands of various *Acacia* spp. The *Juniperus* woodlands are best developed in areas with an annual rainfall of 300-600 mm, as in Jebel Sawdah and Jebel Ashap northwest of Abha, where stands with epiphytic bryophytes and lichens can be observed (Kürschner 1984a, 1984b) (Figure 4.6: 14). In the rain-shadow areas of plateau sites and east-facing slopes the *Juniperus procera* woodlands are often mixed with *Acacia* woodlands dominated by *A. gerrardii* and *A. origena*.

In northwestern Arabia (the Hijaz and Midian mountains) the *Juniperus procera* woodlands are replaced by scattered and isolated stands of the circum-Mediterranean *Juniperus phoenicea*, which reaches its southernmost distribution near Taif. Relict stands on the Sinai Peninsula, fossil finds from the Negev highlands and remnant woodlands in the Edom highlands of Jordan, indicate a formerly wider Mediterranean influence.

Evergreen Needle-leaved Woodlands Dominated by Juniperus phoenicea. Examples of evergreen needle-leaved woodlands dominated by the circum-Mediterranean *Juniperus phoenicea* (Figure 4.7: 11a) are:

Jebel al Lawz (1,300-2,200 m): *Juniperus phoenicea* stands with a dense ground layer of Irano-Turanian taxa such as *Achillea santolinoides, Artemisia sieberi* and *Astracantha echina* subsp. *arabica*.

Jebel Radwa (1,500-1,900 m): scattered stands of *Juniperus phoenicea* in rock desert.

Jebel Subh (above 1,400 m): scattered stands of *Juniperus phoenicea* in rock desert.

Jiddah-Taif transect (1,500-2,000 m): *Acacia gerrardii, Dodonaea viscosa, Euryops arabicus, Juniperus phoenicea,* partly mixed with *J. procera, Psiadia punctulata, Rhamnus staddo* and *Rhus retinorrhoea*.

Evergreen Needle-leaved Woodlands Dominated by Juniperus procera. Examples of the evergreen needle-leaved woodlands dominated by the Afromontane *Juniperus procera* (Figure 4.7: 11b) are:

Jebel Sawdah (above 2,400 m): *Acacia gerrardii, A. origena, Dodonaea viscosa, Erica arborea, Euryops arabicus, Grewia mollis, Hypericum revolutum, Juniperus procera, Nuxia congesta, Rhamnus staddo, Rosa abyssinica* and *Sageretia thea*.

Yemen escarpment mountains (al Luhayya to Ma'rib, 2,200-2,700(-3700) m): Replaced mainly by terraced agriculture and fragments of semi-evergreen sclerophyllous and drought-deciduous thorn woodlands (*Acacia gerrardii, A origena, Buddleja polystachya, Carissa edulis, Cordia abyssinica, Ehretia abyssinica, Grewia mollis, Hypericum revolutum, Juniperus procera, Myrsine africana, Nuxia congesta* and *Rosa abyssinica*).

Evergreen Needle-leaved Woodlands of Juniperus excelsa *subsp.* polycarpos. Examples of evergreen needle-leaved woodlands dominated by the Irano-Turanian *Juniperus excelsa* subsp. *polycarpos* (Figure 4.7: 11c) are:

Jebel al Akhdhar (2,300-3,000 m): *Cotoneaster nummularia, Daphne mucronata, Ephedra pachyclada, Euryops arabicus, Juniperus excelsa* subsp. *polycarpos, Lonicera aucheri, Periploca aphylla* and *Sageretia spiciflora.*

4.3.4.5 Mixed Shrubland and Grassland Formations

These mixed formations consist of drought-deciduous and xeromorphic succulent shrublands and xeromorphic open grasslands. Typical of the dry, interior north- and east-facing slopes of the mountains and often mixed with very open *Acacia* and *Acacia-Commiphora* woodlands are succulent communities dominated by members of the Euphorbiaceae (Figure 4.6: 15-17). They occur as a more or less well developed belt, extending from the southern Asir and Yemeni highlands east to Dhofar in southern Oman (Figures 4.1: 12-13, 4.7: 12-13). Communities with a similar physiognomy but different taxa can be found again on the lower slopes of the Jebel al Akhdhar range, Musandam and Socotra.

Formations Dominated by Boswellia *spp.,* Dracaena *spp. and* Euphorbia *spp.* Examples of mixed formations dominated by *Boswellia* spp., *Dracaena* spp., *Euphorbia ammak* and *E. balsamifera* subsp. *adenensis* (Figures 4.6: 15-17, Figure 4.7: 12a) are:

Jebel Sawdah (east-facing slopes, 2,450-2,200 m): *Acacia oerfota, A. gerrardii, Dracaena ombet, Euphorbia ammak, E. balsamifera* subsp. *adenensis* and *E. schimperiana.*

Yemen escarpment mountains (al Luhayya to Ma'rib, east-facing slopes down to 1,300 m): *Acacia etbaica, A gerrardii, A. oerfota, Dracaena serrulata, Euphorbia balsamifera* subsp. *adenensis* and *Kleinia odora.*

Hadhramaut (al Mukalla to Jol plateau, plateau sites and north-facing slopes): *Acacia asak, A. arabica, A. etbaica, A. mellifera, Boswellia carteri* and *Maerua crassifolia.*

Dhofar mountains (Salalah to Thumrait, north-facing slopes, 900-400 m): *Acacia etbaica, Barleria* spp., *Boswellia sacra, Commiphora habessinica, Dracaena serrulata, Euphorbia balsamifera* subsp. *adenensis* and *Maerua crassifolia.*

Formations Dominated by the Endemic Anogeissus dhofarica. Examples of mixed formations dominated by the endemic *Anogeissus dhofarica* (Figure 4.7: 12b) are:

Dhofar mountains (Salalah to Thumrait, seaward-facing escarpment, 50-500 m): *Acacia senegal, Anogeissus dhofarica, Commiphora* spp., *Croton confertus, Euphorbia smithii, Maytenus dhofarensis* and *Sterculia africana.*

Formations Dominated by the Endemic Dendrosicyos socotranus. Examples of mixed formations dominated by the endemic *Dendrosicyos socotranus* (Figure 4.7: 12c) are:

Socotra, up to 400 m: *Adenium socotranum, Croton socotranus, Cissus subaphylla, Dendrosicyos socotranus, Euphorbia arbuscula, E. septemsulca, E. spiralis, Jatropha unicostata* and *Trichocalyx* spp.

Formations Dominated by the Omano-Makranian Euphorbia larica. Examples of mixed formations dominated by the Omano-Makranian *Euphorbia larica* (Figure 4.7: 12d) are:

Jebel al Akhdhar (450-1,300 m): *Convolvulus acanthocladus, Euphorbia larica, Grewia tenax* subsp. *makranica, Jaubertia aucheri, Maerua crassifolia, Moringa peregrina* and *Pseudogaillonia hymenostephana*.
Musandam (500-1,400 m): *Convolvulus acanthocladus* and *Euphorbia larica*.
Dubai-Musandam mountains (lower slopes up to 1,000 m): *Euphorbia larica, Jaubertia aucheri* and *Pseudogaillonia hymenostephana*.

4.3.4.6 Cold-deciduous Woodlands

Cold-deciduous woodlands dominated by Irano-Turanian *Prunus* species (*P. arabica* and *P. korshinskyi*) and *Pistacia* species are confined almost exclusively to the higher mountains of the northwestern (Jebel al Lawz) and southeastern (Musandam) part of the Arabian Peninsula (Figure 4.7: 14). They probably indicate a former pre-Pleistocene migration of these elements from the Kurdo-Zagrosian mountains into Arabia. Today *Prunus arabica* and *Pistacia khinjuk* form open steppe forests together with *Pistacia atlantica* in the central Negev and highlands of Edom, reaching the southernmost limits of their range in the Midian mountains. Examples are:

Jebel al Lawz (800-1,300 m): *Prunus korshinskyi, Astracantha echina* subsp. *arabica, Pistacia atlantica, P. khinjuk, Retama raetam* and *Rhus tripartita*.
Musandam (1,300-) 1,800-2,000 m: *Prunus arabica, Ephedra pachyclada,* and a dense ground layer of *Artemisia sieberi*.

4.3.5 WADI COMMUNITIES

Seasonal water flows or wadis are one of the most common and important landscape elements of the Arabian Peninsula, draining wide catchment areas and high mountains by well developed tributaries, ravines and runnels. They are bordered by plant communities of an azonal vegetation type, mainly dependent on the water regime, and have played an important role in both the ancient and recent history of tribes and settlements of the Peninsula. The vegetation pattern of wadis is determined by the drainage system, the transport and texture of the sediments, depth of the water table, the frequency of overflows and the variability of rainfall. Although they have not been studied in detail, three main types of wadis can be distinguished on the Arabian Peninsula, indicating different phytogeographical relationships.

4.3.5.1 Wadi Communities Dominated by Taxa of Saharo-Sindian Origin

Wadi communities in arid to semi-arid, extra-tropical climates are dominated by taxa of Saharo-Sindian origin. Such wadis are typical of most of the central part of Arabia, bordering the large wadi systems (e.g. Wadi ar Rimah, Wadi al Batin, Wadi Hanifa, Wadi as Sabha, Wadi ad Dawasir and Wadi al Hamd). They are dominated by *Acacia gerrardii, A. raddiana, A. tortilis* (Figure 4.6: 18), with *Astragalus spinosus, Chrysopogon plumulosus, Cymbopogon commutatus, Lasiurus scindicus, Lycium shawii, Pennisetum* spp., *Pituranthos triradiatus* and *Rhazya stricta* as the main associates.

4.3.5.2 Wadi Communities with Mediterranean and Irano-Turanian Taxa

Wadi communities under a less arid climate are vegetated by taxa of Mediterranean and Irano-Turanian origin. These wadis are mainly restricted to the northwestern, northern and southeastern part of Arabia. Typical of the narrow wadi outlets of the Jebel al Akhdhar range are shrubs such as *Dyerophytum indicum, Nerium oleander* (*N. mascatense*) and *Pteropyrum scoparium.*

4.3.5.3 Wadi Communities with Sudanian and Xero-tropical African Taxa

Wadi communities in subtropical climates are dominated by taxa of Sudanian and xero-tropical African origin, of which five types can be identified:

Acacia ehrenbergiana-Salvadora persica Wadis. Typical of the western coastal plain (Tihamah) with its extensive areas of fluvial deposits and aeolian sands. *Leptadenia pyrotechnica* is an important associate in places where wind plays a role in the deposition of sediments.

Hyphaene thebaica Wadis. Widely distributed in the southwest of the Tihamah (Figure 4.6: 19). Associates are *Acacia* spp., *Jatropha* spp., *Hyparrhenia hirta, Panicum turgidum* and *Tamarix* spp. Often *Cissus* spp., *Cocculus hirsuta* and *Cassythia filiformis* form dense thickets on the wadi border ('Schleier-Synusie').

Prosopis cineraria-Ziziphus spina-christi Wadis. Typical of the coastal lowlands of the Batinah coast of northern Oman. Associates are *Acacia ehrenbergiana* and *A. tortilis.*

Wadis dominated by *Tamarix* spp., *Desmostachya bipinnata* and *Saccharum* spp. Typical of the mountainous Tihamah.

Mesophytic Riverine Forests. Within the mountainous Tihamah and the steep wadis of the southwestern escarpments, mesophytic wadi communities are frequent, dominated by *Breonadia salicina, Berchemia discolor, Combretum molle, Cordia ovalis, Dobera glabra, Ficus lutea, F. cordata* subsp. *salicifolia, F. sycomorus, Moringa peregrina, Nuxia oppositifolia, Phoenix reclinata* and *Tamarindus indica.*

4.4 Summary

The vegetation of the Arabian Peninsula includes more or less dense evergreen and deciduous woodlands, sclerophyllous and xeromorphic shrublands, dwarf shrublands, wadi communities and the sparse vegetation of the rock, gravel and sand deserts. There is a relatively rich montane vegetation which exhibits a distinct altitudinal zonation. These different vegetation types are dominated by taxa of various historical origins, constituting three main phytogeographical units: the Saharo-Sindian regional zone, including an Arabian regional subzone and a Nubo-Sindian local centre of endemism, the Somalia-Masai regional centre of endemism, and the Afromontane archipelago-like centre of endemism. Whereas the major part of extra-tropical Arabia belongs to the Arabian regional subzone, a more or less well defined entity with an autonomous stock of Saharo-Arabian xerophytes was derived from an ancient Mesogean floral stock. The southwestern and southern parts are strongly influenced by a xero-mesic flora of Palaeotropical origin, indicating the strong floristic relationships between Arabia and Africa. Of special interest are the higher parts of the escarpment mountains, which harbour Afromontane vegetation floristically and ecologically comparable to that of the high mountains of East Africa. Towards the north and southeast, marginal intruders of Mediterranean and Irano-Turanian origin occur, indicating former migration routes.

Six main vegetation types can be distinguished on the Arabian Peninsula. These are: 1) The coastal and sabkha vegetation, which consists of *Avicennia marina*, *Arthrocnemum* and *Atriplex* spp. in the intertidal zone, and open drought-deciduous thorn woodlands and shrublands mixed with xeromorphic grasslands on the coastal plains; prominent species are those of *Acacia*, *Commiphora*, *Maerua* and *Ziziphus*. 2) The gravel desert and other scarcely vegetated areas where the vegetation consists of open xeromorphic dwarf shrublands; dominant species are *Haloxylon salicornicum*, *Artemisia* and *Achillea*. 3) The sand deserts, where the vegetation is poor in species and consists of a few common shrubs, notably *Calligonum*, *Cornulaca* and *Artemisia*. 4) The northern plains and the northern coastal lowlands which consist of a very open xeromorphic shrubland, dominated by *Haloxylon salicornicum* and *Rhanterium epapposum*. 5) The montane woodlands and xeromorphic shrublands which occupy the mountainous regions and consists of mixed *Acacia-Commiphora* woodlands at low altitudes and open *Juniperus* woodlands at higher altitudes. Wadi vegetation is an azonal type dependent on the water regime, and is highly influenced by man and his livestock.

Chapter 5

Bryophytes and Lichens

Harald Kürschner[1] *& Shahina A Ghazanfar*[2]

"This is a tale of ignorance; my own and other people's. Knowledge of desert bryophytes is so incomplete, and the publications so fragmentary that I have had to rely largely on my own observations."

Scott (1982)

5.1 Bryophytes

5.1.1 Introduction

Bryophytes are a group of plants that appear to have evolved along a pathway separate from that of the remainder of the plant kingdom, and to have been an early evolutionary 'branch' from ancestral chlorophytes (green algae). These 'primitive' plants have no highly differentiated tissues for water conduction or mineral transport, and they absorb nutrients, water and carbon dioxide directly through their cell walls. They can survive under very varied environmental conditions, and often form a striking part of the vegetation in tundra, forests, wetlands, peatlands and the humid tropics. Some species of mosses and liverworts can recover after many years of nearly complete dehydration, a characteristic that enables them to form an important, though scientifically neglected, part of the vegetation of arid and semiarid environments.

For a long time the Arabian Peninsula was a bryological 'Empty Quarter', with the group being poorly represented in collections from the area. Apart from two

---

Sections [1]5.1 Bryophytes, [2]5.2 Lichens

S. A. Ghazanfar and M. Fisher (eds.), Vegetation of the Arabian Peninsula, 99-124.

collections made by Forsskål (1775) belonging to the genera *Mnium* and *Bryum*, a collection of *Plagiochasma rupestre* from northern Yemen (Savulescu 1928), and a *Splachnobryum arabicum* from Oman (Thériot *et al.* 1934) there appear to have been no early records for the mainland of the Peninsula. The first bryophytes from Saudi Arabia were not recorded until 1982 (Frey & Kürschner 1982). Only the island of Socotra yielded some early records, with several mosses and liverworts recorded by Mitten (1888, collections of the Balfour expedition), and Müller (1900, collections of G. Schweinfurth on the Riebeck expedition).

Extensive bryological exploration of the Arabian Peninsula during the last 15 years has indicated that there are a number of well adapted, xerophytic taxa that appear to have evolved under similar ecological conditions. Today the known number of species consist of one species of hornwort (Anthocerotopsida), 48 liverworts (Hepaticopsida) and 150 mosses (Bryopsida) (Table 5.1). No species have so far been recorded from Bahrain or Qatar.

Table 5.1 Number of bryophyte species in the Arabian Peninsula and Socotra (from Al-Gifri & Kürschner 1996, Al-Gifri *et al.* 1995, Frey & Kürschner 1991a, Kürschner 1996a, 1996b).

Country	Anthocerotopsida (hornworts)	Hepaticopsida (liverworts)	Bryopsida (mosses)
Bahrain	0	0	0
Kuwait	0	0	21
Oman	0	14	30
Qatar	0	0	0
Saudi Arabia	0	27	110
United Arab Emirates	0	2	3
Yemen, north	1	26	49
Yemen, south	0	7	18
Yemen, Socotra	0	11	25

5.1.2 Phytogeography

As a result of varied terrains, habitats, ecological conditions and palaeogeographical history, the bryophyte flora of Arabia is moderately heterogeneous. It consists of six floristic elements (*sensu* Frey & Kürschner 1988) that are derived from different ancestral floral stocks, reflecting earlier processes in palaeogeography and migration.

Several thallose Marchantiales (e.g. *Exormotheca pustulosa,* Figure 5.8: c, *Oxymitria paleacea, Plagiochasma rupestre,* Figure 5.8: b, *Riccia lamellosa* and *Targionia hypophylla*) and well adapted Pottiaceae (*Aloina rigida, Crossidium davidai, Pterygoneurum ovatum, Pseudocrossidium replicatum, Tortula atrovirens* and *Trichostomopsis australasiae*) that originated from an ancient Permo-Triassic floral stock (xerothermic Pangaean ancestors, Frey & Kürschner 1988) are typical of the arid central region and also form the main components of the semiarid and arid bryophyte communities which characterize the limestone cuestas of Jebel Tuwayq

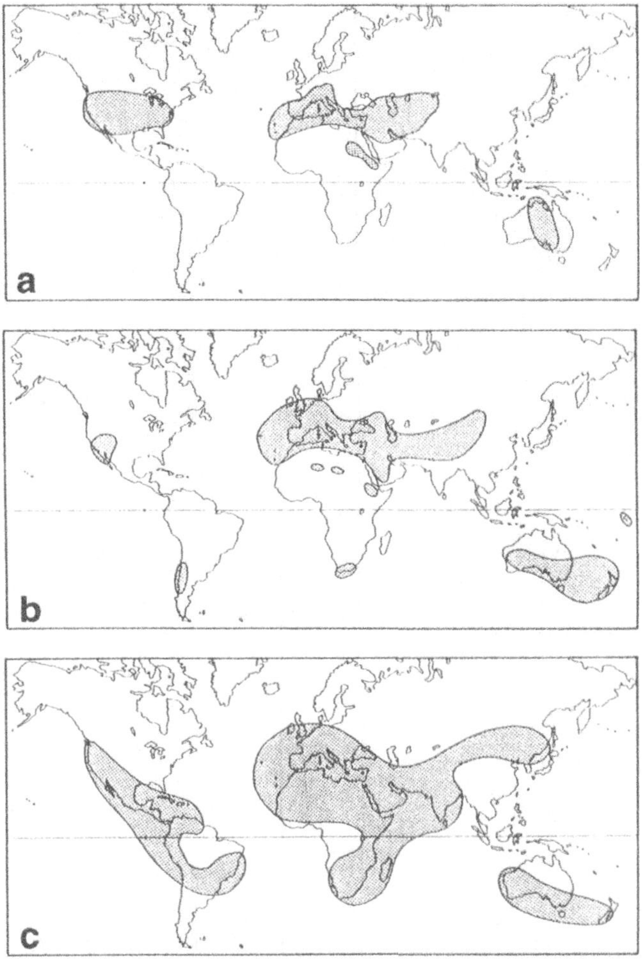

Figure 5.1 Distribution of some xerothermic Pangaean taxa in the Arabian Peninsula. (a) *Riccia lamellosa*. (b) *Tortula atrovirens*. (c) *Targionia hypophylla* (after Frey & Kürschner 1988).

and the Jol plateau in the Hadhramaut. Today, these xerothermic Pangaean elements show a very disjunct distribution in the deserts of the Northern and Southern hemispheres (Figure 5.1), corresponding to that of the Permo-Triassic continental Pangaea. They frequently grow together with other thallose, marchantioid and riccioid liverworts and acrocarpous mosses that are mostly of Mesogean (Old Mediterranean, formerly distributed along the northern and southern coasts of the Tethys sea) or xerothermic Pangaean origin. Today these taxa dominate in the three xerothermic regions of the Holarctic (Mediterranean, Saharo-Sindian and Irano-Turanian region) and the arid northern American region. Taxa of this circum-Tethyan element in Arabia are *Athalamia spathysii, Fossombronia caespitiformis, Mannia androgyna, Riccia atromarginata, R. trabutiana*

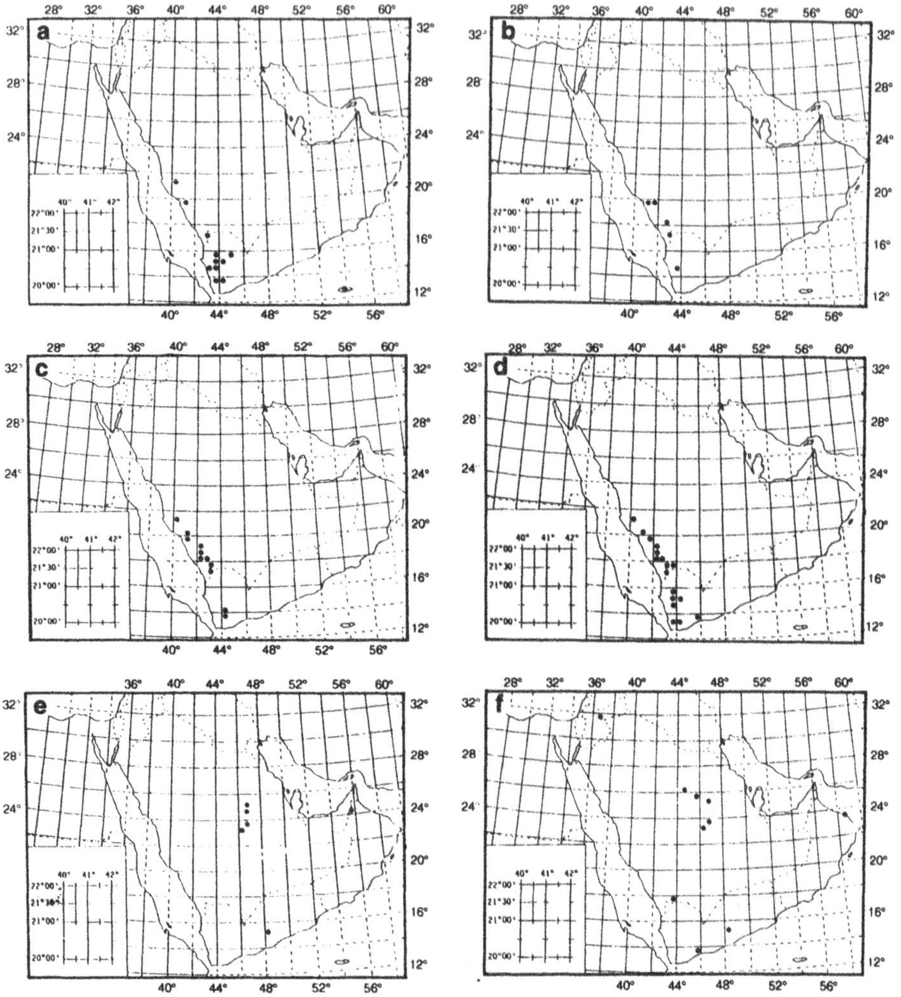

Figure 5.2 Distribution of Palaeotropical (a, b), Afromontane (c, d), and endemic taxa (e, f) on the Arabian Peninsula. (a) *Plagiochasma appendiculatum*. (b) *Tortula porphyreoneura*. (c) *Pseudoleskea leikipiae*. (d) *Tortula fragilis*. (e) *Crossidium laxefilamentosum*. (f) *Tortula mucronifera*.

(Hepaticopsida), *Barbula trifaria* var. *desertorum*, *Crossidium aberrans*, *C. crassinerve*, *C. squamiferum*, *Encalypta intermedia*, *Fissidens arnoldii*, *Hydrogonium ehrenbergii*, *Pleurochaete squarrosa*, *Timmiella barbuloides*, *Tortula brevissima*, *T. inermis* and *Trichostomopsis aaronis* (Bryopsida) and the two pleurocarpous mosses *Fabronia ciliaris* and *Scorpiurium circinatum*.

Within the monsoon-influenced, seasonally more humid western, southwestern and southern escarpment mountains and on Socotra, a conspicuous tropical floral element of xeric and mesic palaeotropical and Afromontane origin occurs. This flora was derived by migration through the East-African mountain chains, thus linking southwest Arabia and Socotra with the African mainland (the northern part of the Somalia-Masai and Afromontane archipelago-like regional centre of endemism *sensu* White & Léonard 1991; see Figure 4.1).

Taxa of the Palaeotropical range in Arabia are *Asterella pappii, Cyathodium africanum, Fossombronia crispa, Frullania ericoides* (pantropical), *Plagiochasma beccarianum, P. eximium, P. microcephalum, Riccia congoana, R. okahandjana* (Hepaticopsida), *Campylopus pilifer* (pantropical), *Fissidens schmidii, F. sylvaticus, Hydrogonium afrofontanum, Hyophila punctulata, Micropoma niloticum, Tortula fragilis* (Figure 5.2: d)and *T. porphyreoneura* (Bryopsida, Figure 5.2). These are typical of the lower elevations of the Tihamah, Hadhramaut and Dhofar, and associated with the drought-deciduous *Acacia-Commiphora* woodlands and the wadis dominated by broadleaved riverine *Combretum molle-Dobera glabra* woodlands.

The strong African character of the southwest Arabian bryophytes is indicated by Afromontane taxa such as *Chonecolea ruwenzorensis, Frullania caffraria, F. obscurifolia, F. trinervis, Lejeunea aethiopica, L. capensis, Plagiochila fusifera, P. squamulosa* (Hepaticopsida), *Fabronia abyssinica, Leucodon dracaenae* var. *schweinfurthii, Macrocoma abyssinicum, Pleurochaete malacophylla, Pseudoleskea leikipiae* (Figure 5.2: c), *P. plagiostoma* and *Racopilum capense* (Bryopsida). Most of these foliose liverworts and pleurocarpous mosses grow as epiphytes in the upper montane forests, and demonstrate both by their life-forms and systematic affinities the relatively young age of this Afromontane element.

The endemic species are of special interest from the systematic and phytogeographical point of view. 18 endemic bryophytes have been recorded so far, belonging to various families (Hepaticopsida: Frullaniaceae, Ricciaceae and Targioniaceae; Bryopsida: Bryaceae, Fabroniaceae, Fissidentaceae, Orthotrichaceae, Pottiaceae, Sematophyllaceae and Splachnaceae). Within the pottioid genera in particular, there is a complex of evolving taxa of circum-Tethyan and xerothermic-Pangaean origin that are restricted to the arid and hyper-arid regions of central Arabia. These central Arabian endemics consist of *Crossidium deserti, C. laxefilamentosum* (Figure 5.2: e), *Tortula mucronifera* (Figure 5.2: f), *Bryum nanoapiculatum, Fabronia socotrana, Fissidens arabicus, F. laxetexturatus,* and *Splachnobryum arabicum* (Bryopsida), and *Riccia crenatodentata* and *Targionia hypophylla* subsp. *linealis* (Hepaticopsida).

As with the angiosperm flora (see Chapters 4 and 12), Socotra has the highest percentage of endemic species (20%), with a number of neoendemics and Afro-Arabian palaeoendemics (i.e. relict species). This is due to Socotra's isolated position and the early Tertiary separation of the island from the African mainland. Taxa restricted to Socotra are *Frullania socotrana* (Hepaticopsida), *Barbula schweinfurthiana, Schlotheimia balfourii,* the recently discovered *Sematophyllum socotrense, Tortella smithii, Weissia artocosana* (=*Symblepharis socotrana*) and *W. socotrana* (Bryopsida).

There is also in Arabia a northern biogeographical element of sub-cosmopolitan bryophyte taxa. Within this element there are taxa of boreal, temperate and sub-Mediterranean - sub-Atlantic distribution that show a former Laurasian distribution pattern. They are concentrated in the north (e.g. Jebel Aja near Hail), in the escarpment woodlands of the northwest, west, and southwest (the Midian, Hijaz and Asir mountains of Saudi Arabia, and the Haraz mountains of Yemen) and in the southeast (Jebel al Akhdhar in Oman). This distribution pattern indicates that there was a former western migration route to the mountains of the Levant and southern Turkey, and an eastern one to the Zagros mountains of Iran (see Chapter 4). Typical taxa of this northern biogeographical element are *Lejeunea cavifolia, Pellia endiviifolia* and *Radula lindenbergiana* (Hepaticopsida), *Amblystegiella subtilis, Anoectangium aestivum, Barbula acuta, B. vinealis, Eucladium verticillatum, Eurhynchium speciosum, Fissidens bryoides, F. exiguus, Grimmia anodon, G. orbicularis, Homalia besseri, Hypnum vaucheri, Orthotrichum diaphanum, Pseudotaxiphyllum elegans, Trichostomum crispulum* and *Weissia tortilis* (Bryopsida).

The distribution of some sub-cosmopolitan taxa has been extended by the activities of man, including *Lunularia cruciata, Marchantia paleacea, Reboulia hemispaerica* and *Riccia sorocarpa* (Hepaticopsida), *Bryum argenteum, B. capillare, Encalypta vulgaris, Fissidens viridulus, Funaria hygrometrica, Hymenostylium recurvirostre, Leptobryum pyriforme, Tortula muralis, Trichostomum brachydontium* and *Schistidium apocarpum* (Bryopsida).

5.1.3 Morphological and Anatomical Adaptations

The mosses of the Arabian Peninsula exhibit several physiological adaptations and survival strategies that enable them to tolerate both heat and water stress and the disturbing and erosive effects of mobile substrates. Desert bryophytes lose water quickly and easily, but are capable of surviving desiccation without damage (Di Nola *et al.* 1983). On re-hydration they are able to resume normal metabolic activity and photosynthesis within a few hours, even after prolonged dehydration, and respiration can be maintained with a very low cell water content. The drought-avoidance strategy, common in desert flowering plants, is rare in desert mosses. Drought avoiding taxa include geophytes and annual or ephemeral species which survive drought periods as subterranean stems or diaspores. The only known geophytic bryophyte in Arabia is *Gigaspermum mouretii* from the Judean desert (Frey & Kürschner 1991b).

The drought-tolerant strategy exhibited by the majority of desert bryophytes involves a variety of structural adaptations for the retention of water and the maintenance of photosynthesis under arid conditions. These adaptive structures, which are derived from highly specialised progenitors of the xerothermic-Pangaean stock, are known as the xerothalloid and xeropottioid life syndromes (Frey & Kürschner 1988), described below. Additionally, most of the drought-tolerant taxa exhibit either a short-lived colonist life-history strategy or the perennial shuttle life-history strategy (*sensu* During 1979 and Frey & Kürschner 1991c). In the latter, taxa are relatively longer-lived and produce large spores which enable them to 'shuttle' to another site within the same community.

5.1.3.1 Xerothalloid Life Syndrome

The xerothalloid life syndrome consists of a variety of morphological and anatomical structures which are typical of most of the highly specialised thallose liverworts growing in deserts (e.g. *Asterella, Exormotheca, Mannia, Plagiochasma, Riccia* and *Targionia*). The most conspicuous characteristics are the rolled-up, shrivelled or folded thalli which expose the intensely anthocyanin-pigmented or hyaline ventral scales, thus protecting the dorsal photosynthetic surface from the sun (Figure 5.3: f). Often the complete plant is sunk into the soil surface, appearing only after a shower of rain or heavy dew ('Trockenschlaf' or 'resurrection plants', Volk 1984; Figure 5.3). Hyaline scales, as in *Riccia lamellosa,* probably help to protect against desiccation. A similar effect is produced by the balloon-like epidermis cells of various *Riccia* spp. (Figure 5.3: a-c), or the chimney-like, hyaline air-chambers of *Exormotheca* spp. ('Fensterthallus', Figure 5.3: d-e), which can be compared with the 'windows' in leaves of some South African Mesembryanthemaceae (*Lithops* spp.), helping to reduce the light intensity reaching the photosynthetically active layer. In cross-section, the thalli are thick, nearly semicircular, and consist of dense, parallel assimilatory pillars in air-chambers (Figure 5.3: a). The thickness of these green pillars is frequently correlated with the degree of insolation.

Several species, such as *Plagiochasma rupestre*, have a non-wetable thallus surface due to the presence of hydrophobic wax globules. These xerophytic Marchantiales are thus not able to take up water with the thallus surface, but do so by pegged rhizoids, with the wax globules helping to direct the flow of water downwards to the substratum for later absorption. These wax globules perhaps also prevent the entry of water into the air-chambers, thus avoiding reduced photosynthesis due to a 'drowning in the desert' effect. In most species this is prevented by hydrophobic cuticular ledges around the pores (Schönherr & Ziegler 1975), but these structures are absent in the completely non-wetable thalli of *P. rupestre*.

5.1.3.2 Xeropottioid Life Syndrome

The xeropottioid life syndrome consists of a variety of morphological and anatomical structures typical of the Pottiaceae which grow in the arid areas of the Arabian Peninsula. These structures are present to delay and reduce water-loss and can be seen in species such as *Aloina, Barbula, Crossidium, Pseudocrossidium, Pterygoneurum, Trichostomopsis* and *Tortula* section *Crassicostatae*.

The most common characteristic of this syndrome are crisp, contorted or spirally-coiled leaves, accompanied by a considerable shrinkage of the lamina and increased rolling-up of the recurved margins on each side. The whole leaf winds helically round the stem, so that the dried leaf is protected both from insolation and desiccation. In *Tortula atrovirens*, which grows in colonies, the leaf movement is so pronounced that the plant often almost disappears in the soil (Figure 5.8: a) and all that can be seen are the exposed bulbiform upper parts with coiled-down

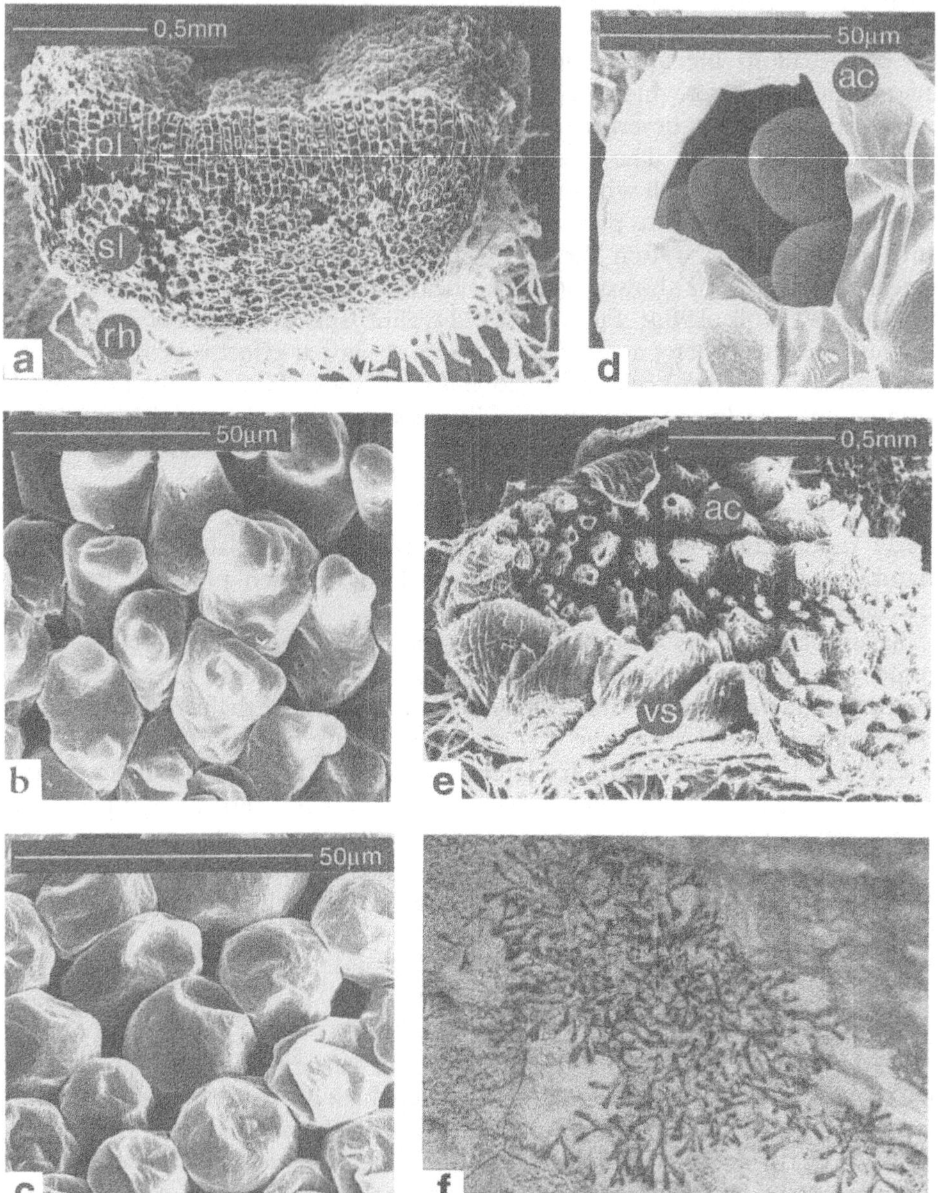

Figure 5.3 Xerothalloid life syndrome. (a) ***Riccia crenatodentata,*** cross-section of the thallus (pl, photosynthetic pillars, sl, storage layer, rh, rhizoids). (b) ***Riccia atromarginata,*** balloon-like epithelial cells (end cells of photosynthetic pillars) with papillae and air pores. (c) ***Riccia lamellosa,*** balloon-like epithelial cells and air pores. (d, e) ***Exormotheca pustulosa***: d, hyaline chimney-like air chambers (ac) inside end cells of photosynthetic pillars; e, thallus with ventral scales (vs) and chimney-like air chambers (ac). (f) ***Targionia hypophylla,*** rolled-up and shrivelled thalli exposing intensely pigmented ventral scales.

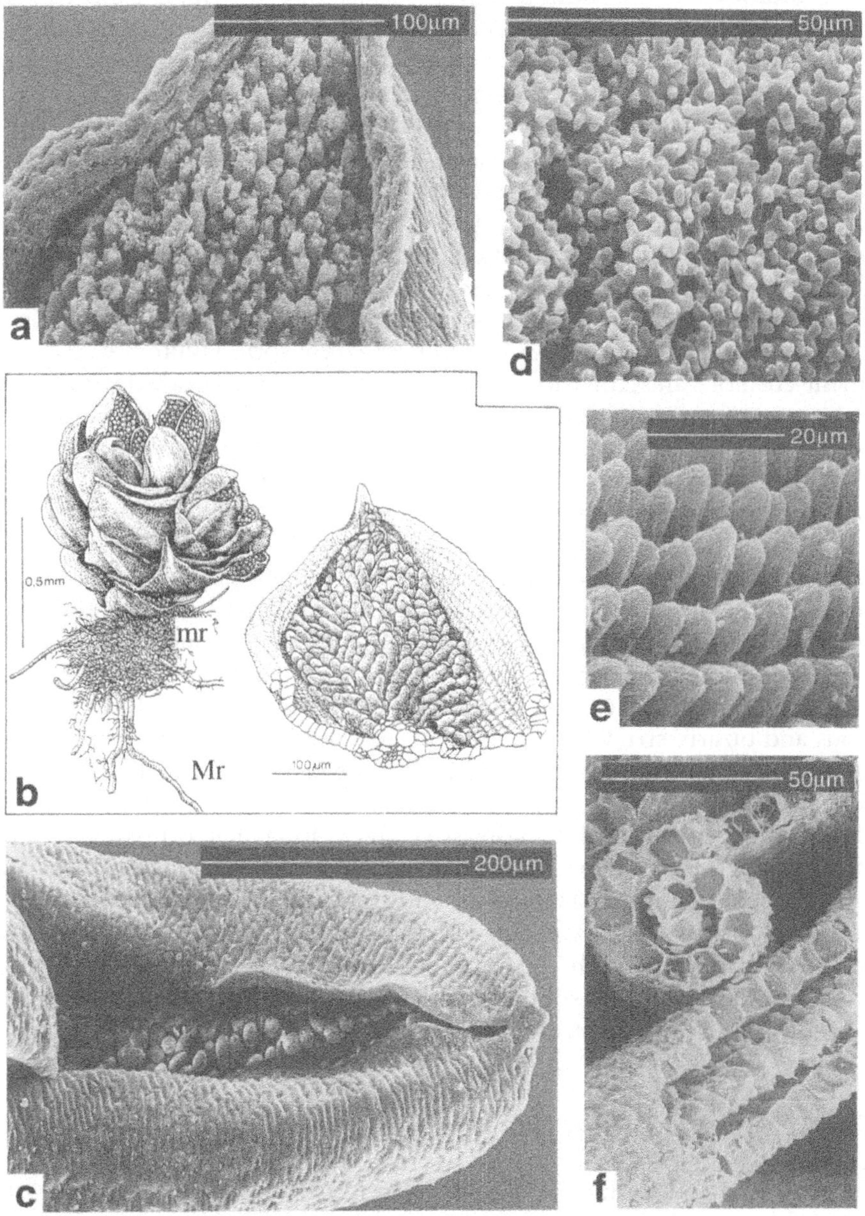

Figure 5.4 Xeropottioid life syndrome. (a) *Crossidium squamiferum* var. *pottioideum,* leaf apex with filament cushion. (b) *Crossidium laevipilum,* bulbiferous plant with dimorphic rhizoids (Mr, macrorhizoids, mr, microrhizoids) and leaf with filament cushion. (c) *Aloina rigida,* cucullate leaf apex with infolded lamina. (d) *Tortula atrovirens,* papillae. (e) *Timmiella barbuloides,* papillae. (f) *Pseudocrossidium hornschuchianum,* spirally-rolled leaf margin (cross-section and longitudinal section of leaf margin), marginal cells differentiated into a specialized photosynthetic region.

leaves. Additionally, the shining abaxial surface of the costa is exposed (e.g. in *Timmiella barbuloides*). This may increase reflection of incoming radiation, and thus reduce evaporation and heat-stress. After rainfall, sand particles and other material are removed by the twisting of leaves (Scott 1982).

A unique feature of several desert mosses is the presence of lamellae, filaments or other outgrowths of the adaxial leaf surface (e.g. *Aloina, Crossidium, Pterygoneurum* and *Tortula* section *Crassicostatae*). This increases the photosynthetic surface of the leaves and acts as a sun-shade, providing a thick and opaque cover (Figure 5.4). In addition, the leaf lamina cells often become hyaline, with strongly infolded or hooded dead cells which protect the filament-cushions and the living parts of the plant. This protects against insolation, desiccation and mechanical damage (e.g. in *Aloina* and *Crossidium* spp., Figure 5.4: a-c). These filament-cushions and lamellae also act as a capillary water conduction system and are able to store water. Besides the increase of the photosynthetic surface by filaments, several taxa (e.g. *Pseudocrossidium* spp. and *Tortula porphyreoneura*) have highly developed chlorophyllous marginal leaf cells which are a special photosynthetic region ('Assimilationsstreifen', Herzog 1926), protected by their location in the spirally-coiled leaf margins (Figure 5.4: f).

Also common among desert mosses are hair-points (glass hairs) formed from dead cells at the tips of leaves (Figure 5.9: c). These reflect sunlight, and may help in the absorption of condensed water vapour from fog and dew, although this has not yet been shown experimentally. In this context, papillae (Figure 5.4: d-e) are also of importance, accelerating water uptake when water is available. These often curious and bizarre structures are present in nearly all pottioid genera, acting as a rapid capillary water movement system (Longton 1988, Proctor 1979), and also reflecting sunlight.

Dimorphic rhizoids, which consist of thick, long macrorhizoids and a fine network of microrhizoids, are very common in desert mosses (Figure 5.4: b). The macrorhizoids anchor the plants in the substrate and the microrhizoids help with the absorption of water by capillary action from the upper soil layer after dew fall. It is possible that, as with root competition in desert flowering plants, the conspicuous occurrence of solitary mosses in various desert habitats is a result of subterranean (i.e. micro-rhizoidal) competition.

5.1.4 Bryophyte Communities

The occurrence and composition of the terrestrial, rock and epiphytic bryophyte communities of Arabia are strongly influenced by climatic, edaphic and topographic factors. The distribution and ecology of each of these communities is described below.

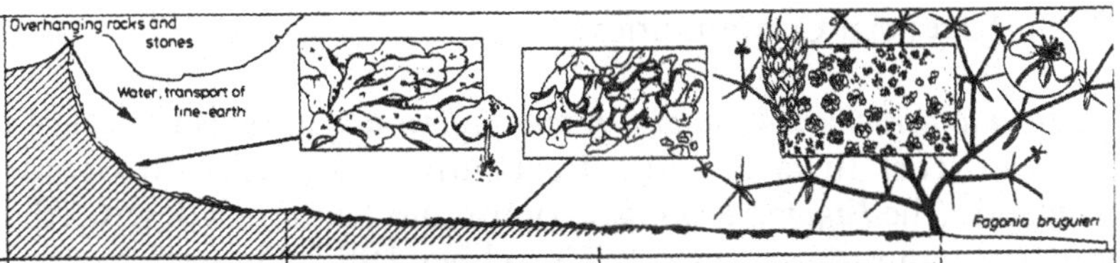

Tortula atrovirens-Crossidium squamiferum synusium

	***Plagiochasma rupestre* micro-zone**	***Riccia trabutiana* micro-zone**	***Tortula atrovirens-Crossidium squamiferum* micro-zone**	**no bryophytes**
Life strategy	Short-lived shuttle species	Perennial/Short-lived shuttle species	Colonists (with a tendency to fertile colonists)	
Life form	Mats	Mats (often overblown with sand)	Short turfs and bulbiformous solitary plants	
Reproductive strategy -sexual reproduction	Frequent within the 1st year	Frequent within the 1st year	Frequent from 2nd-4th year	
-asexual reproduction	Lacking	Rare or lacking	Lacking within fertile Colonists, but frequent within vegetative Colonists (rhizoid gemmae, e.g. *Bryum bicolor*)	
Dispersal strategy	Short-distance dispersal, spores large, 80-100 μm (engychorous tendency)	Short-distance dispersal, spores large, 80-120 μm, capsules cleistocarpous (achorous strategy)	Short- and long-distance dispersal, spores small, 8-20 μm	
Drought-avoidance/ drought-tolerance strategy	← Xerothalloid life-syndromes →		Xeropottioid life-syndromes	

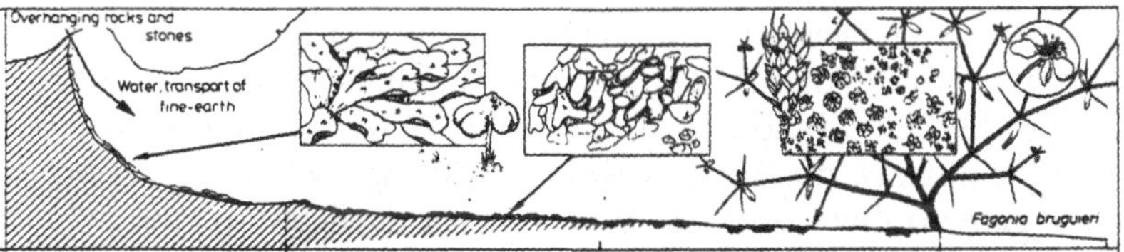

Tortula atrovirens-Crossidium squamiferum synusium

	***Plagiochasma rupestre* micro-zone**	***Riccia trabutiana* micro-zone**	***Tortula atrovirens-Crossidium squamiferum* micro-zone**	**no bryophytes**
Soil				
-Texture	rich in fine earth	± fine-earth	clayey-sandy	sandy
-Moisture	high after rainfall, ± persistent, evaporation low		low, with high evaporation after rainfall	
Microclimate				
-air humidity	c. 30 %, rising up to 50 % after rainfall, ± constant, with low diurnal variation ◄		► 19-50 % with high diurnal variation	
-temperature	with normal diurnal variation, but generally 2-3 °C higher than in open terrain ◄		► with normal diurnal variation	
-illumination	sciophil		photophil	

Figure 5.5 The *Tortula atrovirens-Crossidium squamiferum* synusia of Jebel Tuwayq in central Arabia. Above: Life strategies typical of the different micro-zones. Below: Micro-zonation and environmental conditions (modified after Frey & Kürschner 1987).

5.1.4.1 Terrestrial Communities

The *Tortula atrovirens-Crossidium squamiferum* synusia is typical of the limestone cuestas of Jebel Tuwayq, and is frequently associated with xeromorphic dwarf-shrublands (the *Fagonia bruguieri* community) and rock deserts (the *Gymnocarpos decandrus* community) (Frey & Kürschner 1987; Figure 5.5). This community occurs in regions which receive only 60-100 mm rainfall per year, and grows under boulders where shade and moisture are available and water runoff from rock surfaces can sometimes create a mesic microhabitat.

The *Tortula-Crossidium* synusia is associated with Mesozoic marl sediments. A strong micro-zonation of morphological adaptations and life-history strategies occurs along gradients of soil texture and moisture, temperature and insolation (Figure 5.5). Typical mosses in this synusia are *Barbula trifaria* var. *desertorum, Bryum bicolor, Crossidium davidai* (=*C. asirense*), *C. deserti, C. laxefilamentosum, C. squamiferum* var. *pottioideum, Timmiella barbuloides, Tortula atrovirens* (Figure 5.8: a), *T. mucronifera, Trichostomopsis aaronis* and *T. australasiae* and the dominant liverworts are *Mannia androgyna, Plagiochasma rupestre* (Figure 5.8: b), *Riccia lamellosa* and *R. trabutiana*. This synusia is absent from areas that receive <50 mm of rainfall per year.

The *Plagiochasma rupestre*- and *Riccia trabutiana* - micro-zones consist of short-lived, perennial 'shuttle' species, mainly marchantioid and riccioid liverworts, with drought-resistant gametophytes. These form dense mats and occur only in fissures, rock crevices and on soil beneath overhanging rocks. In this microhabitat, even rain showers of only 1 mm lead to small floods from the smooth-faced rocks (Danin 1972). Obviously in those areas where bryophytes occur, sufficient moisture must be present at least twice during the life cycle, once for the growth of the gametophyte, and a second time for fertilization (the swimming of a male gamete into the archegonium in a film of water). Astonishingly, desert mosses such as *Mannia androgyna, Plagiochasma rupestre* and *Riccia* spp. seem to be distinctly more fertile than the average in more mesic habitats, producing spores as the rule rather than the exception (Scott 1982). In both micro-zones, the incidence of sexual reproduction is therefore rather high. The spores are large (up to 120 μm), indicating short distance dispersal (engychory). Most of the spores remain near the parent plant, and may permit survival of the population if the gametophytes are destroyed by an unusually prolonged drought. This is especially obvious in the *Riccia trabutiana* micro-zone, where capsules are totally embedded in the thallus and are cleistocarpous (achory). Long distance dispersal (telechory) by water after rainfall may occur, but is probably not very effective. Asexual reproduction is rare in both zones. The thalli of all species show morphological adaptations of the xerothalloid life syndrome (i.e. inrolled, shrivelled thalli exposing the ventral scales, non-wetable thalli and inflated epidermis cells), being tolerant of periods of severe drought.

The *Tortula-Crossidium* micro-zone (Figures 5.5) is dominated by acrocarpous mosses with a colonist strategy. Characteristic taxa, mainly of the pottioid genera, have a moderately short life-span (pauciennial, i.e. 1-3 years with a resting stage as spores, to pluriennial, i.e. several years, with seasonality of the gametophyte less distinct) and build predominantly short-turfs or bulbiformous solitary plants on the clayey-sandy crusts. During the dry season they are deeply embedded in the substrate. Colonists have a high rate of both asexual and sexual diaspore production. The spores are small and very persistent in most species, and long distance dispersal is frequent. Two life-history strategies can be observed: fertile colonists with a high degree of sexual reproduction within the first years (e.g. *Crossidium squamiferum* var. *pottioideum* and *Tortula atrovirens*), and vegetative colonists that rely largely on asexual reproduction (e.g. *Bryum bicolor* which reproduces mainly by subterranean rhizoid gemma. The vegetative colonists often appear in primary successional series (microphytic crusts) as pioneers and are important forerunners of the next stages in crust stabilization by other bryophytes and flowering plants. Morphologically, they can be characterized by the xeropottioid life syndrome (i.e. spirally coiled leaves, filament-outgrowths of the lamina, hair-points, papillae and dimorphic rhizoids).

5.1.4.2 Rock Communities

Another habitat for bryophytes in the Arabian Peninsula is that of rocks, rock outcrops and rock crevices with soil pockets. These habitats have diverse water regimes and support many different species. Epilithic communities have so far been studied in three different regions: western (in the Asir mountains of Saudi Arabia and the Haraz mountains of Yemen), central (Jebel Aja near Hail) and southern (Hadhramaut and Jol plateaux in Yemen), but may also occur in the east in Musandam and Jebel al Akhdhar in Oman.

A typical synusia of the Adianto-Primuletum verticillatae *sensu* Deil (1989) is the *Gymnostomum aeruginosum* synusia of calcareous rock crevices, boulders and calcareous sinters with dripping and seeping water. This hydrophytic community is frequent in the montane belt of the Haraz mountains (Jebel Masar, Jebel Shibam and as Zahfah near Menacha) and the Asir mountains (Dahna al Jamal), consisting of typical northern taxa such as *Amblystegiella subtilis, Eucladium verticillatum, Fissidens bambergeri, Gymnostomum aeruginosum, Hymenostylium recurvirostre* and *Plagiochasma rupestre*. Since permanent flowing water is rare on the Arabian Peninsula, the distribution of this synusia is very limited.

A second epilithic community typical on drier rocks has been reported from granite rocks and steep walls of deep gorges and wadis in the Asir mountains of Saudi Arabia (Kürschner 1984a). This community is very poor in species (*Bartramia stricta* (Figure 5.9: a), *Grimmia trichophylla* (Figure 5.9: b), *Pleurochaete squarrosa* (Figure 5.9: a), *Pseudoleskea leikipiae* and *Targionia hypophylla* subsp. *linealis*) but has a high cover and shows a typical successional cycle, dependent on rock inclination and erosion (Figure 5.6). It plays an important role as a pioneer community, collecting fine earth and soil, and thus supporting the establishment of higher chasmophytic plants. The initial phase is a *Grimmia trichophylla-Targionia hypophylla* stage (Figure 5.6: I-II), where pioneer colonization and soil accumulation begins. This is replaced by a *Bartramia stricta-*

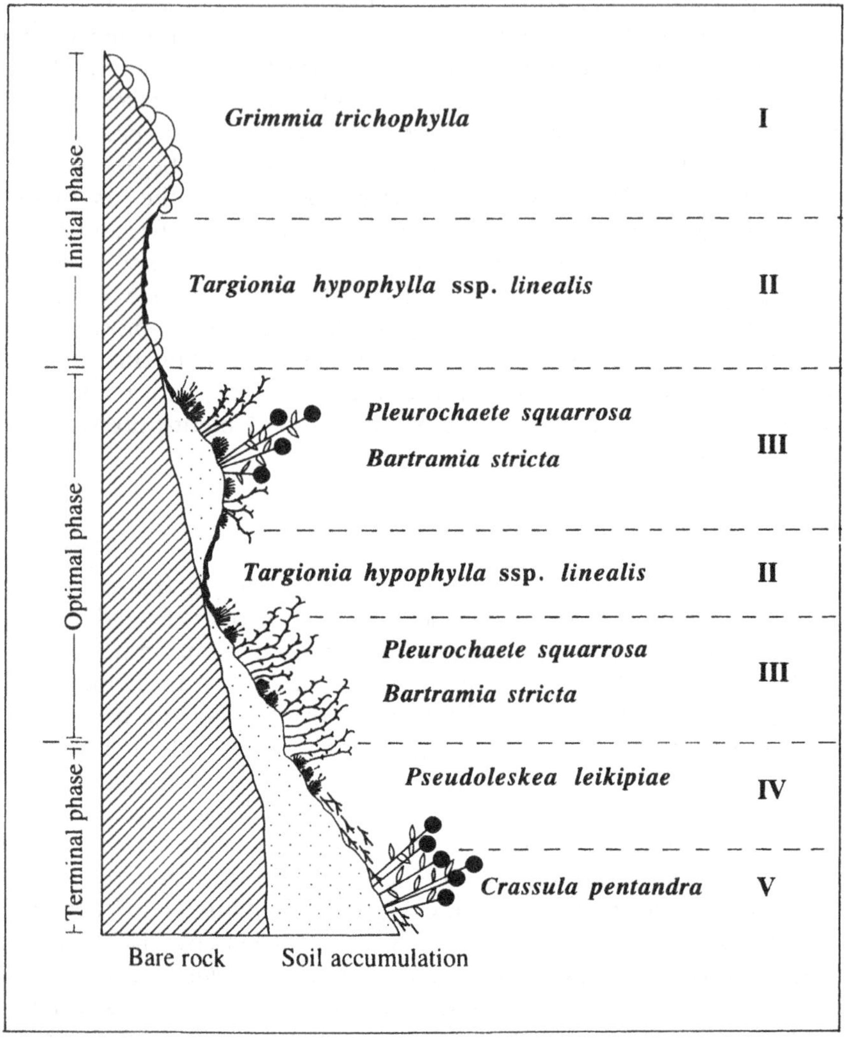

Figure 5.6 Rock community in the Asir mountains of Saudi Arabia, illustrating the relationship between colonization stages (I-V) and soil accumulation (modified after Kürschner 1984a).

Pleurochaete squarrosa stage (Figure 5.6: III) which forms large cushions on bulges of soil reaching up to 10 cm in height, and acts as a germination bed for flowering plants (Figure 5.6: optimal phase). After enough soil has accumulated, these stages, dominated by acrocarpous mosses and thallose liverworts, are occupied by the pleurocarpous, Afromontane *Pseudoleskea leikipiae* and the first flowering plants (*Crassula pentandra* and *Themeda triandra*) which is the terminal phase (Figure 5.6: IV-V). The succession often ends abruptly, when too much soil is accumulated under the cushions. They then become vulnerable to torrential rains which can erode the moss cushions and wash away the accumulated soil particles and weathering products. The succession cycle then starts again.

A third type of rock community, not yet studied in detail, occurs in the central and southern parts of the Arabian Peninsula (Jebel Aja near Hail, the Hadhramaut and Jol plateaux) and in the rocky *harrah* and *hamada* environments. It consists of only a few cushion building acrocarpous mosses such as *Crossidium squamiferum* var. *pottioideum, Grimmia orbicularis* (Figure 5.9: c), *G. pulvinata* var. *africana* and *G. trichophylla* growing in minute fissures on fully exposed rock surfaces. Under boulders and in crevices where water and soil is concentrated, *Aloina rigida, Barbula trifaria, Crossidium crassinerve, C. deserti, Plagiochasma rupestre, Reboulia hemisphaerica, Riccia trabutiana, Timmiella barbuloides* and *Trichostomopsis aaronis* occur, indicating close affinities to the *Tortula atrovirens-Crossidium squamiferum* synusia of the central Arabian Peninsula.

5.1.4.3 Epiphytic Communities

Surprisingly there is a rich epiphytic, bryophyte and lichen flora on the trunks and canopies of *Juniperus* and *Acacia* trees in the Asir mountains of southwest Saudi Arabia (Kürschner 1984b; Figure 5.10). Physiognomically this epiphytic flora is comparable to that of the subtropical cloud forests of the monsoon-influenced high mountains of East Africa. In Arabia they are restricted to altitudes of 2,400-2,900 m, where they are typical of forests and woodlands on west facing slopes. The occurrence of these epiphytes is strongly correlated with three factors: a suitable substrate, i.e. phorophytes in more or less dense stands (*Juniperus procera* and *Acacia origena* woodlands) providing canopy shade and the concentration of rain by rain-tracks down the trunks, an exposed position (escarpments with deep wadi gorges facing into the moist west winds), and a location within the range of the winds coming from the Red Sea (Kürschner 1984b). Depending on the phorophyte, orographic position and exposure, three communities can be observed within this extraordinary epiphytic vegetation.

Typical to the *Juniperus* forests of the humid upper montane region is a Hypno-Leucodontetum dracaenae, which colonizes the middle part of the rough-barked trunks of the junipers. This community consists of the northern *Hypnum vaucheri* and *Tortula laevipila,* the Afromontane taxa *Fabronia abyssinica, Frullania trinervis, Leucodon dracaenae* var *schweinfurthii* (Figure 5.10: b) and *Tortula fragilis,* and the eu-Mediterranean *Leptodon smithii.* Co-dominant with the bryophytic community are various lichens such as *Anaptychia ciliaris,* a typical indicator of xeric sites with frequent high humidity, and *Parmelia flaventior, Physcia adscendens* and *Physconia pulverulenta.* The high humidity in this region is indicated by the occurrence of *Usnea articulata* and *U. bornmuelleri,* the typical canopy epiphytes, with their long hanging beards that give the *Juniperus* forests an 'elfin-like' appearance (Figure 5.10: a,c).

Two communities occur in the more xeric *Acacia origena* woodlands. On the western, northern, and northeastern side of tree trunks, a Fabronietum socotranae can be found. Phytogeographically, this community is a vicariant of the wider

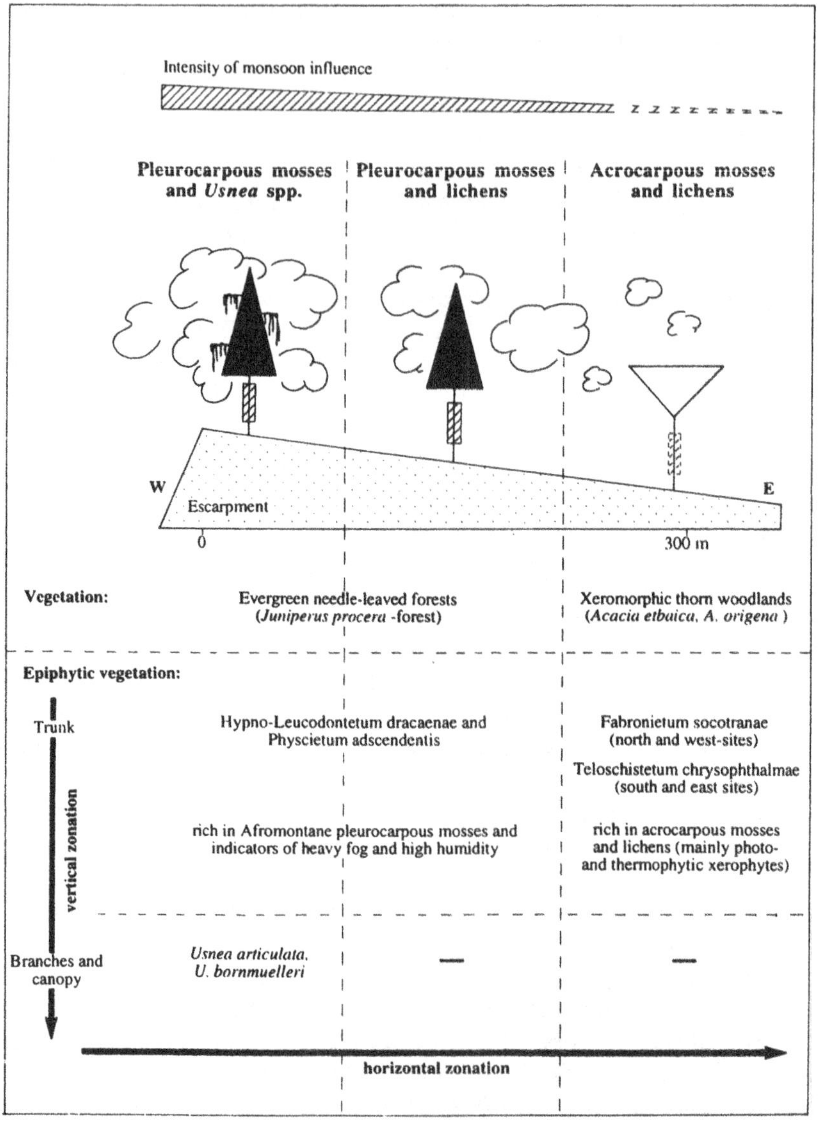

Figure 5.7 Vertical and horizontal zonation in the epiphytic bryophyte communities of the Asir mountains of Saudi Arabia.

distributed Mediterranean Fabronietum pusillae, but is characterized by an Arabian endemic, ***Fabronia socotrana***. Other important species are the cushion forming ***Orthotrichum diaphanum*** and ***Tortula laevipila*** and several lichens such as ***Physcia biziana, Phaeophyscia orbicularis*** and ***Xanthoria parietina***. In the relatively more drier, sun-exposed south- and east-facing sites, the Fabronietum community is replaced by a Teloschistetum chrysophthalmae, a typical Mediterranean,

Figure 5.8 Terrestrial bryophytes of central Arabia. (a) solitary, bulbiferous mosses (*Tortula atrovirens*) on clayey crusts. (b) wet thalli of *Plagiochasma rupestre* under overhanging rocks. (c) wet thalli of *Exormotheca pustulosa*.

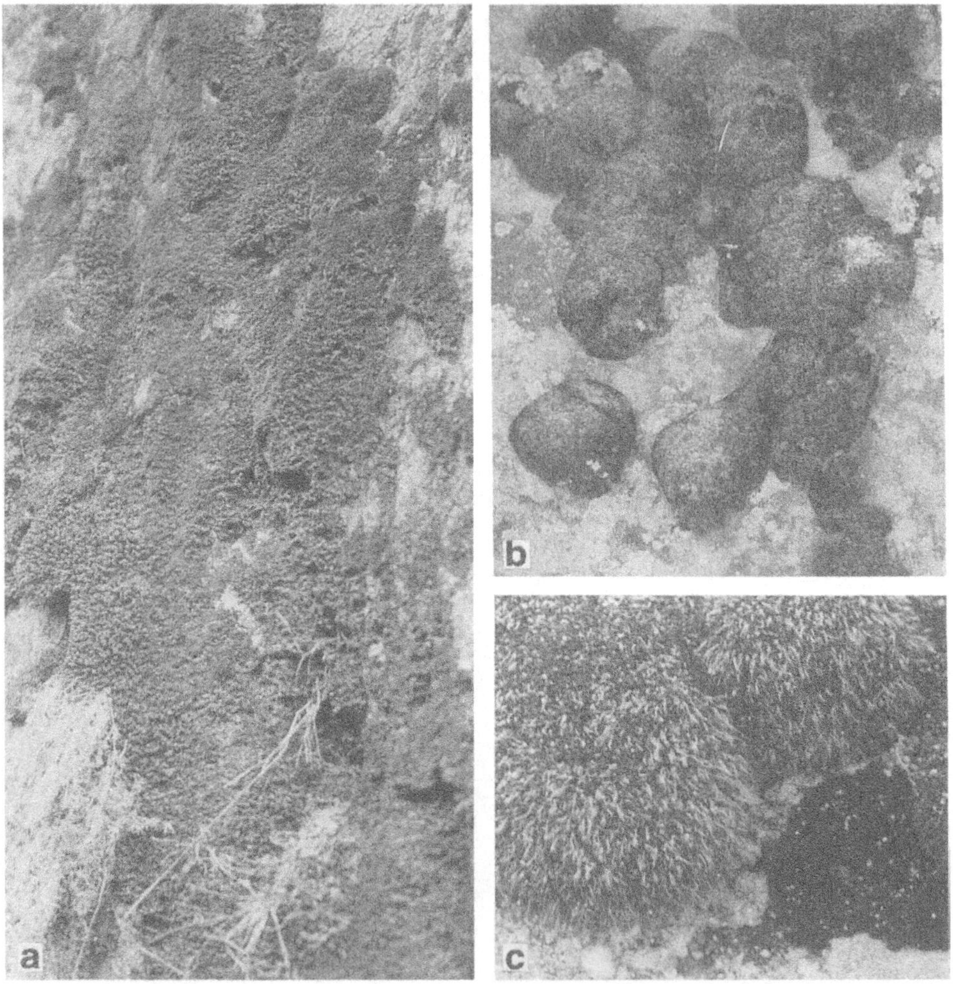

Figure 5.9 Rock communities. (a) large cushions of *Bartramia stricta* and *Pleurochaete squarrosa* on a steep wall in the Asir mountains of Saudi Arabia. (b) cushions of *Grimmia trichophylla* on Jebel Aja near Hail. (c) *Grimmia orbicularis* with hair-points.

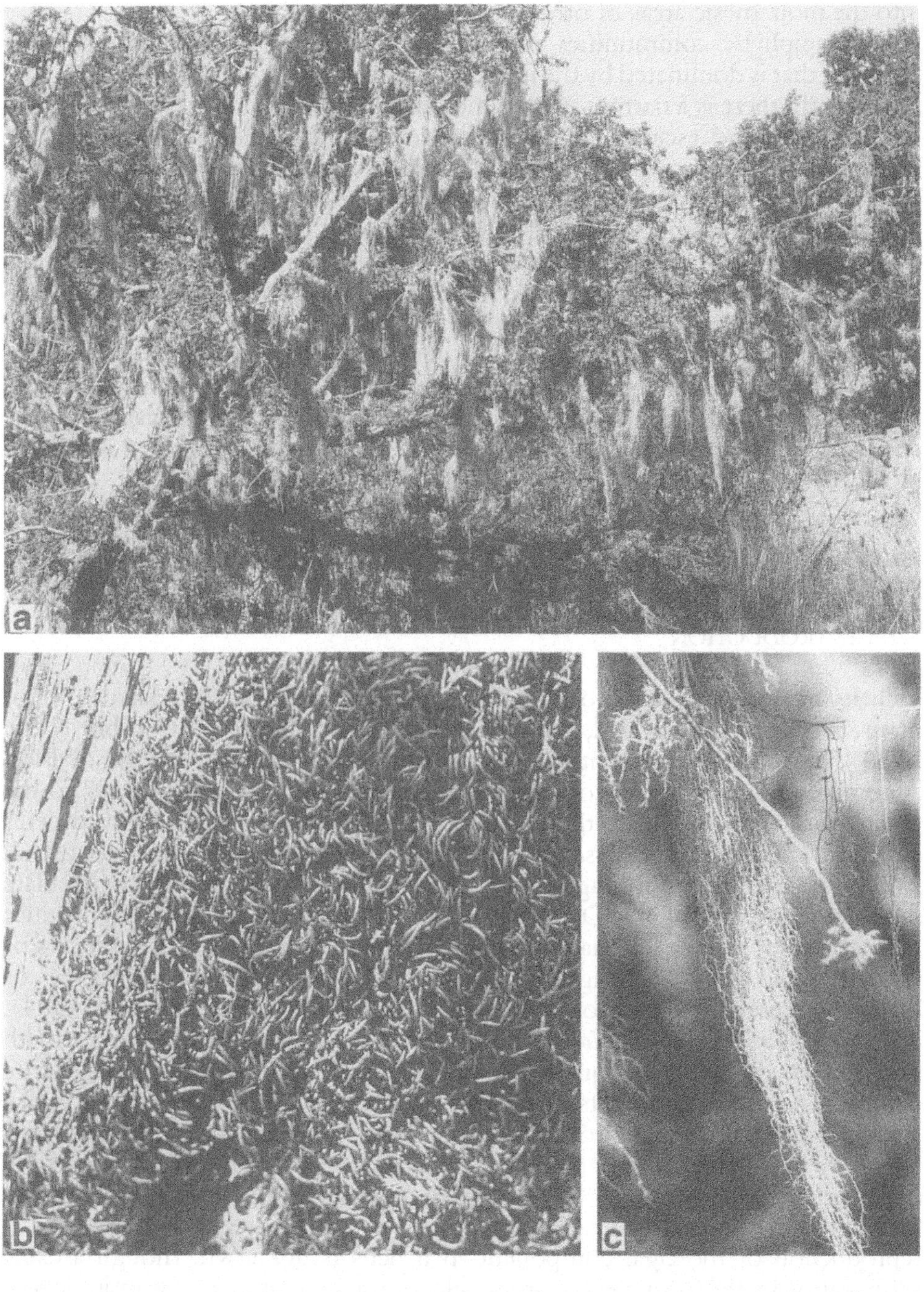

Figure 5.10 Epiphytic communities. (a) beards of *Usnea articulata* and/or *U. bornmuelleri* on *Juniperus procera* in the Asir mountains (Jebel Sawdah) of Saudi Arabia. (b) *Leucodon dracaenae* var. *schweinfurthii*. (c) *Usnea articulata*.

heliophytic (i.e. sun-loving) community, dominated by lichens, that penetrates into the more mesic areas of otherwise largely arid regions.

The epiphytic communities of the Asir mountains show a characteristic zonation that is dominated by the influence of westerly low cloud and fog (Figure 5.7). Firstly, there is a narrow strip of montane juniper cloud forest, restricted to the most exposed escarpment sites, rich in epiphytes and dominated by the Afromontane pleurocarpous mosses, foliose liverworts and lichens. Secondly, there is a drier belt, where the fruticose lichens are absent because of the lower moisture levels. Thirdly, a belt, dominated by xeric *Acacia origena* woodlands, where pleurocarpous mosses are restricted to the escarpment-facing sites, while the xeric south- and east-facing sites of the trunks are dominated by acrocarpous, cushion forming mosses and crustose lichens.

These epiphytic communities are one of the most striking vegetation units in the Arabian Peninsula and, in a physiognomical and phytogeographical sense, exclaves of the Afromontane belt of East Africa. They are indicators of the relatively late separation of the Arabian Peninsula and Eritrean-Somalian highlands.

5.2 Lichens

5.2.1 Introduction

Lichens are a unique group of organisms that consist of two unrelated components, a fungus and an alga, that live in a close symbiotic relationship (Hale 1983). The fungal component, the mycobiont, is usually a member of the division Ascomycetes and the algal component, the photobiont, usually belongs to the unicellular green algae or to the cyanobacteria. The majority of the ascomycetes lichen-forming fungi are Discomycetes (discolichens), and of these almost all lecanoralean (fungi belonging to the order Lecanorales) and some inoperculate discomycetes form lichens. Discomycetes is that group of fungi which produce ascocarps known as apothecia, which often take on the form of cups, saucers or cushions and can be brightly coloured. Species of Basidiomycetes or Deuteromycetes may also form the mycobiont (Hawksworth & Hill 1984). Species of *Trebouxia* or *Nostoc* form the algal component. Lichens with *Trentepohlia* as the photobiont from the Arthoniales and Opegraphales *s.l.* are common in arid environments, since their metabolism is more adapted to salinity.

In a lichen, the fungal associate forms a distinct thallus which has indeterminate growth, and any differentiated cell is usually capable of renewed growth and differentiation (Jahns & Ott 1994). Sexual reproduction is by the formation of asci, with ascospores in cup-shaped apothecia or flask-shaped perithecia. Sexual reproduction of the algal component in lichens is unknown, though asexual propagation by the formation of propagules or fragmentation is common (less so in crustose lichens). Propagules always contains both the fungal and algal components, though this is not the case with the ascospores produced sexually by the fungal component.

Mycobionts produce a number of secondary metabolites that are known as 'lichen substances' or 'lichen acids', which are known for their weathering affect on rocks and soil formation. By studying weathering patterns on rocks, microscopic endolithic lichens have been used as indicators of past climates (Danin 1972).

Lichens are poikilohydric; i.e. they can desiccate completely but resume photosynthesis and growth upon rewetting. In this way they can survive in a variety of habitats and climates and are found on a wide variety of substrates. In deserts, lichens with cynobacterial photobionts are important as they provide most of the fixed nitrogen to the ecosystem.

Information on the lichen flora of the Arabian Peninsula is far from complete, and much exploration remains to be carried out. There has been relatively little work on their ecology, distribution patterns or biogeography, and the information given here is thus based on a limited number of published accounts and checklists and should by no means be considered comprehensive.

More than 230 species of lichens have been recorded from the Arabian Peninsula and Socotra (Abuzinada *et al.* 1986, Ghazanfar & Gallagher in press, Ghazanfar & Rappenhöner 1994, Mies 1994, Mies *et al.* 1995). The highest number of species, *c.* 165, have been recorded from Socotra, including a new genus *Feigeana socotrana* (Mies *et al.* 1995). About 70 and 35 species, respectively, have been recorded from Saudi Arabia and Oman. As yet, no published list exists for Yemen, though the lichen flora is expected to be rich and diverse. No lichen species' lists have been published for the Gulf States of Bahrain, Kuwait or the United Arab Emirates, but a few species occur on stones in Bahrain (pers. obs.) and on stones and tree bark in the United Arab Emirates (Benno Böer, pers. comm.). One species of foliose lichen (*Ramalina lacera*) has been reported from Qatar (Babikir & Kürschner 1992).

5.2.2 Growth-forms

A simple system of lichen classification based on growth-form is commonly followed, with similar morphological forms grouped into three types, though intermediate forms may also occur:

Crustose lichens. In which the thallus is flattened and grows closely appressed to the substrate. The lower cortex and rhizines (rhizoids) are often lacking (Figure 5.11: a).

Foliose lichens. In which the thallus is flattened dorsoventrally, branched and attached to the substrate by rhizines which are produced on the lower surface.

Fruticose lichens. In which the thallus is erect and cylindrical and often branched and bushy, with the lower side of the thallus away from the substrate (Figure 5.10: a,c).

Table 5.2 Number of species of lichens and their growth-form classification in the countries of the Peninsula.

	Species	Crustose	Fruticose	Foliose
Bahrain	1	1	0	0
Kuwait	1?			
Oman	c. 35	c. 20	6	9
Qatar	1	0	0	1
Saudi Arabia	66	38	6	22
United Arab Emirates	c. 3	c. 3	0	0
Yemen (Socotra)	c. 165			

The total number of species and the occurrence of these three growth-forms in the countries of the Peninsula are given in Table 5.2.

5.2.3 DISTRIBUTION

The distribution of lichens in the Arabian Peninsula indicates that they prefer to live in cool, relatively moist locations, areas with frequent dewfall in the morning hours or in areas where cool night temperatures and the availability of moisture results in seasonal fogs (Figure 2.13). Lichens are commonest in the following three regions:

Mountains. Lichens are found at altitudes of 1,300-3,000 m, where temperatures are relatively cool and there is seasonal rain, and occasional low cloud or fog. Lichens occur on shaded rocks, under rock overhangs, on compact soil in rock crevices and in rock shade and on the bark of trees.

Gravel and stony desert plains and limestone coastal hills of central and eastern Oman and the eastern islands of Diimaniyat, Masirah and Halaniyah. A relatively hot climate with usually high day and low night temperatures. Dew is common and a source of available water for plants, and there are seasonal fogs. Lichens occur on the bark of trees, on exposed boulders, limestone rocks and exposed soil (Ghazanfar & Rappenhöner 1994).

Mountainous areas of southwest Arabia and the southern mountains of Oman. Lichens are found above 400 m. These areas are subject to seasonal fogs and low cloud, particularly during the summer monsoon. Lichens occur on the bark of trees, on exposed and shaded rocks and on shaded soil.

Biogeographically the lichen flora of the mountains is a mixture of Mediterrnean and tropical elements. The lichen species of the islands and coastal mountains are however unique in the northwestern Indian Ocean.

5.2.4 LICHEN COMMUNITIES

Favourable substrates for lichen communities in Arabia are tree bark, rock surfaces and crevices, and compacted soil, often cohabiting with bryophyte communities.

5.2.4.1 Corticolous Communities

Corticolous communities of crustose, fruticose and foliose lichens are known from the bark or wood of trees and shrubs in west and southwest Arabia, in southern Oman, and in the woodlands at the eastern edge of Ramlat Wahibah in Oman and in the central desert plain of Oman.

Relatively rich communities of corticolous lichens are common in west and southwest Saudi Arabia, with *Teloschistes villosus*, *T. chrysophtalmus, T. flavicans, Anaptychia kaspica*, *Candelaria concolor*, *Physcia stellaris, Physconia distorta* and *Xanthoria parietina* on *Acacia* and *Juniperus* spp., and the fruticose species *Usnea articulata* and *U. bornmuelleri* (see also section 5.1.4.3 above), extending up to 1.5 m, and *Flavopunctelia flaventior,* on the bark and wood of *Juniperus procera*. The grey, crustose, *Phlyctis argea* occurs on shaded bark. In al Sawdah, an important site of lichen diversity in Saudi Arabia (Abuzinada *et al.* 1986, Bokhary *et al.* 1993), old wooded stumps may bear rich lichen communities of *Amandinea punctata*, *Diplotomma alboatrum*, *Lecanora chlarotera* and *Pertusaria coccodes*.

Communities with *Parmotrema reticulatum*, *P. tinctorum* and several species of crustose lichens (*Opegeopha* spp., *Pyrenmlales*, *Porina* and *Graphidales*) are common on Socotra.

Corticolous lichen communities are found on *Acacia* spp., *Anogeissus dhofarica*, *Boscia arabica* and *Cadaba* spp. in southern Oman. Fruticose, corticolous lichens occur in the monsoon-influenced mountains of southern Oman and Yemen, but not in the northern mountains of Oman, presumably because the available moisture there is insufficient.

In the mist- and fog-affected limestone plateau of central Oman and on the eastern edge of Ramlat Wahibah, well developed communities of *Caloplaca holocarpa*, *Diploicia canescens*, *Dirinaria picta*, *Physcia* spp. *Ramalina duriaei*, *R. lacera* and *Xanthoria parietina* (Figure 5.11: c) occur on *Acacia tortilis*, *A. ehrenbergiana* and *Prosopis cineraria*. In the central desert, the Jiddat al Harasees, the foliose lichen *R. duriaei* (Figure 5.11: b) forms almost pure stands on the lower branches of *A. tortilis*, and in the early hours of the day when dew makes the lichens soft and palatable they are eaten by gazelles (Hawksworth *et al.* 1984).

5.2.4.2 Saxicolous Communities

Saxicolous lichen communities of crustose and foliose growth-forms grow mostly on coarse grained granite, hard igneous rocks and limestone. These communities are usually confined to slopes facing into moisture-laden winds, to rock overhangs and shaded crevices and to exposed rocks in fog-affected areas. In Saudi Arabia *Acarospora lavicola*, *Caloplaca brouardii*, *Dimelaena oreina* and *Parmelia* spp. occur

Figure 5.11 Saxicolous and corticolous lichen communities. (a) ***Buellia subalbula*** and ***Simonyella variegata***. (b) ***Ramalina duriaei***. (c) community of ***Xanthoria parietina***, ***Caloplaca holocarpa*** and ***Diploicia canescens***.

on hard surfaces, whereas the softer sandstones are occupied by *Acarospora strigata*, *Aspicilia* and *Buellia* spp. (Abuzinada *et al.* 1986). More temperate species, such as *Lecanora campestris* and *Rhizocarpon geographicum,* are present in the fog-influenced regions of the western province of Saudi Arabia where temperatures are less extreme (Abuzinada *et al.* 1986).

Saxicolous lichens on limestone are common in restricted locations in central Oman and its offshore islands. There are communities of crustose lichens, such as *Buellia subalbula* (*s.l.*) (Figure 5.11: a) and *Diploicia canescens*, on the fog-affected Huqf escarpment at the eastern edge of the Jiddat al Harasees plateau, and several species of crustose and foliose lichens occur on the islands of Masirah and Halaniyah. Masirah Island has the highest number of lichen species of any location in Oman (Ghazanfar & Gallagher in press, Ghazanfar & Rappenhöner 1994), with all species confined to the limestone cliffs in the southern part of the island. There is a luxuriant growth of lichens on these exposed cliffs, which receive moisture-laden winds from the southwest. These communities include *Arthonia* sp., *Dirina massiliensis*, *Diploicia canescens*, *Buellia* sp., *Ramalina* sp., *Roccella balfourii*, *Simonyella variegata,* and *Roccellographa cretacea.* Two species, *Simonyella variegata* (Figure 5.11: a) and *Roccellographa cretacea,* previously recorded only from Abdalkuri and Socotra, are also found on Masirah. A few locations on the eastern coast of Oman (e.g. Ras al Madrakah and the Barr al Hikman Peninsula) also experience occasional fogs, and in such locations lichens are found restricted to the tops of small limestone hills.

5.2.4.3 Terricolous Communities

Terricolous lichen communities grow in shaded locations on most soil types. The richest communities are found in the mountains on compacted sand in crevices and under rock overhangs. Terricolous lichens, especially species such as *Collema tenax*, *Gloeoheppia turgida*, *Gonohymenia* sp., *Heppia lutosa* and *Peltula radicata*, where the photobiont is a cyanobacteria (Abuzinada *et al.* 1986), are able to withstand desiccation for long periods of time, rehydrating and reproducing when moisture becomes available (Wessels & Büdel 1989). Communities of *Catapyrenium lachneum*, *Fulgensia fulgens, Psora decipiens* and *Toninia diffracta* are commonly found on compact calcareous soil and in the northern mountains of Oman, and in the Asir Region of Saudi Arabia they can cover open expanses of compacted soil (Abuzinada *et al.* 1986). In Oman, terricolous communities are absent from the fog-affected central desert. A few cyanolichens, such as *Gloeoheppia turgida*, are found in the south of Masirah Island.

5.3 Summary

Of all the components of the vegetation of Arabia, the bryophytes and lichens are the least studied, with most of our knowledge coming from work carried out only in the last 15 years. Bryophyte and lichen communities are particularly notable in regions influenced by seasonal fogs and low cloud, such as the western and southwestern mountains and the lower-lying fog deserts such as that of the central desert of Oman. Several species-rich communities occur, of which the most remarkable are the epiphytic bryophytes and lichens on juniper and acacia trees in the high montane woodlands of the Asir and Yemen highlands. Species-rich communities also occur on the islands of Socotra and Masirah. Due to their diversity of both habitat preference and structural form, bryophytes and lichens provide ideal organisms for the study of the ecological, morphological and physiological adaptations that enable organisms to survive and evolve in extreme environments.

Chapter 6

Montane and Wadi vegetation

Ulrich Deil, with Abdul-Nasser al Gifri[1]

"We climbed the mountain for two hours following an extremely steep path and set up camp near the village of Oukend [Wakkan]. Palm trees had by then disappeared and temperate zone plants appeared. I climbed as high as I could, but I could not reach the summit of the mountain. I kept coming upon precipitous rocks which proved to be insurmountable obstacles."

Pierre R. M. Aucher-Éloy, on his excursion to Jebel al Akhdhar, Oman (Jaubert 1843)

[1]Contributed to section 6.3.2

S. A. Ghazanfar and M. Fisher (eds.), Vegetation of the Arabian Peninsula, 125-174.

6.1 Introduction

To Aucher-Éloy and other early travellers, the mountains of the Arabian Peninsula presented a rugged and difficult terrain for passage, and today the more remote regions have yet to yield their final secrets. These mountain chains, lying on the western, southwestern and eastern periphery of the Peninsula, were uplifted during the period of the great Alpine orogeny from the Mid Tertiary onwards, and they were thereafter eroded, often dramatically, during successive pluvial periods (see Chapter 3). Today these mountains exert a strong influence on the climate of the Peninsula (see Chapter 2), and the soil and vegetation structure of their slopes, valleys and summits undoubtedly play a vital, though as yet largely undocumented, role in the hydrogeology of the region. From the Tertiary onwards the mountains were important phytogeographically, providing a relatively equable climatic 'corridor' between Africa and Asia, and *vice versa*, during the Late Tertiary (see Chapter 4), and today the higher altitudes provide a climatic refuge for species that once had much wider distributions.

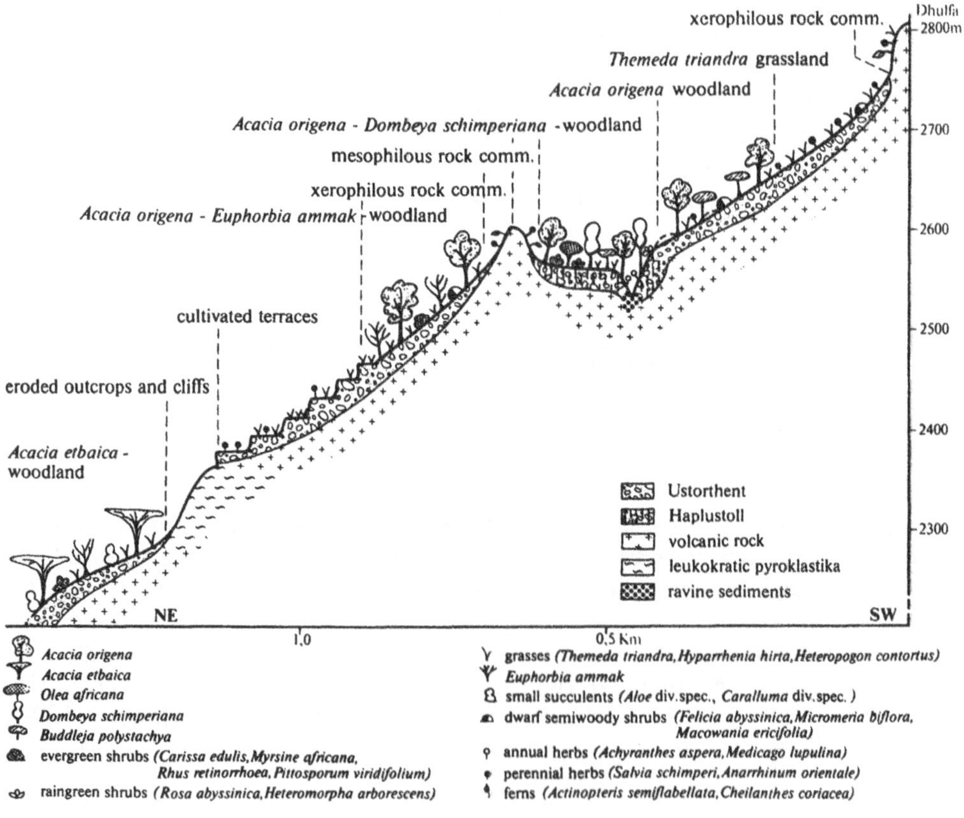

Figure 6.1 Vegetation patterns, soil types and land use on Jebel Shibam in the Haraz mountains of the western Yemeni escarpment (adapted from Grosser 1988).

Attracted by the relatively temperate clime of the mountains, man made a home there for himself and his livestock, and in many areas undertook a dramatic transformation of the natural landscape. This anthropogenic influence has been most marked in southwestern Arabia, where the natural landscape and vegetation have nearly disappeared in many areas, replaced by extensive systems of terraced fields (Figures 6.1, 6.2). Archaeological evidence suggests that these agro-ecosystems date back to the Roman-Nabataean (*c.* 200 BC-100 AD) or early Sabaean (*c.* 950-115 BC) periods.

Over the last three decades, with the opening of the region to international markets and consequent competition with imported products, widespread changes have come to these traditional agricultural systems (Carapico 1985). The shift to a market-orientated production has resulted in a new evaluation of agrarian potential, bringing changes to both the agro-ecosystem and cultural values. Terraces used for grain and legume production have been abandoned, and to a large extent even the ***atar*** of roses has lost its traditional economic and cultural value (Scholz 1984). In the Yemeni highlands, agricultural crops have been replaced by the only lucrative cash crop, ***kat*** (*Catha edulis*) (Figure 6.3), and though cultivated on only 5-8% of the arable land, ***kat*** now contributes about 50% of the total income from agriculture (Kopp 1990, Revri 1983). In addition, severe droughts in the early 1940's and during 1967-1973, combined with the emigration of rural labour, accelerated the disuse and neglect, and consequently the degradation, of the terraced field system. (Deil 1986b, Vogel 1988a, 1988b). Coupled with this there have been considerable reductions in the use of traditional irrigation systems and increases in pump-drawn irrigation, leading to over-exploitation of groundwater reserves in many areas (Kopp 1990).

In general terms, the vegetation of the mountains of the Arabian Peninsula consists of open montane woodlands, deciduous and semi-evergreen shrublands and bushlands and grasslands. Today the most prominent feature of this highly modified vegetation type is the xerophytic succulent plant communities which occupy the rocky outcrops and slopes, the most common landscape of the mountains. In this chapter I firstly discuss the phytogeography and endemism of the montane flora, expanding the relevant sections of Chapter 4 in greater detail for this particular landscape type. I then describe the remnant and relict montane woodlands, including a detailed summary of the information currently available on the apparent dieback in the extant juniper woodlands, followed by a detailed description of the succulent plant communities of the mountain slopes, summits and plateaux, commonly known as 'rock communities'. I conclude with a description of the orophytic grasslands, their importance as rangelands and the vegetation of wadis.

Figure 6.2 (A) The western Yemeni escarpment has been modified by man into an extended sequence of terraces. This is a view from Jebel Shibam towards the Haraz mountains around Manacha. (B) Terraced fields with *Sorghum bicolor* and *Medicago sativa* dominate the landscape near at Tawilah in northern Yemen. Between the boulders and around the fortified houses some remnants of the original *Acacia origena* forests can be seen.

Figure 6.3 *Catha edulis* has replaced *Coffea arabica* as the main cash crop in Yemeni agriculture. In the so-called *sawagi* rainwater harvesting system, the runoff from the rocky surfaces is used for supplementary irrigation of the adjacent terraces (al Maghrabah in the Haraz mountains).

6.2 Phytogeography and Endemism

Three major montane vegetation types can be distinguished in Arabia, the northwestern Mediterranean vegetation type, the southwestern Afromontane type and the eastern Omano-Makranian type (see also section 4.2). These divisions correspond to different climatic regimes and at the same time to the distribution of the three *Juniperus* species, respectively, *J. phoenicea*, *J. procera* and *J. excelsa* subsp. *polycarpos*. The northwest mountains receive the majority of their annual rainfall of 50-150 mm in the winter and early spring, the southwest receives the majority of its 200-500 mm of rainfall in spring and to a lesser extent in summer, and the eastern mountains receive their 150-300 mm of rainfall mostly in the winter and early spring, with an additional summer rainfall peak in some years (see Figure 2.11 and section 2.3.4 for details).

The vegetation of the Eldom mountains in southern Jordan and the northwestern Hijaz mountains in Saudi Arabia is influenced by Mediterranean floristic elements, forming open *Juniperus phoenicea-Pistacia atlantica* woodlands with *Prunus communis, P. korschinskii, Osyris alba* and *Rhamnus dispermus* from 700 to 1,300 m. This community is replaced by a *Quercus calliprinos* forest with *Crataegus aronia, Daphne lineariifolia* and *Calicotome villosa* in the summit zone at 1,300-1,500 m (Baierle 1993, Kürschner 1986a).

The Jebel al Akhdhar range of the northern mountains of Oman represents the Omano-Makranian vegetation of the Nubo-Sindian phytochorion, phytogeographically related to the vegetation of the mountains of southern Iran and Baluchistan (Kürschner 1986a). The summit areas (2,200-3,000 m) are dominated by an open *Juniperus-Olea* woodland, characterized by *Juniperus excelsa* subsp. *polycarpos, Olea europaea* and *Ephedra pachyclada*. Other associated montane species are *Monotheca buxifolia, Teucrium mascatense, Sageretia spiciflora, Helianthemum lippii, Ebenus stellata, Euryops arabicus, Clematis orientalis* and *Cymbopogon schoenanthus* (Ghazanfar 1991b, Mandaville 1977).

The vegetation of the mountainous regions of Dhofar is influenced by the Somalia-Masai phytochorion and constitutes a refuge for mesic Afrotropical floristic elements and some endemic species. The steep southern escarpments are covered with deciduous woodland characterized by *Anogeissus dhofarica, Cadia purpurea, Euclea schimperi, Dodonaea angustifolia, Olea europaea, Maytenus dofarensis, Rhus somalensis, Carissa edulis* and other evergreen shrubs. Several endemic species and the monospecific *Cibirhiza dhofarensis* occur in this monsoon-affected zone (Bruyns 1988). The summit plateau is covered with tall grasslands.

The montane vegetation of southwestern Arabia is influenced by the Eritreo-Ethiopian vegetation type. Afromontane elements and Eritreo-Arabian endemics dominate the evergreen, sclerophyllous shrub communities between 2,000 and 2,800 m (König 1988). These are the northernmost outliers of the Afromontane archipelago-like centre of endemism (see Chapter 4). However, in Arabia, the Afromontane flora is impoverished (White & Léonard 1991) and the floristic diversity is comparable to that of Eritrea (Hedberg 1986) and Jebel Marra in Dharfour (Miehe 1988). The *Hagenia* (*H. abyssinica*) zone and the highly diverse *Helichrysum* shrublands of the East African highlands are absent in Arabia. The only representative of the shrubby *Helichrysum* species in Arabia is the endemic *H. arwae* known from the summit areas around Jiblah in Yemen (Wood 1984).

Characteristic life-forms of the Afro-alpine belt are also present in Arabia. These are the tussock grasses represented by *Festuca cryptantha, Stipa tigrensis* and *Elionurus muticus*, acaulescent rosette plants represented by *Helichrysum forskalii, Salvia schimperi* (Figure 6.4: A) and *Senecio sumarae*, cushion plants represented by *Cichorium bottae* (Figure 6.4: B) and *Dianthus uniflorus* and sclerophyllous shrubs represented by *Macowania ericifolia*. However, the most prominent members of the endemic stock, *Dendrosenecio, Lobelia* and suffrutescent *Alchemilla* are absent in Arabia (Hedberg 1986).

The populations of *Erica arborea* and *Hypericum revolutum* (Figure 6.4: C) in the Jebel Melham and Jebel Burra region of Yemen are fragmentary outposts of the *Erica-Hypericum* belt, which covers large areas in the Bale mountains of southern Ethiopia (Miehe & Miehe 1994) and the Semien mountains (Hurni 1981). Remnant trees of *Cussonia holstii, Cordia abyssinica* and *Pittosporum viridiflorum* suggest that an evergreen montane forest was present in the Yemeni western escarpment prior to human occupation of the region.

Figure 6.4 (A) *Salvia schimperi*, distributed from the Asir to the Hajayrah and in the Ethiopian highlands, is common in the upper montane and subalpine zones.(B) The cushion forming *Cichorium bottae* grows in the montane and alpine zones from the Hijaz mountains to San'a. (C) *Hypericum revolutum*, a widespread shrub in the Afromontane zone, indicates the location of the former *Erica arborea-Hypericum revolutum* horizon in southwest Arabia. (D) The prickly *Acanthus arboreus* colonizes steep grassy slopes on the western escarpment between 1,000 and 3,000 m.

Southern African montane elements are few, represented in Arabia by *Euryops arabicus* and *Kniphofia sumarae* (Lavranos 1983a). Mediterranean and northern temperate floristic elements are well represented, with *Erica arborea*, *Scabiosa*, *Silene*, *Trifolium*, *Arabis alpina*, *Peucedanum* and *Pimpinella*.

Vicariant evolution and species radiation is common in the montane genera and is well illustrated by the distribution patterns of *Acacia*, *Lavandula* and *Teucrium*. In Yemen species of *Acacia* show the following distributional sequence from the Tihamah plains to the top of the Asir escarpment: *Acacia ehrenbergiana*/*A. oerfota*, *A. tortilis*, *A. mellifera*, *A. asak*, *A. hamulosa*, *A. etbaica*, *A. laeta*, *A. origena*, *A. yemenensis* and *A. gerrardii*, each occupying a different bioclimatic niche (Abulfatih 1992, Al-Hubaishi & Müller-Hohenstein 1984, König 1986, 1987, Wissmann 1937). *A. origena* on the seaward side of the Arabian and Eritrean escarpment (Hunde 1982) and *A. gerrardii* in the central plains and eastern plateau in the rain-shadow (Wood 1983b) are a vicarious pair of species. A north-south variation from 1,600-2,200 m in *Acacia yemenensis* is seen with *A. yemenensis* s. str. colonizing steep slopes and abandoned terraces in the Ibb-Taiz area and *A. yemenensis* subsp. *obtusifolia* restricted further to the north from Shahara to Hajjah (Boulos 1994).

The genus *Lavandula*, chiefly distributed in the western Mediterranean-Macaronesian region has a secondary centre of diversity in southwestern Arabia, Somalia and Socotra. 11 species belonging to the sections Subnuda, Pterostoechas and Stoechas occur in Arabia. With the exception of two species, *L. hasikensis* (section Subnuda) and *L. atriplicifolia* (section Stoechas), the taxonomic positions of which are unclear, all other species within the sections are closely related, being separated by the size and shape of the bracts and indumentum (Miller 1985).

Within the genus *Teucrium*, the woody based, cushion-forming species form a vicarious group of Arabian endemics colonizing rocky montane areas: *Teucrium hijazicum* occurs in the Hijaz mountains, *T. yemenense* in the southwestern escarpment, *T. rhodocalyx* in the southern escarpment, *T. nummulariifolium* on the Jol plateau and in Dhofar, *T. mascatense* in the Hajar mountains of Oman and *T. socotranum* and *T. balfourii* in Socotra (King 1988). Within *Nepeta* endemics occur at high altitude along the western escarpment, with *Nepeta sheilae* in northern Hijaz, *N. deflersiana* from the Asir to Jebel Nabi-Shuayb and *N. woodiana* near Ibb (Hedge 1982). The first species is related to the Mediterranean *N. italica*, the latter two to East African taxa.

6.2.1 Arabian-Macaronesian Disjunctions

While studying the succulent vegetation of the Canary Islands, Rivas and Esteve (1965) described the Kleinio-Euphorbietea canariensis communities and outlined a vicarious class, the Kleinio-Euphorbietea eritreo-arabica (now Kleinio-Carallumetea) for the Eritreo-Arabian region, thus demonstrating the close floristic relationship between the *Euphorbia balsamifera* communities in Arabia and Macaronesia (Figure 6.5). Vicarious species found within the genera *Euphorbia*, *Ceropegia*, *Campylanthus*, *Lavandula*, *Caralluma*, *Kleinia* and *Aloe* all belong to the mid-Tertiary African stock. Within the genus *Euphorbia*, the evolutionary lines of

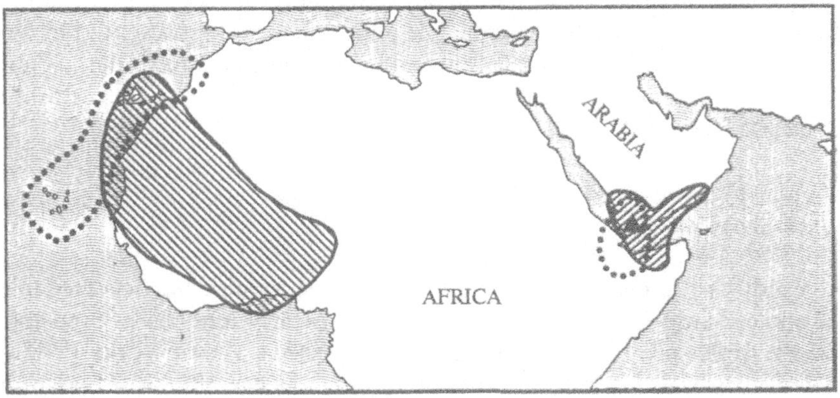

Figure 6.5 Taxa with an Eritreo/Arabian-Macaronesian distribution. Δ *Ceropegia* section *Sarcodactylus* and the pseudovicarious ▲ *Ceropegia rupicola*. .... *Aeonium*. ▧ *Euphorbia balsamifera* s. str. ▨ *Euphorbia balsamifera* subsp. *adenensis*.

the sections *Diacanthium* (now subgenus *Euphorbia,* Carter 1985), *Pachycladae* and *Tirucalli* had apparently already split up before geological events and climatic changes in North Africa severed the connections between the Eritreo-Arabian and Macaronesian populations (Deil & Müller-Hohenstein 1984).

Similar disjunctions can be seen in *Aeonium* (Crassulaceae). The only remnant species from this genus present in Arabia is *Aeonium leucoblepharum* (section Holochrysa; Figure 6.21d). It has primitive characters both in growth-form (Lems 1960) and photosynthetic pathway (Lösch 1987). Its next relative (in morphological similarity and CAM type) is *A. gorgoneum* from the Cape Verde Islands. Species of *Ceropegia* in sections Sarcodactylus and Phalaena also show the same Arabian-Macaronesian disjunct distribution. Species in section Sarcodactylus in the Canary Islands and the pseudo-vicarious *C. rupicola* (section Phalaena) in Yemen and the Asir mountains have succulent stems, showing adaptative convergence in vegetative characters (Bruyns 1989).

6.2.2 Palaeoendemics

The rocky habitats in southern Arabia shelter many specialized and endemic species, including several palaeoendemics such as *Aeonium leucoblepharum*, *Delosperma harazianum* (Poppendiek & Ihlenfeldt 1978), *Psilotum nudum* (Miller 1984), the monospecific genus *Dhofarica* (Miller 1988) and *Centaurothamnus maximus* (Wagenitz *et al.* 1982). *Primula verticillata*, another such species (Primulae section Sphondylia, a group with primitive characters), has survived on the Arabian Peninsula and its surroundings in relict stands with the maidenhair fern since the Tertiary. The section is present in western Saudi Arabia and Yemen (the Adianto-Primuletum verticillatae on Jebel Fayfa in the Asir, König 1987), in Ethiopia (the Adianto-Primuletum simensis) and on Jebel Catharina in Sinai (the Adianto-Primuletum boveanae). The genus *Dionysia*, represented by a single species, *Dionysia mira*, distributed on the Hajar mountains in Oman, is another element of the Archaeprimulae, indicating a migration route with the Zagros mountains of Iran.

6.2.3 NEOENDEMICS

Adaptive radiation and speciation by geographical separation has occurred in many genera, and each mountain range in the western escarpment has its own endemic species. Examples are found in the genera *Euphorbia*, *Aloe* (Wood 1983a) and *Phagnalon* (Qaiser & Lack 1985). In the Asclepiadaceae, the most endemic-rich family, the genus *Caralluma* has its centre of radiation in the Eritreo-Arabian lowlands (Albers 1983) and has *c.* 30 endemic species distributed in the mountains of western and southern Arabia (Miller & Nyberg 1991). Only the polyploid species, *C. europaea*, is distributed further north in Sinai (Bruyns 1987a).

A high degree of endemism is present amongst the succulent genera. These are *Caralluma* with *c.* 21 endemic species (not following Plowes 1995 re- classification of *Caralluma*; Bruyns 1987b, Bruyns & Jonkers 1994, Lavranos 1962, 1963a, 1979, Rauh & Wertel 1965), *Ceropegia* with 10 endemic species (Bruyns 1989, Chaudhary & Lavranos 1985, Field & Collenette 1984), *Sarcostemma* with 5 endemic species (Bruyns & Forster 1991, Lavranos 1974), *Duvalia* with 4 endemic species (Lavranos 1983b), *Echidnopsis* with 4 endemic species (Lavranos 1993), *Rhytidocaulon* with 4 endemic species (Field 1981b) and *Huernia* with 6 endemic species (Field 1981a, Lavranos 1963b, Newton & Lavranos 1993). Neoendemics in the families Euphorbiaceae, Asteraceae, Moraceae, Liliaceae and Crassulaceae occur in the genera *Euphorbia* (Carter 1982, 1985, Ghazanfar 1993b), *Kleinia* (Halliday 1983), *Senecio* (Boulos & Wood 1983), *Dorstenia* (Friis 1983), *Aloe* (Lavranos 1965, 1967, 1985, 1995, Lavranos & Newton 1977, Wood 1983a) and *Kalanchoe* (Lavranos 1965, 1967, 1985, 1995, Raadts 1981, Raadts 1995, Wood 1983a).

6.2.4 TAXA OF HOLARCTIC ORIGIN

The herbaceous vegetation between 2,500 and 3,500 m in the western escarpment of Yemen contains the highest percentage of taxa of Holarctic origin in the whole Peninsula, an observation originally made by Wissmann (1972) when discussing the potential distribution of *Juniperus* forests. The taxa are: *Asplenium trichomanes*, *A. adiantum-nigrum*, *Minuartia filifolia*, *Silene macrosolen*, *S. yemenensis*, *Gypsophila umbricola*, *Dianthus uniflorus*, *D. longiglumis*, *Cardamine hirsuta*, *Arabis alpina*, *Potentilla dentata*, *P. reptans*, *Alchemilla cryptantha*, *Trifolium semipilosum*, *Lotus schimperi*, *Salvia schimperi*, *Sanicula europaea*, *Galium yemenense*, *Micromeria biflora*, *Teucrium yemenense*, *Scabiosa columbaria*, *Swertia polynectaria*, *Campanula*

Campanula edulis, Anthemis yemenensis, Felicia dentata, Helichrysum forskahlii, H. nudiflorum, H. splendidum, Senecio schimperi, S. harazianum, S. sumarae, Crepis rueppelii and *C. nabi-shuaybii*. Some of the species are Arabian endemics, others occur also in East Africa or are widespread in the Holarctic region.

6.3 Montane Woodlands

The montane woodlands and shrublands of the Arabian Peninsula belong to the Afromontane classes Hyperico-Rhamnetea, predominantly of the order Pistacio-Eucleetalia, and the Juniperetea procerae (Bussmann & Beck 1995), chiefly of the order Oleo-Juniperetalia (Knapp 1968).These classes are identical to Oleetea africanae and Juniperetea procerae *sensu* Zohary (1973). Most of the *Juniperus* communities can be included in the Myrsino africanae-Juniperetum procerae, found from the Asir to southern Kenya (Bussmann & Beck 1995), and the montane bushland included in the *Olea-Tarchonanthus* community *sensu* Vesey-Fitzgerald (1957b). The bushland associations are the impoverished northernmost outliers of the Hypericetum revolutae (Bussmann & Beck 1995).

Perhaps due to the attractions of the relatively cooler and moister montane climate, the vegetation of the mountains has been described in some detail, especially in the west. Work on the higher regions of the Asir mountains includes that of Abulfatih (1984, 1992), Brooks and Mandil (1983), El-Karemy and Zayed (1996), Fayed and Zayed (1989), Fisher (1997), König (1986, 1987) and Kürschner (1984b). The vegetation of the Hijaz mountains has been studied by Abd El-Ghani (1994) and by Vesey-Fitzgerald (1957b), and that of the southern Jordan mountains by Kürschner (1986b) and Baierle (1993). In Yemen the vegetation has largely been studied in the central part of the western escarpment, and ranges from checklists and floristic overviews (Al-Hubaishi & Müller-Hohenstein 1984, Scott 1942, Wissmann 1937) to descriptions and vegetation maps (Deil & Müller-Hohenstein 1985, Grosser 1988, Hepper 1977, Hepper & Wood 1979, Scholte *et al.* 1991, Wood 1983b). The vegetation of the Dhofar mountains has been investigated by Radcliffe-Smith (1980), Sale (1980) and Miller and Morris (1988), and that of the northern Oman mountains and Musandam by Fisher and Gardner (1995), Gardner and Fisher (1996), Ghazanfar (1991b) and Mandaville (1977, 1985).

6.3.1 Evergreen Juniper Woodlands

Three species of *Juniperus*, *J. phoenicea*, *J. procera* and *J. excelsa* subsp. *polycarpos*, occur in the semi-arid montane regions of the Peninsula. *J. phoenicea* is found from the Mediterranean along the Hijaz mountains to *c.* 21°N at Taif in western Saudi Arabia. *J. procera* occurs in Zimbabwe and northwards through the highlands of East Africa to the southwest and west of the Arabian Peninsula in Yemen and Saudi Arabia (Hall 1984), and is sympatric with *J. phoenicea* for about

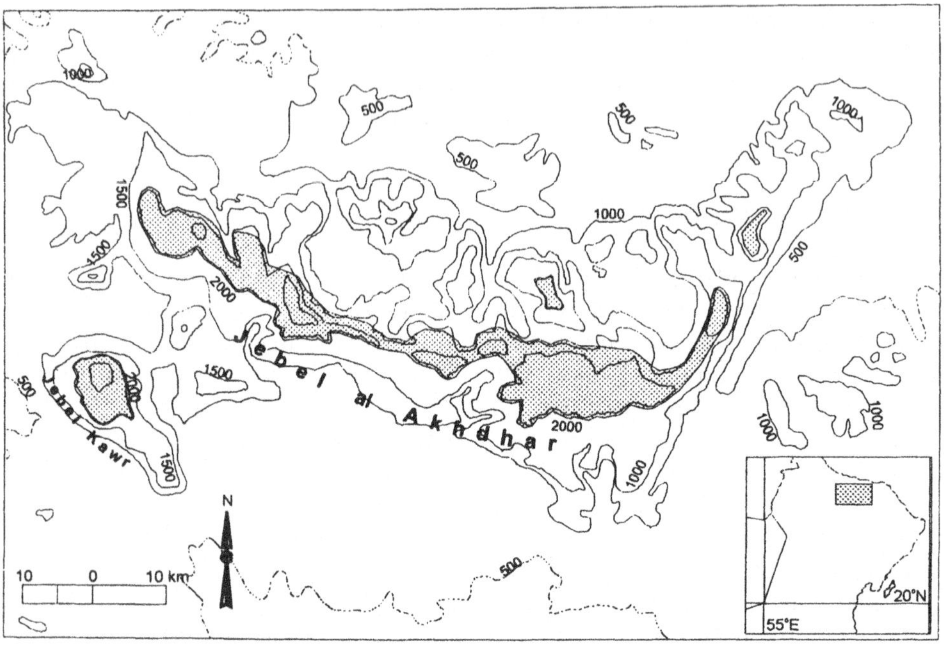

Figure 6.6 The distribution of *Juniperus excelsa* subsp. *polycarpos* (shaded area) in Jebel al Akhdhar; the inset indicates the position of the main map in northern Oman. After Gardner and Fisher (1996), with amendments.

30 km along the escarpment mountains in the vicinity of Taif (Asmodé 1989). The area of sympatry is approximately coincident with the transition from a Mediterranean to a subtropical precipitation regime on the western mountains (see section 2.3.4 and Kerfoot 1961). *J. procera* appears to have originated from the Irano-Turanian *J. excelsa* during the Miocene-Pliocene (see section 4.2 and Kerfoot 1961). *J. excelsa* subsp. *polycarpos* has an Irano-Turanian distribution, occurring from eastern Turkey through the Caucasus and the Kopet mountains into Afghanistan, reaching the Tien Shan and mountains of Kirgizstan in the northeast, the mountainous regions of Baluchistan in Pakistan and Himachal Pradesh in India in the southeast, and with its most southerly population by some four degrees of latitude, isolated from other populations by the Arabian Gulf and the deserts of eastern Arabia, in the western Hajar range of the northern mountains of Oman (Farjon 1992).

The montane *Juniperus* woodlands of the Peninsula are found above altitudes of *c.* 1,400 m, though below 2,000 m they are uncommon and sparse, and largely restricted to slopes with sheltered northerly aspects. At the highest altitudes, generally above *c.* 2,500 m, the woodlands are largely monospecific. In some areas, *Juniperus* forms mixed stands with *Acacia* spp. A second vegetation layer is often dominated by evergreen shrubs. In the Jebel al Akhdhar of Oman, for example, *J. excelsa* subsp. *polycarpos* is associated with *Olea europaea* and *Monotheca buxifolia*, with *Sageretia spiciflora* and *Dodonaea viscosa* growing in the shrub layer,

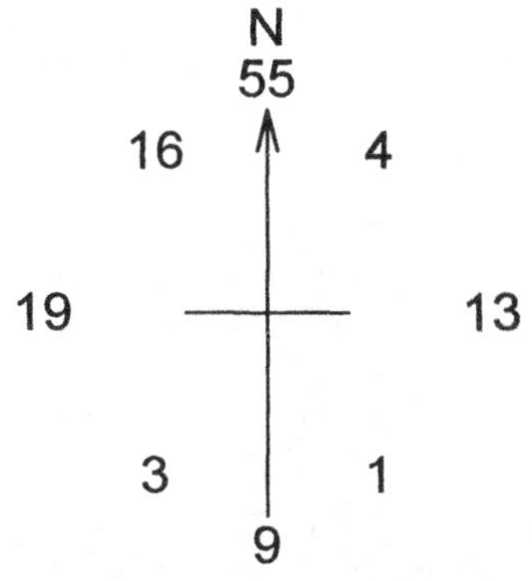

Figure 6.7 The compass position of the 120 *Juniperus* trees ≤2 m in height growing next to or beneath their 'nurse' trees in 32 hectares of Hayl Juwari, a woodland in the northern mountains of Oman. After Fisher and Gardner (1995).

and *Ebenus stellata, Helianthemum lipii* and *Cymbopogon schoenanthus* dominating the ground cover (Ghazanfar 1991b, Mandaville 1977).

In the northern mountains of Oman *J. excelsa* subsp. *polycarpos* is restricted to the highest areas of the central massif of Jebel al Akhdhar and the outlying mountains of Jebel Qubal and Jebel Kawr (Figure 6.6), where it generally forms open woodlands with a maximum density of *c.* 100 trees per hectare (Fisher & Gardner unpublished). On exposed slopes juniper is distributed from 2,100 m to the highest summit at 3,009 m, with no upper tree line, whilst on well-shaded north-facing slopes trees grow down to 1,375 m; elsewhere within its range this species grows at altitudes of up to 4,000 m, and only the lower part of the potential altitudinal range is therefore represented in Oman (Gardner & Fisher 1996). The low densities of these juniper woodlands compared, for example, to the much higher densities of the *J. excelsa* subsp. *polycarpos* woodlands of Baluchistan (M. Fisher pers. comm.), and their restriction to the relatively higher elevations on well-shaded slopes, is due to the fact that the more southerly Hajar mountains are considerably drier than other areas within the species' range.

The vital importance of moisture for this species can be seen in its use of 'nurse' plants (older *J. excelsa* subsp. *polycarpos, Olea europaea, Daphne mucronata, Lonicera aucheri, Monotheca buxifolia* and *Berberis* sp.) for shelter when young, and by the fact that seedlings are mostly found underneath the northerly side of the canopy of such plants (Figure 6.7). The further importance of moisture in what is probably in general a marginal environment for this species can be seen in the clustering of larger, better condition trees within and around low-lying areas of seasonal water flow (Figure 6.8).

Of recent cause for concern is the observation of widespread decline in both the *J. excelsa* subsp. *polycarpos* woodlands of Oman and the *J. phoenicea* and *J. procera* woodlands of Saudi Arabia (Figure 6.9; Asmodé 1989, Fisher 1997, Fisher & Gardner 1995, Gardner & Fisher 1994, 1996, Hajar *et al.* 1991). This

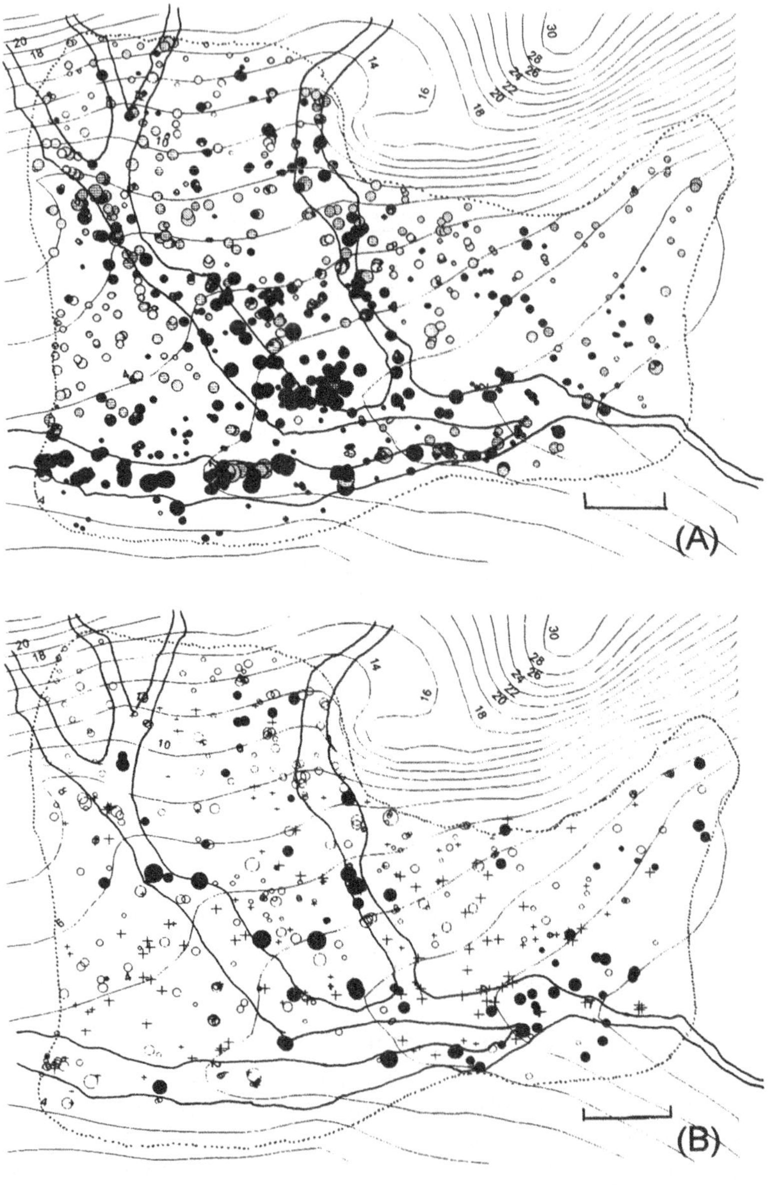

Figure 6.8 Map of individual *Juniperus excelsa* subsp. *polycarpos* in the valley of Hayl Juwari in the northern mountains of Oman. Limits of the surveyed area are indicated by a dashed line and wadis by a solid line. Contour intervals are at 2 m with 0 m passing through 2,250 m. Scale bar represents 100 m. (A) All living trees, with circle diameter proportional to tree height and degree of shading indicating tree condition, with darker being better. (B) All dead trees, with circle diameter proportional to tree height and dark shading representing death apparently due to burning and no shading representing death due to other causes; trees of unknown height are represented by crosses (large = burnt, small = cause of death unknown). Modified after Fisher and Gardner (1995).

Figure 6.9 (A) Stand of very poor condition juniper trees on the exposed slopes above Dar Sawda, Jebel al Akhdhar at 2,250 m. (B) Juniper trees in good condition growing on the summit plateau at 2,600 m above Hayl Juwari at the western end of Jebel al Akhdhar. After Fisher and Gardner (1995).

decline appears to be largely occurring at lower altitudes, and is most marked on exposed slopes, with some stands consisting of mostly dead or dying trees (Figure 6.10). Mean tree condition, proportion of trees bearing berries and male cones and proportion of living trees decline with decreasing altitude in a similar way in the open *J. excelsa* subsp. *polycarpos* woodlands of Oman and the denser *J. procera* woodlands of Raydah Reserve in the Asir highlands of Saudi Arabia (Figure 6.11). The importance of water balance for *J. excelsa* subsp. *polycarpos*, as indicated by the critical influence of macro- and micro-topography on local patterns of

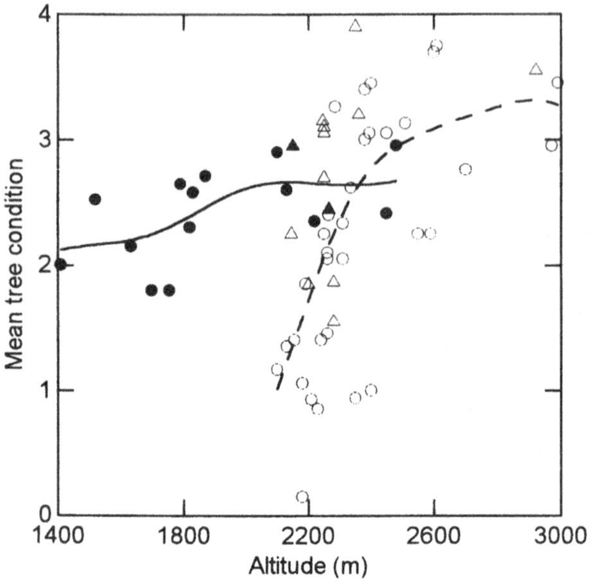

Figure 6.10 Variation with altitude of the mean tree condition (where 0 = dead, and 1-4 represent 25, 50, 75 and 100% respectively of tree alive) of sixty juniper woodland sites in the northern mountains of Oman; filled symbols indicate sites that receive some degree of shade during the day and unfilled symbols sites that are in the open; triangles indicate sites within wadis (i.e. relatively more mesic sites) and circles non-wadi sites (i.e. relatively more xeric sites). Smoothed lines (with distance-weighted least squares smoothing) are through shaded (continuous) and unshaded (dashed) sites. After Fisher and Gardner (1995).

distribution, and the coincidence of the transition zone between healthy and unhealthy *J. procera* woodland (*c.* 2,400-2,500 m, Figure 6.11) with the lower limit of moisture-dependent fruticose lichens (see also Chapter 5), suggests that decline may be related in some way to climate (Fisher 1997).

In this context Fisher (1997) proposed four hypotheses for the decline of the juniper woodlands of the Arabian Peninsula, all involving climate changes, though operative at different spatial and temporal scales: (1) Overgrazing by domestic livestock has altered local vegetation structure, causing decline at lower altitudes through effects on microclimate. (2) The global temperature rise of the twentieth century (Houghton *et al.* 1990), with elevated spring temperatures in the Middle East (Nasrallah & Balling 1993), is causing woodland decline through temperature-induced dieback at the lower juniper ecotone. (3) Dieback is caused by periodic droughts combined with long regeneration cycles, the effects of which are more marked at the lower, hotter elevations. (4) The present arid phase in the climate of Arabia, which began between 4,000 and 6,500 years ago (McClure 1984, Sanlaville 1992), is still developing, causing woodland dieback through a long-term increase in aridity. Given the slow growth of these *Juniperus* spp., the investigation of these hypotheses is a major conceptual challenge. The most

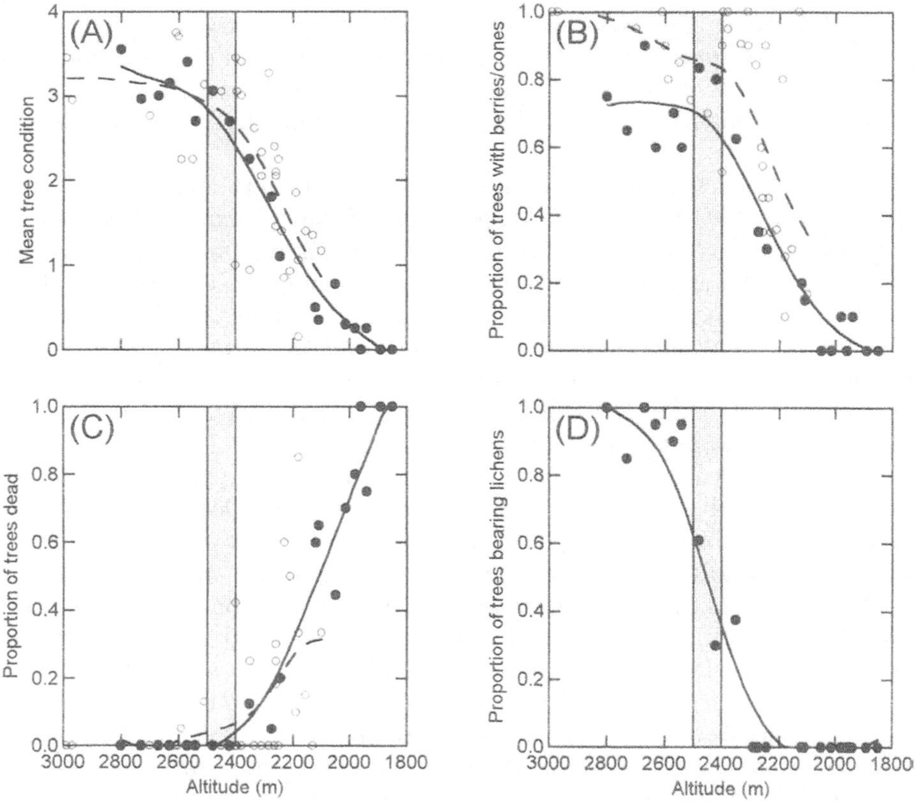

Figure 6.11 Variation with altitude of (A) mean tree condition (see legend to Figure 6.10 for definition), (B) proportion of trees with berries and/or male cones, (C) proportion of trees dead, and (D) proportion of trees bearing fruticose lichens in Raydah (see section 12.4.5.3) in the Asir of Saudi Arabia (•, —) and in the northern mountains of Oman (○, ----) from Figure 5 of Gardner & Fisher, 1996). Fruticose lichens do not occur in the northern Oman mountains. Curves were produced with distance-weighted least squares smoothing using the same smoothing parameters for both data sets. The altitude axis is in descending order to emphasise the decline in tree condition with decreasing altitude. The shaded area indicates the zone of transition from a healthy to an unhealthy woodland. Modified after Fisher (1997).

important tool in this investigation will probably be that of dendroecology (Fisher 1995b, Fisher & Gardner in press).

6.3.2 Drought-deciduous and Semi-evergreen Woodlands

Drought-deciduous *Acacia* and semi-evergreen *Olea europaea* open woodlands are the dominant vegetation over the lower altitudes of most of the western, southwestern and eastern mountains of the Peninsula. *Acacia asak*, *A. etbaica* and

A. *tortilis* occur up to *c*.1,500 m, associated with *Maerua crassifolia* in the Hijaz and *Grewia villosa*, *G. mollis*, *G. tembensis*, *Barleria trispinosa*, *Maytenus* spp. *Pyrostria* sp., *Adenium obesum* and *Anisotes trisulcus* in the Asir (König 1986, 1988, Miller & Cope 1996, Vesey-Fitzgerald 1957b). The genus *Maytenus* is very diverse in the montane zone of Arabia, being represented by 17 species (Sebsebe 1985).

In the Yemen escarpment mountains *Acacia* (*A. asak* and *A. mellifera*) is associated with *Commiphora* (*Commiphora gileadensis*, *C. kataf*, *C. myrrha* and *C. abyssinica*) and other tree species such as *Terminalia brownii*, *Cordia abyssinica*, *Adina microcephala* and *Ficus* spp. to form the dominant woody vegetation. In the eastern desert of the Yemen escarpment a mosaic of open *Acacia-Commiphora* shrubland with *Euphorbia balsamifera*, *Maerua crassifolia*, *Ziziphus spina-christi* and occasional *Dracaena serrulata* occur down to 1,300 m (Al-Hubaishi & Müller-Hohenstein 1984, Miller & Cope 1996, Scholte *et al.* 1991).

The *Acacia-Commiphora* shrubland also occurs in the Dhofar mountains, dominated on the seaward-facing escarpments by the endemic *Anogeissus dhofarica*, which is associated with *Carissa edulis*, *Croton confertus*, *Dodonaea angustifolia*, *Euclea racemosa* subsp. *schimperi*, *Euphorbia smithii*, *Jatropha dhofarica*, *Maytenus dhofarensis*, *Olea europaea*, *Rhus somalensis* and *Sterculia africana*. *Anogeissus dhofarica* is replaced by *A. benthii* in southern Yemen. On the drier slopes *Acacia etbaica* occurs with *Boswellia sacra*, *Maerua crassifolia* and *Dracaena serrulata* (Miller & Morris 1988, Radcliffe-Smith 1980). *Acacia tortilis* and *A. gerardii*, associated with *Maerua crassifolia*, *Acridocarpus orientalis* and *Moringa peregrina* occur on the slopes of the Hajar and Musandam mountains of northern Oman. Other woody associates are *Prunus arabica* and the endemic *Ceratonia oreothauma* subsp. *oreothauma*, the former found in Musandam and the eastern Hajar and the latter only in the eastern Hajar mountains.

In the mountains of northern Oman four plant communities can be distinguished from 650 to 2,800 m (Ghazanfar 1991b). Altitude has a strong influence on the structure of the vegetation, with species richness greatest from 1,000 to 1,480 m (Figure 6.12). There is a high diversity of growth-forms on the mountains, but chamaephytes and hemicryptophytes are the most common.

In the mountains of western Arabia from 1,500 to 1,800 m the *Acacia asak*-*A. etbaica* woodland is replaced by *Acacia origena* woodland. *Acacia origena* and *A. gerardii* replace *A. tortilis* and the other *Acacia* species dominant on the lower slopes and form open semi-evergreen bushland or open woodlands with *Barbeya oleoides*, *Celtis africana*, *Ehretia abyssinica*, *Monotheca buxifolia*, *Olea europaea*, *Pistacia falcata* and *Juniperus*.

In most mountainous areas, and particularly in Yemen, the tree cover has been destroyed and replaced by agricultural terraces, although an evergreen sclerophyllous shrubland has survived on steep slopes. Its characteristic species in southwest Arabia are *Barbeya oleoides*, *Berberis holstii*, *Buddleja polystachya*, *Cadia purpurea*, *Carissa edulis*, *Catha edulis*, *Debregeasia saeneb*, *Dodonaea viscosa*, *Dombeya torrida*, *Ehretia cymosa*, *Heteromorpha arborescens*, *Hypericum revolutum*, *Ochna inermis*, *Olea europaea*, *Osyris abyssinica*, *Pittosporum viridiflorum*, *Psiadia arabica*,

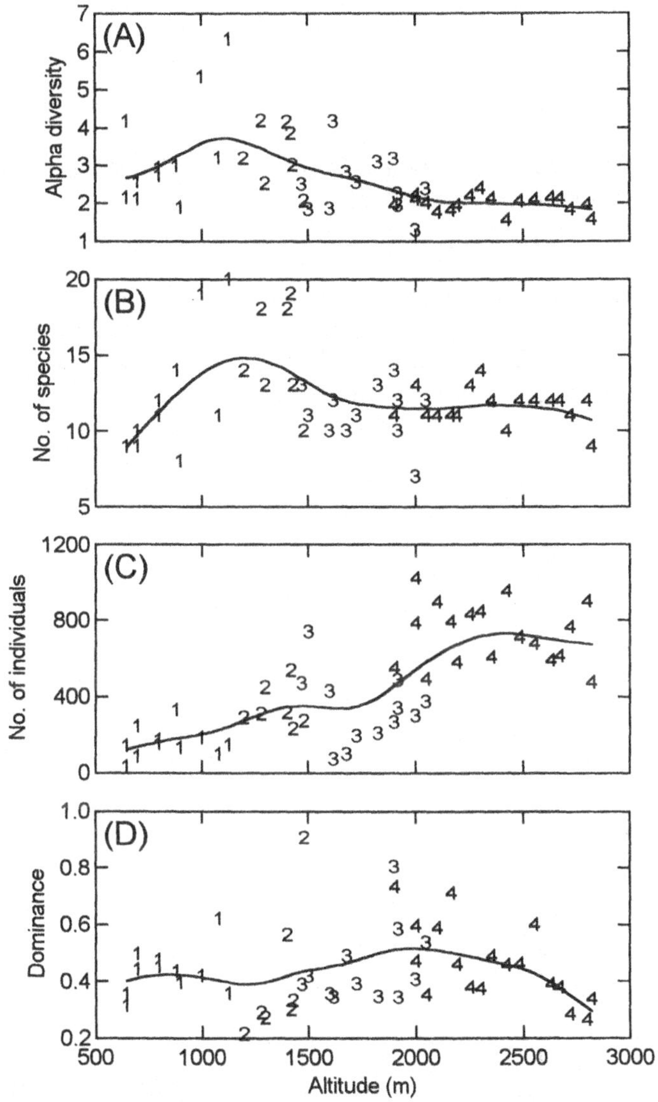

Figure 6.12 Relationship of α-diversity (A), species richness (B), number of individuals (C) and dominance (D) with altitude in the four plant communities (1-4) on Jebel al Akhdhar in the northern mountains of Oman. 1, *Acacia tortilis-Rhazya-Fagonia* (650-1,130 m); 2, *Euphorbia larica* (1,130-1,480 m); 3, *Monotheca-Olea-Dodonaea* (1,480-2,050 m); 4, *Juniperus-Ephedra-Teucrium* (2,050-3,000 m). After Ghazanfar (1991b).

P. punctulata, Pouzolzia mixta, Rhamnus staddo, Sageretia spiciflora, Maesa lanceolata, Maytenus spp., Monotheca buxifolia, Myrica salicifolia, Myrsine africana, Nuxia congesta, Tarchonanthus camphoratus, Teclea nobilis, Rhamnus staddo, Rhus retinorrhoea and *Rosa abyssinica*. Most of these species are Afromontane or Eritreo-Arabian in distribution.

6.4 Rock Communities

"We often thought that the Romans should have named this part of the Peninsula Arabia Spinosa rather than Arabia Felix."

Scott (1942)

Plant communities of rocky habitats are characteristic of mountainous regions throughout the Arabian Peninsula. They have been studied on the western escarpments from the Asir mountains through to the Haraz mountains in the Yemeni highlands, on the southern limestone plateaux from the Jol plateau to the Dhofar mountains, on the Haggier mountains of Socotra, on the eastern escarpment of the Hajar mountains from Jebel al Akhdhar to Musandam, in the deeply incised wadis Bana and Hadhramaut, on the flat plains of the Audhali

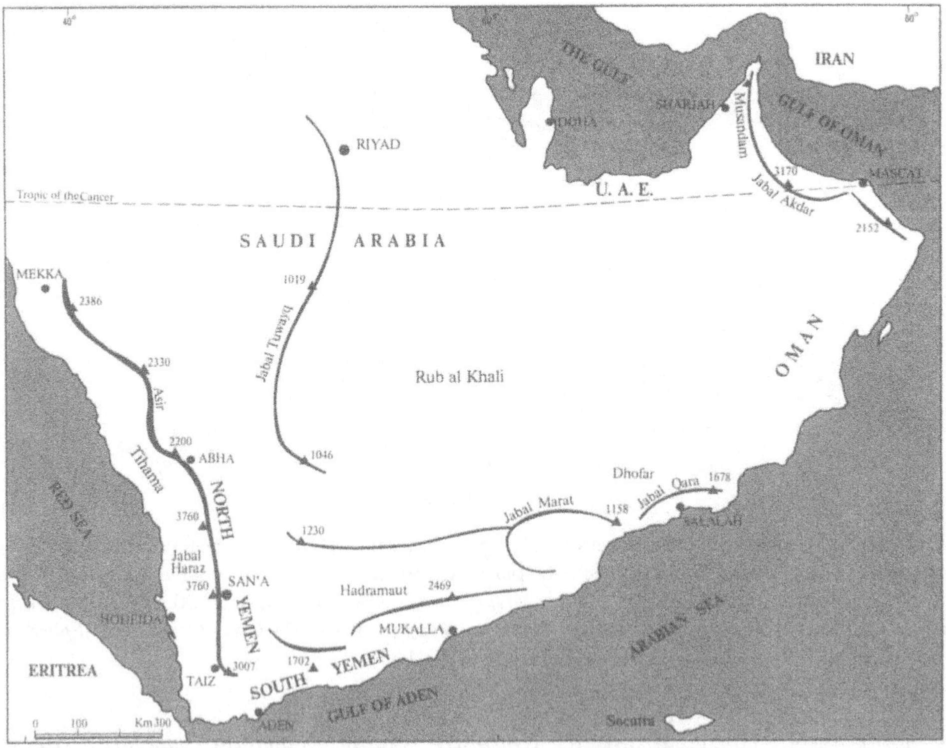

Figure 6.13 The location of the main areas where the plant communities of rocky habitats ('rock communities') have been studied.

plateau in the rain-shadow of the surrounding mountain ridges, and on Jebel Tuwayq in central Saudi Arabia (Deil 1988, 1989, Deil & Müller-Hohenstein 1984, 1985, 1996, Ghazanfar 1991b, Grosser 1988, König 1987, Kürschner 1986c, Mandaville 1985). The description of the vegetation of rocky habitats is based largely on work that has been carried out on the plant communities of southern Arabia (Figure 6.13). This region contains most of the rocky habitat types of the Peninsula and covers a wide range of bioclimatic zones. The area falls within the transition zone from the Somalia-Masai regional centre of endemism at low and middle altitudes to the Afromontane archipelago-like centre of endemism at high altitudes, and above the tree line even taxa of the Mediterranean region of the Irano-Turanian phytochorion are present (see Figure 4.2).

Rock vegetation is relatively diverse, and can be grouped into several phytosociological classes based on the syntaxonomic treatment proposed by Knapp (1968) for East Africa (for syntaxonomic details see Table 1 in Deil 1991). Each class, characterized by its particular combination of species and dominant life-forms, is distinguished by its position on gradients of temperature and humidity (Figure 6.14).

Dry rocks at lower and middle altitudes (1,000-1,500 m) are colonized by succulent *Euphorbia* spp., *Aloe* spp., Asclepiadaceae taxa from the Stapelieae group and *Kleinia* spp. of the Kleinio-Carallumetea class. On wet rocks at higher altitudes in northern Yemen and the Asir mountains, other succulent species (*Aloe* spp. and Crassulaceae) and candelabrous shrubs form the Crassulo-Aloetea communities. Communities with small poikilohydric ferns occur on terraced farm walls and in small rock fissures. At middle altitudes (1,500-2,000 m) these communities belong to the Cheilanthi-Actiniopteridetea class and at higher altitudes (2,000-3,000 m) to the Asplenion aethiopicum group. Rock cavities

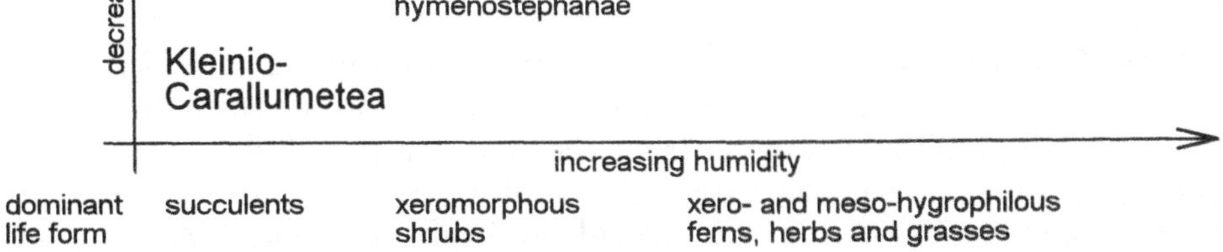

Figure 6.14 Rock vegetation community classes, characterized by particular combinations of species and dominant life-forms, can be distinguished by their positions on gradients of temperature and humidity.

with dripping water contain the hygrophilous Adiantetea capilli-veneris communities.

Although flat, stony habitats colonized by shrub communities have not yet been studied in detail, the Carallumo petraeae-Euphorbietum adenensis community is found on the plateau between San'a and Ma'rib (Deil & Müller-Hohenstein 1984), the Elionuro-Macowanietum in Yemen and the Pseudogaillonio hymenostephanae-Euphorbietum laricae communities in the United Arab Emirates and Oman (Deil & Müller-Hohenstein 1996). The succulent communities are described below, followed by the fern-rich and shrub communities.

6.4.1 Dry Rock Communities Rich in Succulents

Succulent communities are largely restricted to two types of rocky habitat. Firstly, areas with steep slopes, shallow soils and rocky surfaces, in which the plants are rooted in rock fissures and runoff is considerable. Secondly, inhabited and abandoned man-made habitats surrounding settlements and cemeteries, where succulent communities have replaced the natural vegetation (Deil 1988). Succulents, protected by their spines or by chemical repellents (such as the necrotic effects of *Caralluma penicillata* when browsed by sheep, Mossa *et al.* 1983), are especially abundant in ruderal places. The high nitrogen content of soils near villages also favours the presence of succulents. The introduced cactus species *Opuntia dillenii* and *O. ficus-indica,* which often occur on dry rocky areas, have, to a large extent, invaded the endemic succulent communities (Ellenberg 1989), facilitated by the dispersal of their seeds by baboons. Quasi-natural habitats, such as the rocks and rock surfaces which supply some of the water by runoff to the sophisticated rainwater-harvesting system of terraced agriculture, are often colonized by succulents (Deil 1988).

Only in the subtropical Yemen highlands, the northern slopes of Jebel Qara and Jebel Qamr in Dhofar and in the arid coastal plains along the Gulf of Aden, (see the distribution of drought-deciduous/semi-succulent shrublands in Arabia, Figure 4.1) do succulents occur on flat rocky areas, forming communities with *Euphorbia balsamifera* subsp. *adenensis* (Figure 6.15: A) and *Dracaena*.

6.4.1.1 Kleinio-Carallumetea Communities

Species of *Euphorbia* form the main components of this community, which occurs at the frost-free lower and middle altitudes. The distribution and altitudinal ranges of *Euphorbia* spp. along a transect from the Red Sea near Luhayyah over the Haraz mountains and the San'a basin to Ma'rib is given in Figure 6.16. These *Euphorbia* spp. are associated with succulents from other families. These are Asclepiadaceae (*Sarcostemma, Ceropegia, Caralluma, Huernia* and *Duvalia*), stem-succulent Compositae (*Kleinia* and *Senecio*), leaf-succulent Liliaceae (*Aloe*) and Agavaceae (*Sansevieria*), succulent climbers in Vitaceae (*Cissus*) and pachycaulous

A

B

Figure 6.15 (A) The limestone plateau between San'a and Ma'rib, covered by a *Euphorbia balsamifera* community. (B) Fortified villages and abandoned fortresses are secondary habitats for chasmophytic succulents; this photograph illustrates a *Euphorbia cactus* community near Mabyan on the Yemeni escarpment.

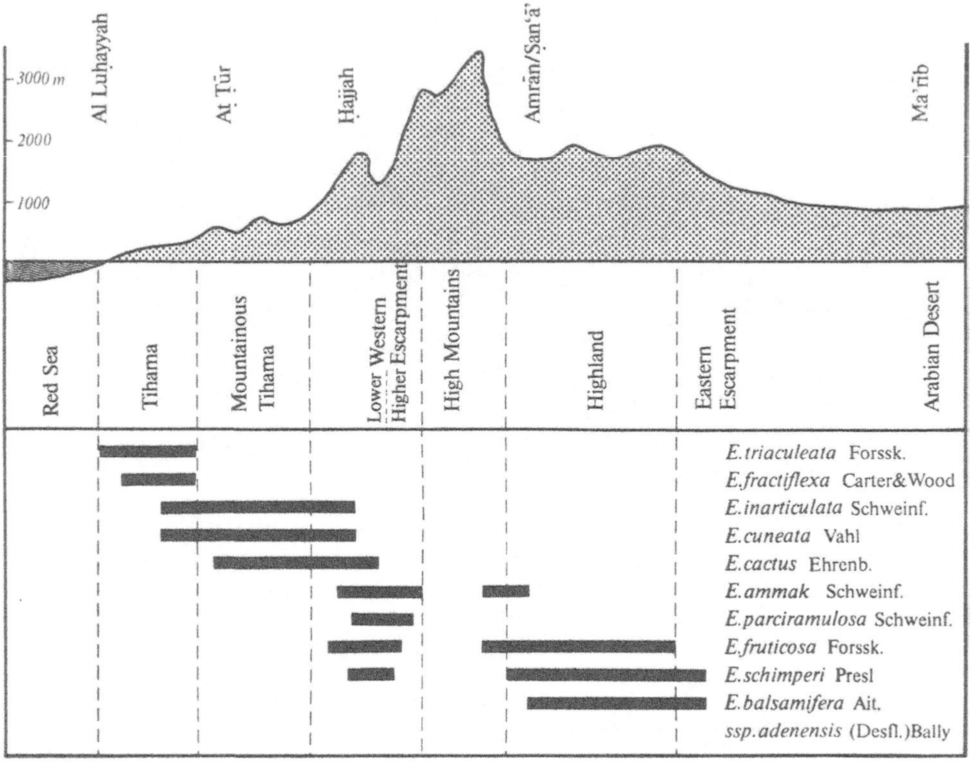

Figure 6.16 Distribution of succulent *Euphorbia* spp. on a transect through the central parts of the western Yemeni escarpment.

trees such as *Adenium obesum* (Apocynaceae) and *Adenia venenata* (Passifloraceae). These associations are described below.

Rocky outcrops and granite domes in the Yemeni foothills are colonized by the Aloe sabaeae-Euphorbietum inarticulatae community in which *Aloe sabaea* and *Euphorbia inarticulata* are the dominant species (Figure 6.17, left). The Carallumosubulatae-Euphorbietum inarticulatae can be found in ruderal locations around villages and on abandoned settlements (Figure 6.17, right). Both associations have close floristic affinities to similar stands in Eritrea and northern Somalia and belong to the Euphorbion inarticulatae community-group.

Three associations have been studied at middle altitudes (1,600-2,000 m) in Yemen. These are, from north to south, the Aloe yemenicae-Euphorbietum cacti in the surroundings of Hajjah at Jebel an Nasirah, the Aloo menachensis-Euphorbietum cacti near Manacha in the Haraz mountains and the Ceropegio rupicolae-Euphorbietum parciramulosae on rocky outcrops near Taiz. *Euphorbia cactus* (Figure 6.15: B) is the dominant species in all three associations. Other characteristic associated species are *Plectranthus* spp., *Caralluma cicatricosa* and *Ceropegia rupicola*. Vicarious stands occur in Dhofar (Miller & Morris 1988,

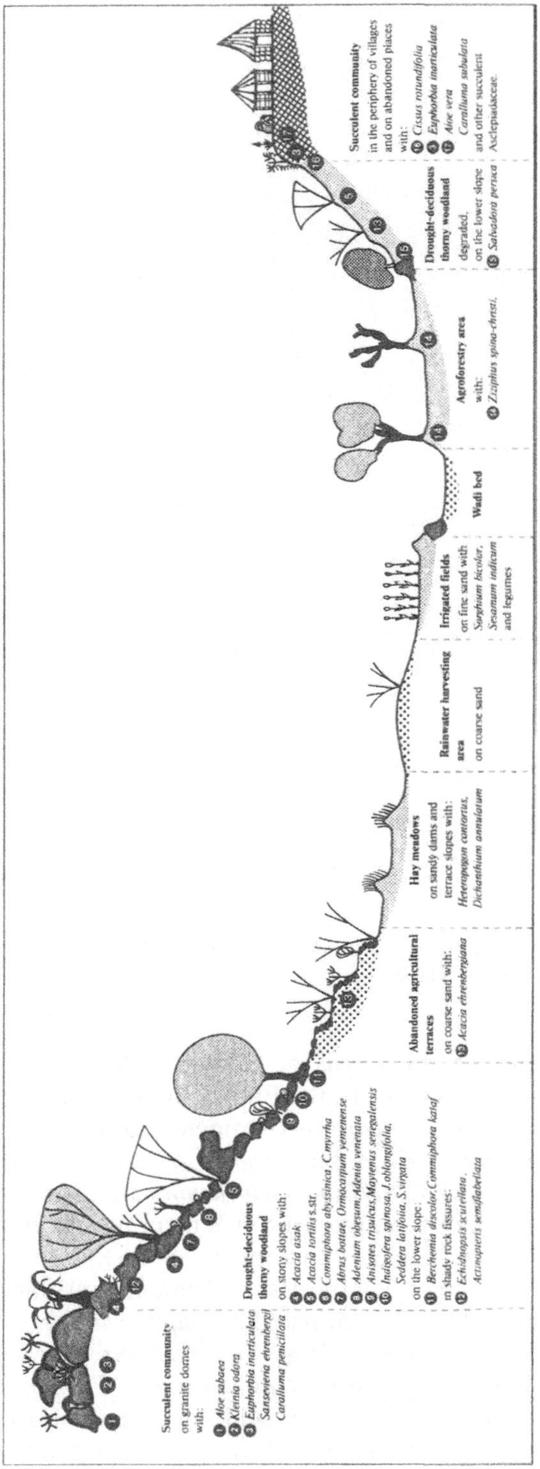
Succulent community
on granite domes
with:
1 Aloe sabaea
2 Kleinia odora
3 Euphorbia inarticulata
Sansevieria ehrenbergii
Caralluma penicillata
Drought-deciduous
thorny woodland
on stony slopes with:
4 Acacia asak
5 Acacia tortilis s.str.
6 Commiphora abyssinica, C.myrrha
7 Abrus bottae, Ormocarpum yemenense
8 Adenium obesum, Adenia venenata
9 Anisotes trisulcus, Maytenus senegalensis
10 Indigofera spinosa, J.oblongifolia,
Seddera latifolia, S.virgata
on the lower slope:
11 Berchemia discolor, Commiphora kataf
in shady rock fissures:
12 Echidnopsis scutellata,
Actinopteris semiflabellata
Abandoned agricultural
terraces
on coarse sand with:
13 Acacia ehrenbergiana
Hay meadows
on sandy dams and
terrace slopes with:
Heteropogon contortus,
Dichanthium annulatum
Rainwater harvesting
area
on coarse sand
Irrigated fields
on fine sand with
Sorghium bicolor,
Sesamum indicum
and legumes
Wadi bed
Agroforestry area
with:
14 Ziziphus spina-christi,
Drought-deciduous
thorny woodland
degraded,
on the lower slope
15 Salvadora persica
Succulent community
in the periphery of villages
and on abandoned places
with:
16 Cissus rotundifolia
3 Euphorbia inarticulata
17 Aloe vera
Caralluma subulata
and other succulent
Asclepiadaceae.

Figure 6.18 *Euphorbia fruticosa*, a member of the stem succulent section *Diacanthium*, scattered over an *hamada* plain near Huth in Yemen.

Radcliffe-Smith 1980), in the mist oasis of Erkwit in Sudan and along the Eritrean escarpment. On superficial soils, the Euphorbion cacti communities are replaced by evergreen shrubland (Hyperico-Rhamnetea) and on deeper soils, by drought-deciduous woodland with *Acacia negrii, A. origena* and *A. hockii*. *Euphorbia fruticosa* (Figure 6.18), *Caralluma meintjesiana* and *Aloe vacillans* grow on the leeward side of the western escarpment in Yemen, in rock fissures and in the basins. Studies in this region are incomplete, but the succulent plant communities of the semidesert region of the eastern Yemeni escarpment are better known. The open shrubland there is dominated by *Euphorbia balsamifera* subsp. *adenensis*, with associated succulents such as *Aloe vacillans, Caralluma edulis, C. hexagona* (Figure 6.21: C), *C. petraea, C. quadrangula, Euphorbia schimperi, Kleinia odora* and *Sarcostemma viminale* (Deil & Müller-Hohenstein 1984). On the Audhali plateau in southern Yemen *Euphorbia balsamifera, Caralluma penicillata,*

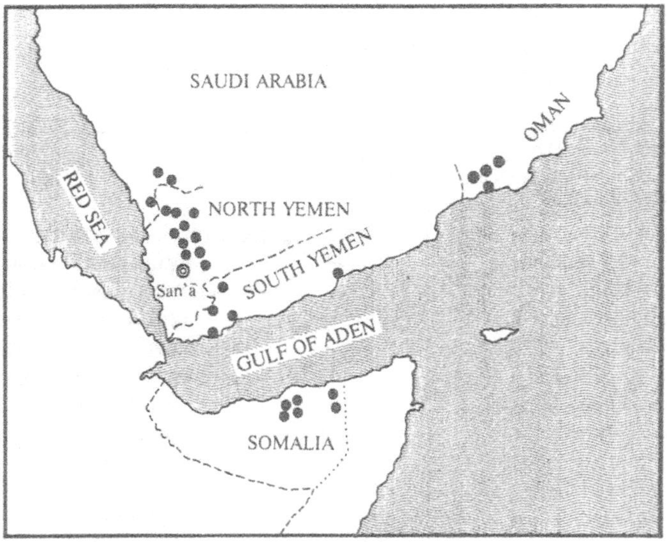

Figure 6.19 Distribution of *Euphorbia balsamifera* subsp. *adenensis* in the southwest of the Arabian Peninsula, in Somalia and on Jebel Qara in Dhofar.

C. plicatiloba, C. quadrangula, Dorstenia foetida and *Echidnopsis squamata* occur (Rauh 1966), and associations of *Euphorbia applanata* with *E. hadramautica, Dorstenia foetida, Endostemon tereticaulis* and *Haplophyllum amoenum* occur on cliffs in the Hadhramaut (Thulin & Al-Gifri 1995).

Euphorbia balsamifera subsp. *adenensis* is also common on the limestone escarpment mountains of Jebel Qara in Dhofar and is locally abundant on the coastal plains of southern Yemen and northern Somalia (Figure 6.19). The effect of strong daytime insolation on *Euphorbia balsamifera* is reduced by the high albedo of its whitish branches. The plant creates a microclimate within its canopy which is favourably different from the surrounding climate (Mies & Aschan 1995).

Rock communities with *Dracaena serrulata* have not yet been studied in detail on the Arabian mainland. They occur from the Asir to Dhofar, and in Socotra they are characterized by *Dracaena cinnabari*. Vicarious communities can be found in Djibouti (with *Dracaena ombet, Aloe trichosantha* and *Caralluma priogonum*) and in Northern Somalia (the *Acocanthera-Euphorbia abyssinica-Dracaena ombet*-belt with *Aloe somalensis, A. trichosantha, Caralluma speciosa, C. dicaputae* and *Euphorbia turbiniformis*).

Cliff vegetation on Socotra is also dominated by succulents (Boulos *et al.* 1994, Gwynne 1968, Mies 1995, Mies & Zimmer 1994, Popov 1957, Vierhapper 1907) and is high in endemic species, even at the generic level, with monospecific genera such as *Duvaliandra* and *Socotra* (Asclepiadaceae). Most of the species (such as *Dracaena cinnabari, Kleinia scotii, Cissus subaphylla, Euphorbia septemsulcata,*

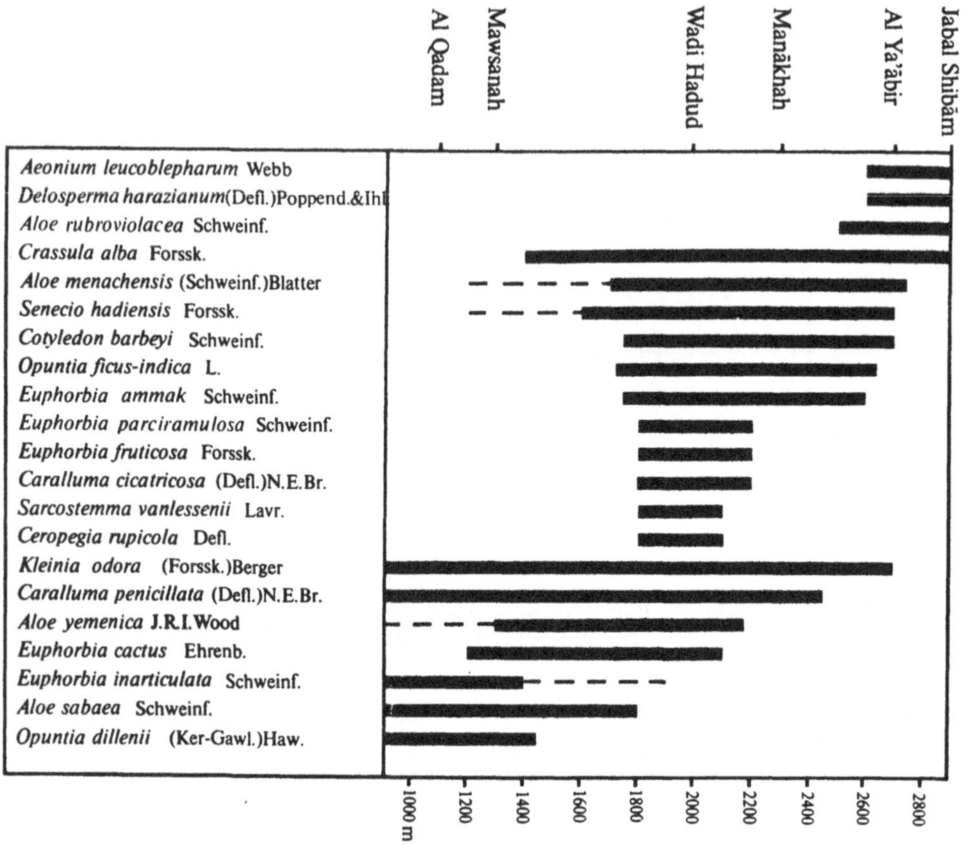

Figure 6.20 Altitudinal range of succulents in the Haraz mountains of Yemen.

E. spiralis and *Aloe perryi*) have vicarious relatives on the mainland. Common species of the mainland communities are rare on Socotra.

6.4.1.2 Crassulo-Aloetea Communities

In the upper montane belt between 1,800 and 2,000 m the succulent Euphorbiaceae and Asclepiadaceae are replaced by the succulent Crassulaceae and frost resistant *Aloe* spp. (Figure 6.20). These succulent communities have close floristic similarities with the Afromontane zone of East Africa and can therefore be included in the Crassulo-Aloetea class proposed by Knapp (1968) for the Ethiopian highlands. *Aeonium leucoblepharum, Cotyledon barbeyi, Crassula* spp. and *Kalanchoe* spp. are always present in these communities in Arabia. These are the characteristic species of the Centaurothamno-Aeonion leucoblephari, an alliance which is named after the montane southwest Arabian *Centaurothamnus maximus* and the montane Abyssinian-Eritreo-Arabian *Aeonium leucoblepharum* (Figure 6.21: D). Other characteristic species are *Delosperma harazianum, Crassula schimperi, C. alba* and the endemic *Phagnalon woodii, P. harazianum* and *P. stenolepis*

(Qaiser & Lack 1985). Two associations, the Aeonio-Aloetum tomentosae in semi-humid regions of the Sumarah Pass and Jebel Raymah at 2,700-3,700 m on the Yemeni escarpment, and the Aeonio-Aloetum rubroviolaceae (Figure 6.21: A, D) on the more humid slopes of Jebel Haraz at 2,200-2,800 m have been studied. Northern outliers of the latter group have been noted in the Asir (König 1987).

6.4.2 Humid Rock Communities Rich in Ferns and Bryophytes

6.4.2.1 Cheilanthi-Actiniopteridetea Communities

At lower and middle altitudes in Yemen and Dhofar (300-1,800 m), communities with small, xeromorphic ferns of the family Sinopteridaceae (*Cheilanthes catanensis, C. fragrans* and *C. coriacea*), *Actiniopteris semiflabellata* and *A. radiata* (Pichi-Sermolli 1962) and the cryptogam *Selaginella imbricata* grow on terrace walls and in rock fissures. These exhibit floristic and ecological similarities with rock communities in the thermo-Mediterranean area (Perez *et al.* 1989), in southwestern Australia (Pignatti & Pignatti 1995) and in the high Andes of Peru (Gutte 1986). In the *Combretum molle* ravine forests of the Asir these ferns are associated with the endemic bryophyte *Fissidens arabicus* (Pursell & Kürschner 1987), and on Socotra with *Adiantum balfourii, Begonia socotrana* and *Exacum affine* (Popov 1957).

Fern communities in the *Monotheca-Olea* zone of the northern mountains of Oman may belong to the same class, associated with *Dionysia mira, Cheilanthes fragrans, Viola cinerea, Phagnalon viridiflorum* and *Umbilicus botryoides* (Mandaville 1977). In Jebel Qara in Dhofar the humus-rich shaded cliffs are colonized by *Remusatia vivipara, Hypodematium crenatum* and *Adiantum incisum* (Radcliffe-Smith 1980). In drier rock fissures, small succulents such as *Dorstenia foetida* (Figure 6.21: B) and *Talinum portulacifolium* are associated with poikilohydric ferns.

6.4.2.2 Asplenietea trichomanis Communities

Fern communities at high altitudes (1,800-3,000 m) are dominated by *Asplenium aethiopicum, A. trichomanes, A. ceterach* and *Cheilanthes farinosa. Asplenium aethiopicum* may be the characteristic species for this Sudanian-Arabian group of fern communities. This group is the most southerly and most impoverished outlier of the Mediterranean class Asplenietea trichomanis. It is also recorded from Jebel Marra in Sudan (Miehe 1988) and from Jebel Aswad in the eastern Hajar mountains of Oman (Mandaville 1977).

A B C D

Figure 6.21 (A) *Aloe rubroviolacea*, endemic on basalt cliffs in the Yemeni escarpment between 2,500 and 3,000m. (B) *Dorstenia foetida* (Moraceae) in rock fissures in the Haraz mountains. (C) Ascepiadaceae from the tribe Stapelieae occupy the dwarf succulent niche in rock habitats at lower and middle altitudes; the most species rich genus is *Caralluma*, here represented by *C. hexagona*. (D) *Aeonium leucoblepharum*, a Crassulaceae endemic to the high Yemen and Abyssinian highlands.

6.4.2.3 Adiantetea capilli-veneris Communities

Hydrophilous plant communities, dominated by the maidenhair fern *Adiantum capillus-veneris* and chalk incrusted mosses occur throughout the subtropical regions of the world. They belong to the class Adiantetea (Deil 1996). Two associations occur in Arabia: the Adianto-Epipactidetum veratrifoliae community which grows along permanent water courses in the Hajar and Musandam mountains of Oman (Deil & Müller-Hohenstein 1984, Ghazanfar 1992a, Mandaville 1977) and in the Jordan valley (Baierle 1993). Rock cavities with dripping water in the mountain belt of the escarpment mountains of southwest Arabia are colonized by the Adianto-Primuletum verticillatae, which consists of *Adiantum capillus-veneris* and *Primula verticillata.* In Wadi Sarfait in Dhofar the Adiantetea community includes *Pteris vittata* and *Samolus valerandi* (Radcliffe-Smith 1980).

6.4.3 Rock Vegetation Dominated by Shrubs and Herbs

6.4.3.1 Euphorbietea laricae Communities

These communities consist of shrubs dominated by *Euphorbia larica,* a species with an Omano-Makranian distribution from Baluchistan and southwest Iran through to northern Oman. On rock outcrops in the southeastern mountains of Arabia the open *Acacia tortilis-A. ehrenbergiana* scrub of the plains is replaced by these communities. Further to the west. *E. larica*, is replaced by other species from the section Tirucalli, *viz. Euphorbia masirahensis* on Masirah Island off the eastern coast of Oman, *E. dhofarica* in Dhofar and *E. schimperi* in southwest Arabia (Ghazanfar 1993a).

On the rocky outcrops of the Hajar mountains two species of the Rubiaceae, *Pseudogaillonia hymenostephana* and *Jaubertia aucheri* colonize fissures in the limestone and are associated with *Euphorbia larica.* This Pseudogaillonio hymenostephanae-Euphorbietum laricae community (Deil & Müller-Hohenstein 1996) can be included in the Euphorbietea laricae, a class originally described by Zohary (1963) for southern Iran. Other *E. larica* associations are with *Barleria, Commiphora, Convolvulus* and *Iphiona* in the foothills of the Hajar and Musandam mountains (Frey & Kürschner 1986, Ghazanfar 1991a, Mandaville 1985).

6.4.4 Arabo-Alpine Shrub Communities

Chasmophytic vegetation, dominated by chamaephytes, occurs at high altitudes in the mountains of the Peninsula. This vegetation is very different from the lowland, shrubby rock communities and has no species in common with it. The Arabo-Alpine vegetation is related more to the dwarf shrub communities of the Ethiopian highlands (= Helichryso-Micromerietea *sensu* Knapp 1968). Members of this community include the vicarious *Phagnalon* spp., which are distributed along the Yemen escarpment (Qaiser & Lack 1985). The genus shows a disjunct distribution not uncommon in the high montane vegetation of the Arabian Peninsula (see also Chapter 4). The north-facing basaltic cliffs near Jiblah in northern Yemen are colonized by *Helichrysum arwae* (Wood 1984), a species with its closest relatives in the Drakensberg mountains of South Africa. A similar Arabian-South African disjunction can be found in the genus *Delosperma* (Aizoaceae).

6.4.4.1 Elionuretea mutici Communities

Grass communities colonizing flat rocky ground with sparse soil cover, dominated by *Elionurus muticus,* were described from the pampa in northern Argentina by Eskuche (1992) and Fontana (1996) as the Elionuretea mutici class. An *Elionurus muticus-Themeda triandra* grassland on rocky outcrops is also described from South Africa (Eckhardt *et al.* 1995). Observations on similar plant communities in Arabia are fragmentary, although the Elionuro mutici-Macowanietum ericifoliae community, dominated by the dwarf shrub *Macowania ericifolia* (Asteraceae), occurs on flat, stony basalt layers in the Haraz mountains of central Yemen.

6.4.5 Other Herbaceous Rock Communities

Plant communities on wet rocks, dominated by hygrophilous herbs such as *Minuartia filifolia, Osteospermum vailantii, Felicia dentata, Cineraria abyssinica, C. schimperi, Arabis alpina, Silene burchelii* and *Saponaria umbricola*, occur in the Haraz mountains and near Hajjah in Yemen. *Albuca abyssinica* in Yemen and *A. pendula* in the Asir (Mathew & Collenette 1994) cover cliffs in the *Juniperus* zone, *Pelargonium* spp. in the *Olea-Barbeya* zone, and *Peperomia* spp. at the Sumarah Pass and on Jebel Qara in Dhofar (Radcliffe-Smith 1980).

6.4.6 Rock Vegetation Mosaics

Cliffs and rocky escarpments offer a great diversity of habitats and microhabitats according to their degree of exposure and incline, drainage and rock type. Within a specific altitudinal belt and climatic zone the pattern of the physical properties of a cliff results in a pattern of associated plant communities or a vegetation mosaic. Vegetation in Arabia has generally not yet been studied at this level of complexity, but two examples from the Yemeni escarpment can be given.

A vegetation mosaic on a basaltic cliff on an exposed humid western slope of Jebel Shibam near Manacha in the Haraz mountains is illustrated in Figure 6.22. At an altitude of 2,700 m the following associations occur, from top to bottom: (A) The Elionuro-Macowanietum in rock fissures on the flatter areas near the top

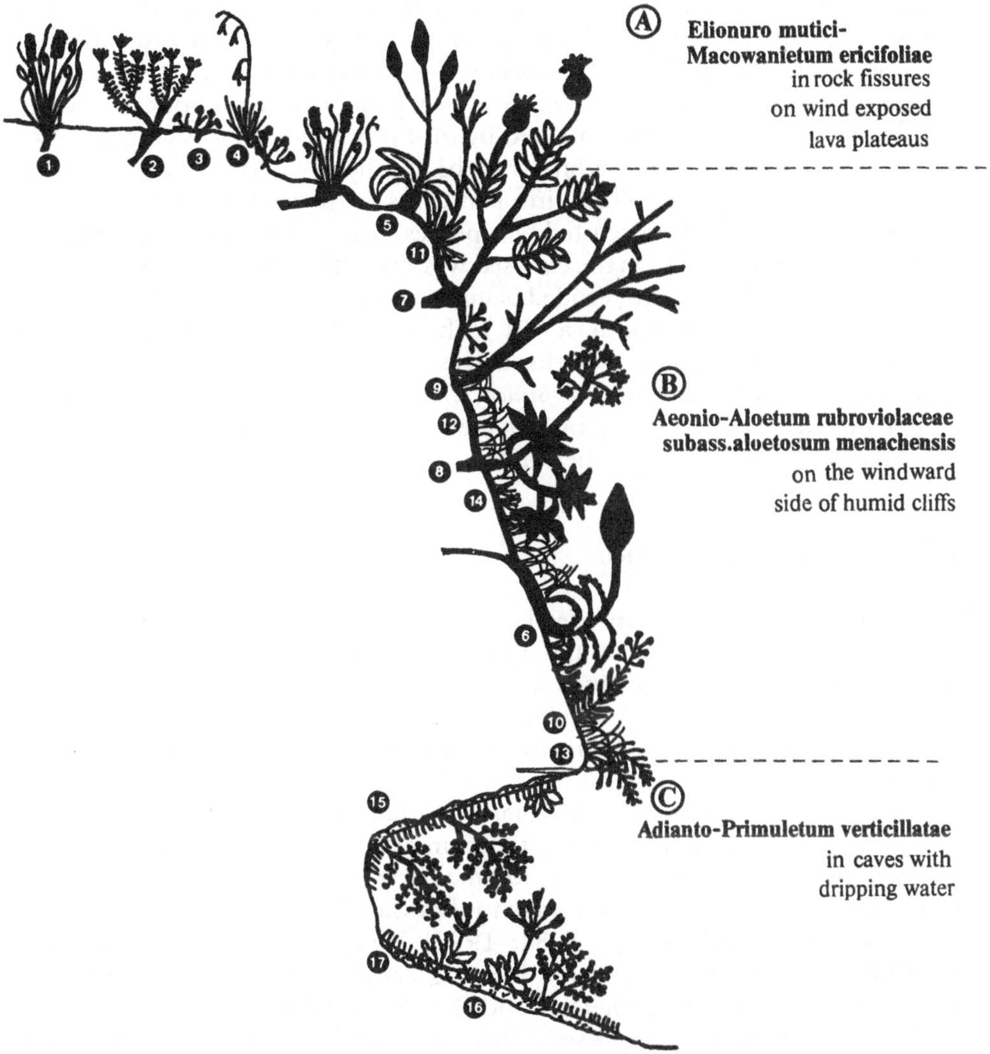

Figure 6.22 Vegetation mosaic on cliffs in the Haraz mountains of the Yemeni escarpment. (A) 1, *Elionurus muticus*; 2, *Macowania ericifolia*; 3, *Felicia dentata*, *Micromeria biflora*, *Scutellaria rubicunda* subsp. *arabica* and *Phagnalon* div. spec.; 4, *Hyparrhenia hirta*, *Aristida adscensionis* and other drought resistant grasses. (B) 5, *Aloe menachensis*; 6, *Aloe rubroviolacea*; 7, *Centaurothamnus maximus* and other shrubs and tall herbs such as *Acanthus arboreus*, *Vernonia leopoldii*, *Rumex limoniastrum* and *Plectranthus barbatus*. 8, *Aeonium leucoblepharum*; 9, evergreen xeromorphic shrubs such as *Rhamnus staddo*, *Maytenus parviflora*, *Hypericum revolutum* and *Myrsine africana*; 10, dwarf succulents such as *Crassula alba* and *C. schimperi*; 11, hygrophilous grasses such as *Phaenanthoecium moestlinii*, *Oryzopsis holciformis*, *Themeda triandra* and *Bromus leptoclados*; 12, low herbaceous plants such as *Minuartia filifolia*, *Silene burchelii* and *Campanula edulis*; 13, small ferns such as *Asplenium aethiopicum* and *Ceterach officinarum*; 14, bryophytes such as *Hypnum vaucheri*. (C) 15, *Adiantum capillus-veneris*; 16, *Primula verticilliata* s. str.; 17, *Gymnostomum aeruginosum*.

of the cliff, where water supply for this xeromorphic dwarf shrub community is provided by runoff from the bare rock surface. (B) The Aeonio-Aloetum rubroviolaceae, the endemic sub-association with *Aloe menachensis* (Figure 6.22) on steep cliffs, where water supply is relatively greater since the rock fissures are broader and deeper and supplemented by mist precipitation; this community is a mixture of succulents, mesophilous herbs and hygrophilous grasses. (C) The Adianto-Primuletum verticillatae in open caves at the foot of the cliff, with a continuous layer of chalk-incrusted bryophytes irrigated by dripping water.

A similar catena occurs on the exposed western cliffs at the Sumarah Pass near Ibb at 2,900 m in Yemen. An eroded plateau with *Macowania ericifolia, Dianthus uniflorus* and *Thymus laevigatus,* a cliff with an Aeonium leucoblepharum-Aloe tomentosa community, and tall herbs such as *Acanthus arboreus, Vernonia leopoldii, Plectranthus barbatus, Artemisia abyssinica, Thalictrum minus, Hypericum revolutum* and *Myrsine africana* growing on boulders, screes and man-made terraces in ruderal areas at the bottom of the cliff.

6.5 Orophytic Grasslands

The majority of the Arabian grasslands are secondary in nature. They have replaced the original *Juniperus* and *Acacia* woodlands and the evergreen *Olea-Barbeya-Tarchonanthus* woodlands which were destroyed by felling and burning. Plant species which indicate the former distribution of *Juniperus* woodlands include *Carex negrii* and *C. brunnaea*, which are also associated with *Juniperus procera* in East Africa (Hooper 1984).

The only remaining primary grassland vegetation in Arabia is now restricted to small Arabo-Alpine islands on the summits of the highest peaks in the southwest of the Peninsula above 2,500 m. These grasslands include tufted, low growing, perennial endemic species such as *Festuca cryptantha, Andropogon crossotos* and *Tripogon oliganthos* (Cope 1984, 1992). These are associated with cold resistant *Stipa* species and cushion-forming species such as *Cichorium bottae* (Figure 6.4: B), *Anthemis yemenensis* and *Dianthus uniflorus*.

Our knowledge of the grasses of the Arabian Peninsula has considerably improved during the last 15 years with studies on *Stipagrostis* (Freitag 1989b, Scholz 1969, 1983), *Oropetium* (Scholz & König 1984) and *Aristida* (Scholz & König 1988) and with the checklist of the Arabian genera of Poaceae by Cope (1985). Knowledge of the distribution patterns of *Oropetium africanum* (from northwestern Sahara to the central Arabian plains with an outpost in the Tibesti), *Stipagrostis socotrana* (from the Jol area to Socotra), *Chrysopogon macleishii* (endemic to the monsoon-affected mountains of Dhofar, with its closest relatives on the Indian subcontinent, Cope 1993) has contributed greatly to the understanding of the phytogeographical relationships of the Arabian flora.

Vicariant montane grass species occur in the genus *Stipa*. *Stipa tigrensis*, a dense tufted grass of open habitats grows from the upper montane to the subalpine zones in the *Juniperus* woodlands of the Asir and Yemen (Freitag 1989b). Its eastern vicariant is *Stipa mandavillei* found at 1,300-2,500 m on Jebel al Akhdhar in northern Oman. *S. mandavillei* grows with *Tripogon purpurascens* and *Cymbopogon schoenanthus*, a grass community which occurs in the *Monotheca buxifolia-Olea europaea-Dodonaea viscosa* zone of the northern mountains of Oman and the Zagros mountains of Iran.

The montane grasslands of western Arabia harbour a number of endemic species such as *Striga yemenica* at *c*. 2,600 m near Jiblah (Musselman & Hepper 1988) and *Habenaria aphylla* near Ibb (Cribb 1987). Characteristic members of montane grassland communities in the humid southwest corner of Arabia are Umbelliferae of Holarctic origin. These are *Peucedanum inaccessum,* endemic to the west-facing humid slopes of the escarpment in Jebel Melhan and Mahabishah (Townsend 1983), *Pimpinella woodii* on Jebel Badaan in the Ibb area (Townsend 1986a), and *Oreoschimperella arabiae-felicis* at 2,300 m near Jiblahand northwest of Abha on hillsides in the Asir at 2,600 m (Townsend 1986b).

6.5.1 Grass Communities

The communities of the orophytic grasslands of the Peninsula await detailed study, though some information is available from the Haraz mountains and Dhofar. In the Haraz mountains the grasslands can be classified according to altitude, habitat, slope and frequency of disturbance (Table 6.1). At 2,000-3,000 m in the Haraz mountains of Yemen *Themeda triandra, Heteropogon contortus, Andropogon distachyos* and *Hyparrhenia hirta* form the dominant grassland vegetation. They are all indicator species of the Themedo-Hyparrhenietea, a widespread tall grassland community in eastern and southern Africa (Knapp 1968). 'Dry' grassland types are characterized by *Tetrapogon villosus, Elionurus muticus, Stipa tigrensis* and *Pennisetum setaceum,* and moist habitats by hygrophilous grasses such as *Agrostis viridis,* associated with tall herbs such as *Acanthus arboreus* (Figure 6.4: D) and geophytes such as *Crinum yemenense, Scadoxus multiflorus* and *Kniphofia sumarae*.

In Dhofar the tall *Themeda* grassland on the southern slopes and summit plateaux (above 750 m) exposed to the southwest monsoon are dominated by *Themeda quadrivalvis, Setaria pumila, Heteropogon contortus, Sporobolus spicatus* and *Arthraxon pusillus* (Radcliffe-Smith 1980).

6.5.2 Grasslands as Rangelands

Annual vegetation is the most important source of fodder for livestock in Arabia. At lower and middle altitudes fodder is usually tree and shrub browse, whilst at higher altitudes grasses and herbs are the dominant fodder vegetation (Thalen & Kessler 1988). Grasses are also dried and used as hay, and natural fodder resources are often completely utilized and stocking rates too high to support all of the livestock during droughts.

Table 6.1 Characteristic grass species of different habitats and successional stages of differing moisture conditions in the submontane and montane belts of the Haraz mountains of Yemen.

Habitat or successional stage	Submontane belt		Montane belt			
		1,700 m	1,800 m	2,400 m	2,500 m	
	Dry	Moist	Dry	Moist	Dry	Moist
Subclimax grassland		*Cenchrus ciliaris* *Cenchrus setigerus* *Panicum maximum* *Bothriochloa insculpta*	*Themeda triandra*	*Andropogon distachyos* *Bothriochloa insculpta*	*Themeda triandra* *Helictotrichon elongatum* *Sporobolus africanus*	*Hyparrhenia quarrei* *Andropogon distachyos* *Bromus leptoclados* *Brachypodium sp.* *Festuca caprina*
Fallow grassland			*Aristida adscensionis* *Heteropogon contortus* *Hyparrhenia hirta* *Themeda triandra* *Aristida congesta* *Eragrostis papposa* *Microchloa kunthii* *Pennisetum villosum* *Cynodon dactylon* *Eleusine floccifolia* *Eragrostis braunii*	*Digitaria abyssinica*	*Airstida adscensionis* *Heteropogon contortus* *Hyparrhenia hirta* *Themeda triandra* *Aristida congesta* *Eragrostis papposa* *Microchloa kunthii* *Pennisetum villosum* *Cynodon dactylon* *Eleusine floccifolia* *Eragrostis braunii*	*Digitaria abyssinica*
Rocky slopes	*Tetrapogon villosus* *Cenchrus ciliaris* *Heteropogon contortus* *Aristida adscensionis* *Hyparrhenia hirta* *Tragus racemosus* *Leptothrium senegalense* *Tetrapogon cenchriformis* *Digitaria nodosa*		*Tetrapogon villosus* *Cenchrus ciliaris* *Heteropogon contortus* *Themeda triandra* *Aristida adscensionis* *Hyparrhenia hirta* *Tragus racemosus* *Panicum maximum* *Eustqachys paspaloides* *Cynodon dactylon*		*Heteropogon contortus* *Cenchrus ciliaris* *Themeda triandra* *Aristida adscensionis* *Hyparrhenia hirta* *Cynodon dactylon*	*Arthraxon prionodes*

Table 6.1 continued.

Habitat or successional stage	Submontane belt		Montane belt			
		1,700 m	1,800 m	2,400 m	2,500 m	
	Dry	Moist	Dry	Moist	Dry	Moist
Pioneer stages	*Pennisetum setaceum* *Tricholaena teneriffae* *Tetrapogon villosus* *Aristida adscensionis* *Hyparrhenia hirta* *Heteropogon contortus* *Rhynchelytrum repens* *Cynodon dactylon* *Tetrapogon cenchriformis*	*Digitaria abyssinica* *Arthraxon prionodes* *Enneapogon cenchriformis*	*Pennisetum setaceum* *Tricholaena teneriffae* *Tetrapogon villosus* *Aristida adscensionis* *Hyparrhenia hirta* *Rhynchelytrum repens* *Cynodon dactylon*		*Pennisetum setaceum* *Aristida adscensionis* *Hyparrhenia hirta* *Heteropogon contortus* *Rhynchelytrum repens* *Cynodon dactylon*	*Digitaria abyssinica* *arthraxon prionodes*
Outcrops and terrace walls	*Pennisetum setaceum* *Heteropogon contortus* *Aristida adscensionis* *Hyparrhenia hirta*	*Arthraxon prionodes* *Pennisetum macrourum*	*Pennisetum setaceum* *Stipa tigrensis* *Heteropogon contortus* *Aristida adscensionis* *Aristda aloensis* *Hyparrhenia hirta*	*Arthraxon prionodes* *Pennisetum macrourum*	*Pennisetum setaceum* *Elionrus muticus* *Stipa tigrensis* *Heteropogon contortus* *Aristida adscensionis* *Aristda aloensis* *Hyparrhenia hirta*	*Arthraxon prionodes* *Phaenanthoecium moeslinii*
Weeds in cultivated fields	*Eragrostis papposa* *Cynodon dactylon*	*Echinochoa colona* *Eragrostis cilianensis* *Eragrostis barrelieri* *Chloris pycnothrix* *Digitaria abyssinica* *Setaria verticillata*	*Eragrostis papposa* *Digitaria abyssinica* *Cynodon dactylon*	*Chloris pycnothrix* *Eragrostis cilianensis* *Eragrostis barrelieri* *Setaria verticillata*	*Eragrostis papposa* *Cynodon dactylon*	*Digitaria abyssinica* *Eragrostis cilianensis*
Moist places		*Echinochoa colona* *Arundo donax* *Dactyloctenium aegyptium* *Bothriochloa insculpta*		*Agrostis viridis* *Andropogon distachyos* *Bothriochloa insculpta* *Pennisetum thunbergii*		*Agrostis viridis* *Andropogon distachyos* *Pennisetum thunbergii* *Bromus leptoclados* *Brachypodium sp..*

In the drought-deciduous *Acacia-Commiphora* woodlands of the at Tur basin total standing crop is closely related to the density of the tree and shrub layer. The biomass accessible to small ruminants (below the browse line) ranges from 11 kg dry matter ha^{-1} on the stony pediplains to 250 kg ha^{-1} on wadi terraces and 390 kg ha^{-1} in dense shrubland on steep slopes. Most of the mapped area in the at Tur basin has a carrying capacity of less than one sheep ha^{-1}. Only valley forests offer more that 400 kg dry matter ha^{-1} (Deil & Rappenhöner 1989, Müller-Hohenstein & Rappenhöner 1991).

Stubble pasturing and weed clearing provides about 35% of the total feed demand in Dhamar (Kessler 1987) and 30% in the Tihamah foothills (Deil & Rappenhöner 1989, Müller-Hohenstein & Rappenhöner 1991). Annual production of the herbaceous layer in Dhamar ranges from 120 to 500 kg dry matter ha^{-1} yr^{-1} (Kessler 1987). Lopping of *Ziziphus spina-christi* and *Dobera glabra*, which are scattered over the agricultural area, contributes 5-10% of the feed supply. The mean crude protein of the herbaceous layer in Dhamar is maximum in March and August (Kessler 1989). *Themeda triandra* offers fodder of a high quality at the beginning of the rainy season and protein content declines from 8% to 2% between March and July (Grosser 1988). Amongst the three indigenous grasses of high elevations studied for their response to grazing, *Tetrapogon villosus* is one of the few species which can survive heavy grazing pressure and tolerate defoliation in combination with water stress. It is suited to drier habitats due to its ability to adjust its water use efficiency. *Themeda triandra*, which occurs mainly on slopes with seasonal rain and is normally lightly grazed is the least tolerant while *Anthropogon distachyos* is intermediate between the two (Fennema & Briede 1990).

Traditional rangeland management in Arabia includes forage reserves on common ground or in private ownership called *hima* in Oman and Saudi Arabia and *mahjur* in Yemen. They are used for grazing at the end of the dry season and serve as fodder reserves in which the grass is generally cut by hand. Kessler (1995) studied the floristic composition, vegetation cover and the standing biomass of 26 such forage reserves in the Dhamar plains of Yemen and found a higher vegetation cover and percentage of perennial grasses than on grazed areas. Crude protein contents were low at the end of the dry season, but covered the minimal feeding requirements for sheep. Tree cover was lower <1% within the protected areas since firewood collecting was not forbidden. See also section 12.3 for further discussion of this topic.

6.6 Wadi Vegetation

6.6.1 Physical Characters and Irrigation Practices

A prominent feature of the wadi ecosystems of the Arabian Peninsula is the occurrence of periodic runoff following rain. Although there are no permanent rivers, larger wadis such as Wadi Mawr, Wadi Surdud, Wadi Siham, Wadi Rhima' and Wadi Zabid which originate in the southwestern mountains, Wadi Bani Awf and Wadi Sahtan which originate in the northern mountains of Oman, and Wadi Hadhramaut, Wadi Jawf and Wadi Bana, have permanent water along parts of their length. In the lower wadi courses water flowing into the alluvial fans recharges the groundwater. Only during episodic floods brought by heavy rain does surface water reach the sea. Increase in surface water flow is dramatic during floods; for example during the hydrological year 1965/66, surface flow in Wadi Shiham in the central Yemeni Tihamah ranged from 0.1 to 350 $m^3 s^{-1}$ (Kraft *et al.* 1971). Permanent settlements and agriculture along wadi courses rely on both water supply and the large quantities of sediment transported to the plains during floods.

Depending on altitude and periodicity of surface runoff, two wadi irrigation systems are used (Dequin 1977): (1) *sayl* irrigation, first described by Niebuhr (1772), in which water is diverted from permanent streams to terraced fields, as practised in the middle escarpment and Tihamah foothills in Yemen and, (2) *oqam* irrigation, where flood water is diverted by dykes to flood fields. These two irrigation systems date from at least two millennia (Kopp 1975) and reached perfection during the Sabaean period. A series of flood-oases were the basis for the main caravan route along the eastern Yemeni escarpment, and in Ma'rib 9,300 ha were irrigated at that time.

6.6.2 Wadi Flora

The middle altitudes (500-2,000 m) of the deeply incised valleys of the western escarpment shelter the richest Sudanian vegetation type found in Arabia (Wood 1983b). Characteristic trees of these riparian forests are *Breonadia salicina*, *Celtis toka*, *Combretum molle*, *Cordia africana*, *Ficus cordata* subsp. *salicifolia*, *F. glumosa*, *F. lutea*, *F. populifolia*, *F. pseudosycomorus*, *F. sur*, *F. sycomorus*, *F. vasta*, *Mimusops laurifolia*, *Phyllogeiton discolor* (= *Berchemia discolor*), *Tamarindus indica*, *Terminalia brownii*, *Trichilia emetica*, *Ziziphus mucronata* and *Z. spina-christi*. Low growing trees and shrubs are *Phoenix reclinata*, *Nuxia congesta*, *N. oppositifolia*, *Maesa lanceolata*, *Myrica salicifolia* and the introduced species *Pandanus odoratissima* and *Jatropha curcas*. Since this habitat offers the best location for irrigated fields, these riparian forests have now been reduced to small fragmented riverside galleries. Native trees are mostly replaced by cultivated species such as *Ceiba pentandra*, *Melia azedarach*, *Tamarindus indica*, *Terminalia brownii* and *Cordia africana*.

Wadis with periodical or intermittent water flow are characterized by the genus *Tamarix*. The Arabian Peninsula is the centre of diversity of this genus, with members of the *Tamarix nilotica* group (*T. nilotica* s. str., *T. arabica*, *T. mascatensis* and *T. aucheriana*) and *T. aphylla* (Baum 1978, 1989). *T. nilotica* s. str., like *Halopeplis perfoliata* (Figure 9.3), has a circum-Arabian distribution (Figure 6.23). Species of *Tamarix* have an interesting adaptation to desert conditions. Their salt glands not only excrete salt, but in the early hours of the morning the

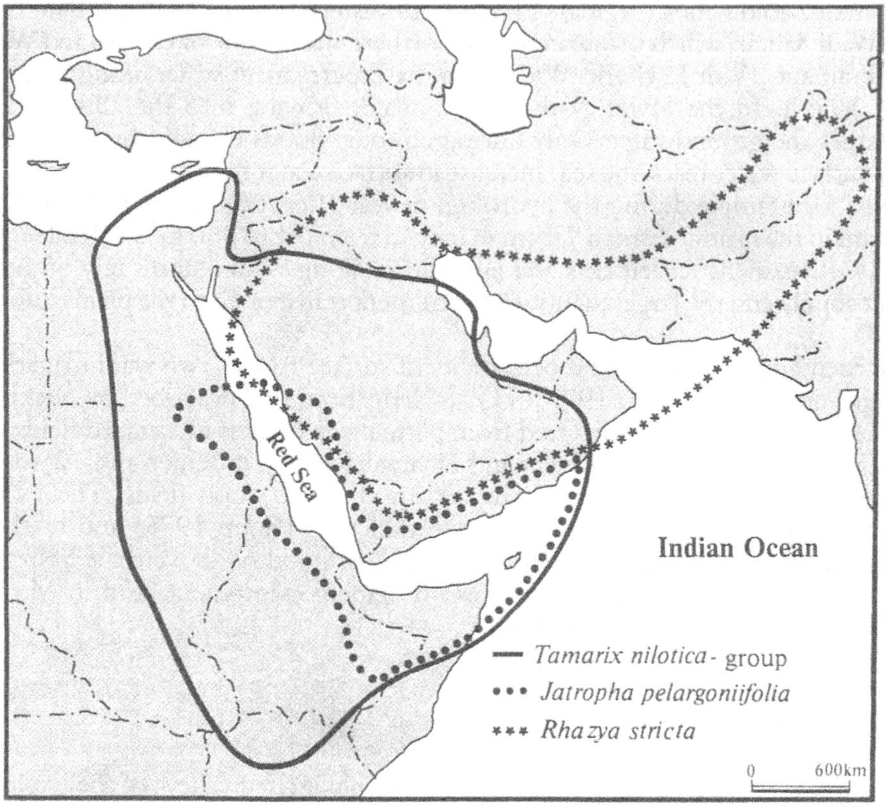

Figure 6.23 Distribution of the *Tamarix nilotica* group, *Jatropha pelargonifolia* and *Rhazya stricta,* common wadi species, on the Arabian Peninsula and its surroundings.

twigs are covered by an alkaline solution, absorbing moisture hygroscopically and also providing an extra source of CO_2 (Waisel 1991).

There are only a few endemic species in the wadi flora. The fan palm, *Livistonia carinensis* (= *Wissmania carinensis*) is regionally endemic to Wadi Hadjer in the Jol of southern Yemen, Djibouti and northern Somalia (Bazara *et al.* 1990, Dransfield & Uhl 1938), and the dwarf palm *Nannorrhops ritchieana* to Oman and Baluchistan. The fan palm is associated with *Conocarpus lancifolius, Desmostachya bipinnata, Hyphaene thebaica* and *Pluchea dioscoridis. Nogalia drepanophylla* (Boraginaceae) is regionally endemic to the stony wadis of Dhofar, Jol and Somalia (Verdcourt 1989). *Acacia johnwoodii* (=*A. abyssinica* var. *macroloba*) is endemic to the western escarpment from Taif to Taiz, and is characteristic of the gallery forest community. Its closest relative is *A. elatior*, a riparian tree in East Africa (Boulos 1994). These wadi endemics illustrate the Dhofar-Jol-Socotra-northern Somalia and montane Eritreo-Arabian pattern of endemism (Miller & Nyberg 1991, see Chapter 4 for further details).

The distribution of some wadi species, such as *Rhazya stricta* and *Jatropha pelargoniifolia*, reflect the transition from a winter to a summer precipitation regime and from a climate with winter frost to one with a frost-free winter (Figure 6.23). *Rhazya stricta* is a typical wadi species north of Jeddah (Vesey-Fitzgerald 1957b), in central Arabia (Baierle *et al.* 1985, Vesey-Fitzgerald 1957a), and in the high Yemen, Hadhramaut, northern Oman and the United Arab Emirates (pers. obs.). *Jatropha pelargoniifolia* (*sensu* Hemming & Radcliffe-Smith 1987) occurs in wadis on the Tihamah, from Jeddah to Bab al Mandab, and from Aden to Dhofar. In less arid conditions, *R. stricta* is replaced by the Mediterranean-Irano-Turanian *Nerium oleander*. *Rhazya stricta* is protected from foraging animals by its high alkaloid content (El-Tawil 1983) and can become dominant in wadis near settlements.

6.6.3 Wadi Vegetation Sequences

The pattern of vegetation along wadi courses is determined by the nature of the drainage system (i.e. riverbed gradient and stream velocity), sediment grain size, groundwater level, frequency of overflow, distance from the wadi bed with respect to the water table and differences in the catchment area. Two types of wadis can be distinguished: autochthonous wadis which start and end in the same bioclimatic belt, and allochthonous wadis which traverse several regions. In allochthonous wadis, surface runoff, water supply and plant phenology are largely independent of the precipitation regime of the area itself. A vegetation pattern referred to as a 'wadi sequence' can be observed along the courses of such wadis.

To demonstrate regional differences, wadi vegetation is described here by representative wadi sequences, beginning with the extra-tropical vegetation types of central Arabia and ending with the tropical vegetation types of southwestern Arabia.

6.6.3.1 Extra-tropical Wadi Vegetation

Central Arabia. Wadis in the central Arabian desert have been studied by long distance transects from Taif to Riyadh (Baierle & Frey 1986). The vegetation belongs to the Saharo-Arabian pseudo-savanna wadi type, with *Zilla spinosa, Astragalus spinosus, Rhazya stricta* and *Anvillea garcinii* present under a scattered tree layer of acacias (*Acacia tortilis, A. ehrenbergiana, A. raddiana* and *A. gerrardii*). The herb and shrub layer is extremely poor in species, with *Panicum turgidum, Salsola cyclophylla* and *Indigofera spinosa* being the main species. The eastern part is relatively rich in species and differentiated according to the substrate: *Cymbopogon commutatus, Chrysopogon plumosus* and *Hyparrhenia hirta* characterize small runnels with sparse fine soil, *Rhanterium epapposum* the medium size wadis with a thicker layer of sandy and silty material and *Haloxylon salicornicum* the moderately saline sandy soils.

Eastern Saudi Arabia. The vegetation sequence in Wadi as Sahba in eastern Saudi Arabia was described by Mandaville (1965). Wadi bottoms are dominated by *Haloxylon salicornicum* and by scattered trees of *Acacia flava*. Pseudo-savannas occur on wadi slopes, with *Panicum turgidum, Lasiurus scindicus, Lycium barbatum* and *Ochradenus baccatus,* and the steep wadi valleys are characterized by the *Rhanterium epapposum-Anvillea garcinii* shrub community. The gravel plains and side channels are dominated by *Rhanterium epapposum* and *Rhazya stricta*.

UAE and the Northern Oman Mountains. The canyon-like wadis of the mountains of the eastern UAE and northern Oman shelter the Saccharo-Nerietum oleandri (Deil & Müller-Hohenstein 1996). Wadi beds are strewn with boulders and gravel, and stream velocity during episodic flooding is high. On fine sand between the boulders a short-lived synusia comprising the hygrophilous herb *Parietaria lusitanica* and the fern *Onychium melanolepis* develops after run-off. Where the wadi bed is wide and filled with water-storing gravel and sandy sediments the areas are intensely utilized; *Phoenix dactylifera, Mangifera indica, Punica granatum, Citrus* spp. and other fruit trees are commonly cultivated in these wadi oases. However, remnants of the original vegetation can still be seen, particularly *Nerium oleander, Ficus cordata* subsp. *salicifolia* and *Acacia nilotica*. In the wadi alluvium emerging from the mountains *Physorrhynchus chamaerapistrum* and *Ochradenus aucheri* form a community, and runnels in the foreland with fine deposits support *Tamarix* spp. and *Calotropis procera*, and those with gravel, *Acacia tortilis* and *Rhazya stricta*.

Dhofar Mountains. Wadis on the humid southern slope of the Dhofar mountains shelter gallery forests with *Ficus sycomorus, F. vasta, F. lutea, Tamarindus indica* and a liana veil of *Cissus quadrangularis, Cyphostemma ternatum, Luffa acutangla* and other species. In the dry forelands the wadis are characterized by the dwarf palm *Nannorrhops ritchieana* (Radcliffe-Smith 1980), *Dyerophytum indicum* and *Tamarix* spp. *(T. aphylla, T. aucheriana, T. arabica* and *T. mascatensis*; Miller & Morris 1988).

6.6.3.2 Tropical Wadi Vegetation

The tropical wadi sequence in the Asir was studied by König (1986), in the western Yemeni escarpment by Deil (1986a) and in Wadi Dhar near San'a by El-Monayri *et al.* (1991) and Dubaie *et al.* (1993).

Asir Mountains. In the Asir the wadi vegetation sequence starts in the *Juniperus procera-Acacia origena* belt at 2,600 m with a *Maesa lanceolata-Nuxia congesta-Hypericum revolutum* shrub community in ravines, and with *Nuxia oppositifolia, N. congesta, Buddleja polystachya, Myrica salicifolia* and *Maytenus undulatus* in the *Olea-Juniperus* belt near 2,300 m. The sequence continues with *Breonadia salicina-Ficus sycomorus-F. palmata-Combretum molle* gallery forests at mid-altitudes (1,400 m), in contact with the *Acacia asak-A. etbaica* woodland on slopes which

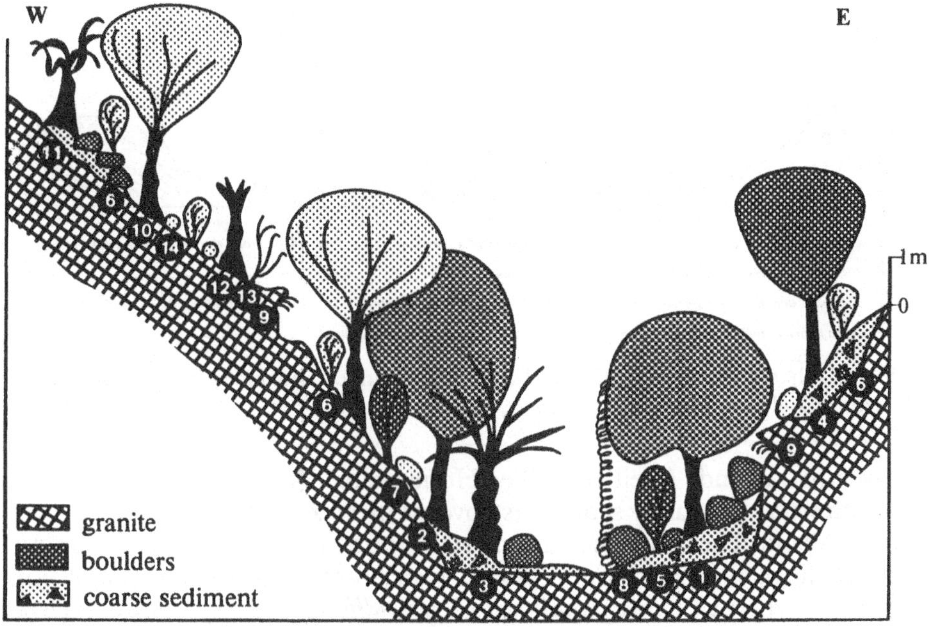

Ravine forest with (1) *Combretum molle, Berchemia discolor* (2) *Tamarindus indica* (3) *Ficus* div.spec. (4) *Boscia angustifolia,* seldom *Dobera glabra, Mimusops laurifolia* (5) *Ochna inermis, Tarenna graveolens arabica, Ehretia obtusifolia, Dodonaea viscosa, Maytenus* div.spec., *Carissa edulis* (6) *Premna resinosa, Ecbolium linnaeanum, Grewia* div.spec., *Anisotes trisulcus* (7) *Barleria* div.spec. (8) *Cissus rotundifolia, Cocculus hirsutus , Cassytha filiformis* (9) poikilohydric ferns
Dry woodland with (10) *Acacia asak* (11) *Commiphora abyssinica, C.myrrha* (12) *Adenium obesum* (13) *Abrus bottae, Ormocarpum yemenense* (14) *Seddera* div.spec.

Figure 6.24 Life-forms, species composition, microhabitats and substrata in a ravine of Wadi al Gubaiqi in the Tihamah foothills of Hajjah province in Yemen.

are not influenced by groundwater. Sandy sediments in the draining runnels and sandy terraces of larger wadis in the Jizan Tihamah are colonized by *Acacia ehrenbergiana, A. johnwoodii, Ficus cordata* subsp. *salicifolia* and *Ziziphus spina-christi*, and the wadi bed itself by *Tamarix nilotica, T. aphylla* and *Salvadora persica*. In areas of the plains where episodic floods deposit silt with a sandy overlay, *Hyphaene thebaica* is present, indicating a high water table (König 1986).

The Yemeni Tihamah. Along autochthonous watercourses in the mountainous Tihamah, riparian forests start at high elevations with *Combretum molle-Berchemia discolor* forests in deeply incised ravines (Figures 6.24). This is a vegetation type very rich in species and life-forms, with broad-leaved mesomorphic trees, evergreen xeromorphic trees and shrubs and poikilohydric ferns. On lower slopes between 450 and 600 m they are replaced by the drought-deciduous *Acacia*

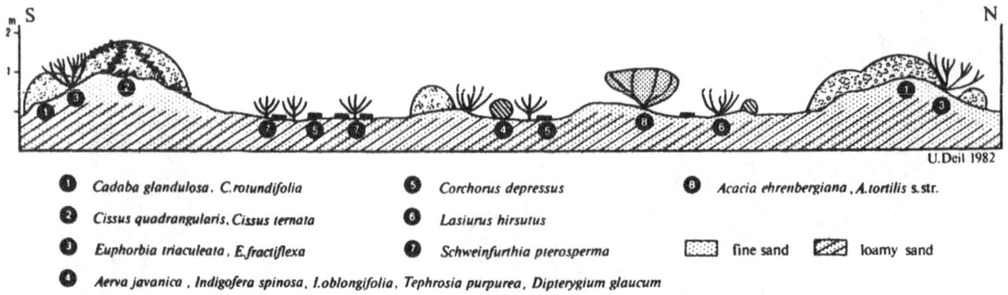

Figure 6.25 Characteristic vegetation and substrata in runnels and dune valleys in the Tihamah plain near az Zuhra in Hodeidah province, Yemen.

johnwoodii woodlands (Deil & Müller-Hohenstein 1985). The wadi bottom is filled with stone and sandy sediments covered by an *Acacia ehrenbergiana-Cassia senna* woodland. The latter association reaches northwards to Jeddah (Batanouny 1979). In the plains there is only episodic water-flow in the shallow hollows and dune valleys, where a *Cadaba glandulosa-C. rotundifolia-Schweinfurthia pterosperma* community occurs, mixed with elements of the widespread *Acacia tortilis-A. ehrenbergiana-Lasiurus scindicus* pseudo-savanna (Figure 6.25).

The vegetation sequence along large allochthonous wadis is illustrated by that of Wadi Mawr (Figures 6.26-6.29). The catchment area starts near Sadah in the highland plains, runs through the basin of at Tur where it is joined by two other large wadis, Wadi Ayyan and Wadi La'ah, and terminates in the Tihamah near az Zuhra. Mesophytic riverine forests of the lower escarpments and mountainous Tihamah consist, respectively, of *Breonadia salicina-Ficus* and *Berchemia discolor-Combretum molle-Trichilia emetica* communities. These are replaced by the *Tamarix-Desmostachya bipinnata* community on the lower inland slopes. A *Tamarix-Saccharum spontaneum* community characterises the central Tihamah plain, and in coastal areas a *Tamarix-Zygophyllum simplex* community occurs on saline clay in the wadi bottom and a *Hyphaene thebaica* community on silty plains that have a high water table.

Within the mountainous Tihamah water-flow is permanent and floods frequent, with wadi inclinations of 7-15% and sediments of gravel and coarse sand. A riverside catena in the mountainous Tihamah is given in Figure 6.27 and the actual vegetation pattern at the confluence of Wadi Mawr and Wadi La'ah is illustrated in Figure 6.28. On small sandy islands in the wadi bed young *Tamarix nilotica* and the ephemeral *Bacopa monnieri* community are pioneers. The sandy terrace is stabilized on the riverside by a dense, tall layer of *Desmostachya bipinnata* with some *Jatropha curcas* shrubs and scattered *Tamarix* trees. Adjacent to the riverside vegetation there is an *Acacia ehrenbergiana-Cassia senna* community. Degraded drought-deciduous *Acacia-Commiphora* woodlands grow on the slopes, with *Salvadora persica* fringing the base. The pediplain is utilised for agriculture

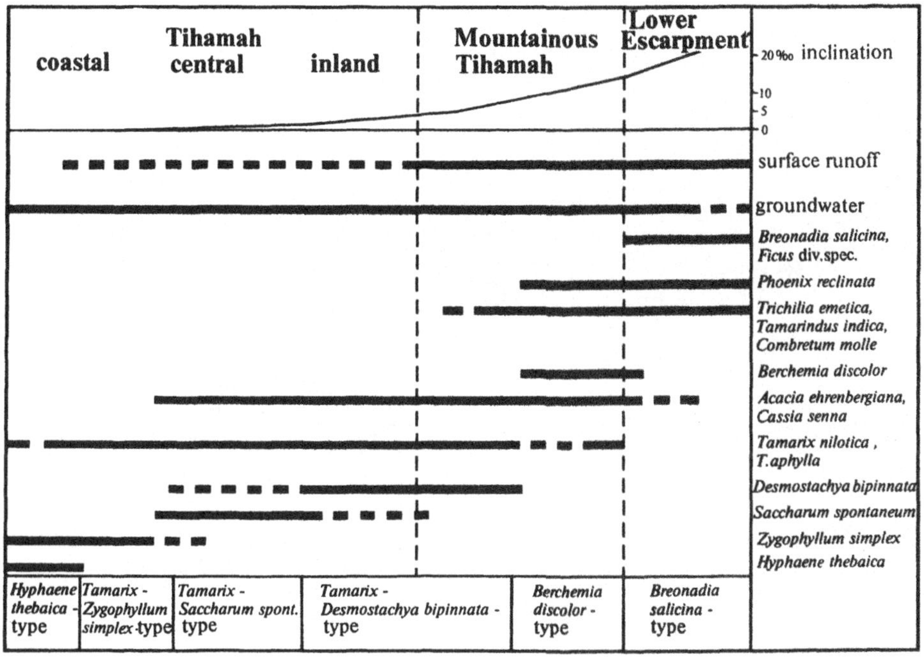

Figure 6.26 Sequence of vegetation types, the distribution of characteristic species and the influence of surface runoff and groundwater along allochthonous wadis such as Wadi Mawr in the western escarpment and on the Tihamah plain.

and agroforestry, with small settlements, fringed by a succulent Carallumo-Euphorbietum inarticulatae community, predominantly situated between the pediplain and the surrounding slopes. The latter shelter remnants of a species rich *Acacia-Cadaba-Commiphora* woodland.

The flow of water in the Tihamah foothills is intermittent, wadi inclinations are <3% (with a consequent reduction in stream velocity relative to the mountainous Tihamah) and sediments vary from fine sand to silt. The wadi bottom is colonized by the tall tufted grass *Saccharum spontaneum* and *Tamarix* trees (Figure 6.29). The high river bank is stabilized by an evergreen *Tamarix nilotica-T. aphylla-Salvadora persica* woodland with a *Desmostachya* undergrowth. *Desmostachya* also penetrates into irrigation canals and fields and has to be removed from time to time. The floods are diverted by a dyke, *oqam,* to the dammed fields, where *Sorghum* and *Pennisetum* are cultivated. A *Senra incana-Cassia-Heliotropium-Aerva javanica* community is common in uncultivated fields.

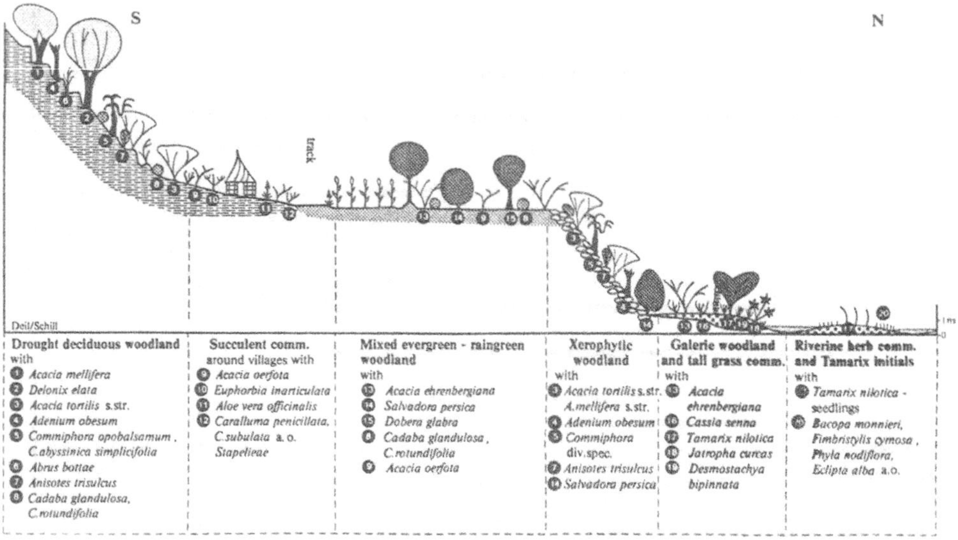

Figure 6.27 Plant communities on a riverside catena in Wadi Mawr, an allochthonous wadi, in the at Tur basin of the Tihamah foothills of Hajjah Province, Yemen.

Haraz mountains. In the ravines and valleys of the Western escarpment of the Haraz mountains, the following vegetation sequence occurs (Müller-Hohenstein *et al.* 1987):

2,500-1,500 m: ***Ficus palmata, F. vasta, Cordia africana, Dombeya schimperiana, Maesa lanceolata*** and ***Ehretia cymosa***.
2,000-1,000 m: ***Teclea nobilis, Ficus lutea*** and ***Oncopa spinosa***.
1,500-500 m: ***Ficus cordata*** subsp. ***salicifolia, F. populifolia, F. sycomorus, Breonadia salicina, Trichilia emetica, Phoenix reclinata*** and ***Tamarix aphylla***.
1,000-500 m: ***Acacia johnwoodii, Tamarindus indica, Terminalia brownii, Mimusops laurifolia, Ficus glumosa, Nuxia oppositifolia, Berchemia discolor, Celtis integrifolia*** and ***Tamarix nilotica***.

Ibb. Gullies and steep sided valleys in the high rainfall area around Ibb have remnants of a gallery forest of ***Ficus vasta, F. sycomorus, Cordia africana*** and ***Nuxia oppositifolia***. ***Myrica salicifolia, Debregeasia bicolor*** and ***Erythrococca abyssinica*** form the shrub layer (Hepper & Wood 1979).

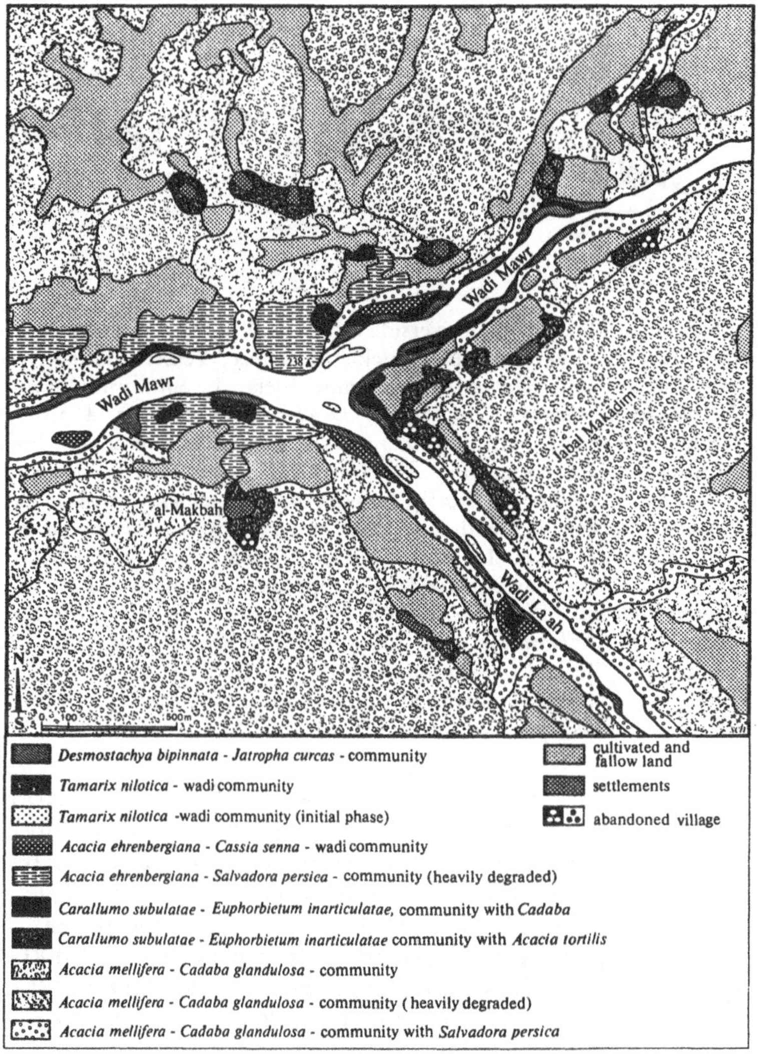

Figure 6.28 Vegetation patterns at the confluence of Wadi Mawr, an allochthonous wadi, and Wadi La'ah in the at Tur basin of Hajjah Province, Yemen.

6.6.4 Herbaceous Wadi Communities

Intermittently flooded low terraces along large wadis with fine grained sediments are covered with a layer of herbaceous species. Dominant members of this community are *Bacopa monnieri*, *Eclipta alba* and *Phyla nodiflora*. Associated species occurring in temporarily inundated areas are *Exaculum pusillum* and *Corchorus depressus*, and *Fimbristylis complanata*, *F. cymosa*, *F. sieberiana* and *Mimulus gracilis* in habitats with stagnant water and permanently wet areas. This

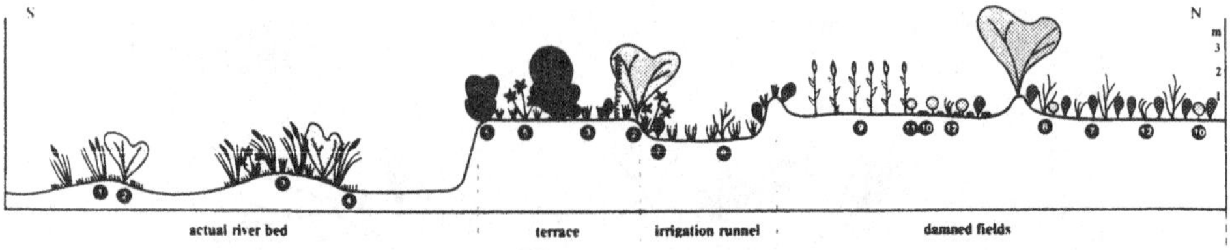

Figure 6.29 Characteristic species on a riverside catena in Wadi Mawr, an allochthonous wadi, in the Tihamah plains near az-Zuhra in Hodeidah Province, Yemen. Actual river bed with *Tamarix* riverine shrubs and *Saccharum* on sandy islets: 1, *Saccharum spontaneum* var. *aegypticum*; 2, *Tamarix nilotica* and *T. aphylla*; 3, *Desmostachya bipinnata*; 4, a shortlived herb community following flooding with *Cyperus rotundus, Fimbristyis cymosa, Brachiaria reptans, Dactyloctenium scindicum, Eragrostis minor, Corchorus* sp., *Eclipta alba, Flaveria trinervia* and *Glinus lotoides*. Terrace: 5, *Salvadora persica*; 6, *Jatropha curcas*; 2, *Tamarix nilotica* and *T. aphylla*; 3, *Desmostachya bipinnata* liana veil with *Leptadenia arborea*, *Cissus quadrangularis*, *Momordica balsamina* and *Cocculus pendulus*. Irrigation runnel: 7, *Indigofera oblongifolia*; 8, *Abutilon pannosum* and *Senra incana*. Dammed fields: 9, crops such as *Sorghum bicolour*, *Pennisetum glaucum*, *Sesamum indicum*, *Vigna aconitifolia*, with weeds such as *Pulicaria jaubertii*, *Heliotropium longiflorum*, *Corchorus* div. spec., *Cleome* sp., *Dactyloctenium aegyptium*, *Paspalidium desertorum*, *Cynodon dactylon* and *Cyperus rotundus*. Fallow land: 10, *Zygophyllum simplex*; 11, *Schouwia purpurea* and *Tephrosia purpurea*; 12, shrubs such as *Aerva javanica*, *Cassia italica*, *C. holosericea*, *Jatropha lobata*, *Solanum dubium* and *Heliotropium* spp.

vegetation type, first reported by Vierhapper (1907) from Socotra, is of pantropical distribution. It is recorded from wadis in the Andean highlands of Peru (Samolo floribundi-Bacopetum monnieri, Eleocharido caribaeae-Bacopetum monnieri; Müller & Gutte 1985) and from the French West Indies (Eleocharido-Bacopetum monnieri Fimbristylo cymosae-Lippietum nodiflorae; Foucault 1978). In Arabia *Bacopa monnieri* is found along streams below 2,000 m, and is replaced at higher altitudes by *Apium nodiflorum, Eleocharis palustris, Ranunculus multifidus* and other elements of Holarctic origin. *B. monnieri* also colonizes man-made habitats such as irrigation channels.

6.7 Summary

This chapter provides an overview of the phytogeography, endemism and vegetation of the montane woodlands, rock communities, orophytic grasslands and wadis of the mountainous regions of the Arabian Peninsula. Between 1,500 and 2,000 m the drought-deciduous *Acacia-Commiphora* woodland of the lowlands is replaced by evergreen forests and sclerophyllous shrubs. A northwestern *Juniperus phoenicea-Pistacia atlantica* community type with a Mediterranean character can be distinguished, a southeastern *Juniperus excelsa-Ephedra pachyclada* type with species of Omano-Makranian distribution and a southwestern *Juniperus procera-Myrsine africana* type with Afromontane and Eritreo-Arabian elements. In the man-made terrace landscape of the highlands of Yemen the montane forests have largely disappeared and have been replaced by the *Olea-Tarchonanthus* and *Hypericum-Rhamnus-Sageretia* shrublands.

Woodlands of all three species of *Juniperus* on the Peninsula appear to be exhibiting extensive dieback at the lower altitudes of their range. Whilst this phenomenon appears to be related in some way to moisture stress, the exact nature of the cause is still under investigation.

Rock communities are generally concentrated on the escarpments and have a high degree of endemism. Adaptive radiation and speciation by geographical isolation has resulted in vicarious neoendemics within *Euphorbia, Aloe, Phagnalon* and many genera of the Asclepiadaceae. The rocky habitat also supports palaeoendemics and isolated taxa such as *Dhofarica, Centaurothamnus* and *Delosperma*. The vegetation in dry rocky habitats is rich in succulents, with the class Kleinio-Euphorbietea occurring in the frost-free lower regions and the Crassulo-Aloetea in the upper montane belt. The associations in relatively more humid rocky habitats are rich in ferns and bryophytes and belong to the syntaxa Cheilanthi-Actiniopteridetea, Asplenion aethiopicum and Adiantetea capilli-veneris.

The orophytic grasslands are dominated by species of northern temperate origin mixed with Mediterranean and Afro-alpine elements. The existence of a *Themeda triandra* grassland between 2,000 and 2,800 m is largely due to anthropogenic influences. Primary grassland communities of *Elionurus muticus* occur on shallow soil, with *Festuca cryptanta* and *Stipa tigrensis* on the highest peaks.

The characteristic vegetation of wadis is determined by the nature of the drainage system, the groundwater level, the sediment type and overflow frequency. The patterns of plant communities along the wadi courses are described as wadi vegetation sequences. Although these are in general dominated by *Tamarix* spp. and are of azonal character, a Mediterranean-SubSaharan type with *Retama raetam, Nerium oleander* and *Pistacia atlantica*, a north Arabian *Acacia* pseudo-savanna with Saharan shrubs in the understorey, a southeastern community type with *Saccharum griffithii, Nerium oleander* and *Dyerophytum indicum* and a southwestern riparian forest of tropical character can be identified.

The latter is the richest Sudanian vegetation type in Arabia and is a northern outlier of the gallery forests of East Africa.

Chapter 7

Vegetation of the Plains

Shahina A Ghazanfar

"Our road lay over a gradually declining sandy plain, sprinkled with dwarfish acacias, brushwood, and desert herbage ... The cool breath of the morning was delicious, and as we sped along, the desert air, and the consciousness of being on the edge of the great wilderness, with a boundless expanse of open country unrolling gradually in front of us, caused an exhilaration of spirits which made the ride an extremely pleasant and enjoyable one."

SB Miles, on his journey to Adam in Oman (Miles 1901)

7.1 Introduction

The almost boundless expanse of open country which Miles referred to in his travelogue comprises the vast rock and gravel deserts of the Arabian Peninsula. These plains consist of relatively flat and featureless coastal belts, wadi fans, low rocky plateaus and lava fields (Figure 7.1). Soil cover is usually shallow and overlain with large and small rocks, stones, gravel and sand. Alluvial deposits and silts are present where the plains lie adjacent to mountains. Lacking the species diversity of the mountains and the drama of the sand deserts, the vegetation of the plains is probably the least studied of that of all the landforms of the Arabian Peninsula.

The terms *hamada* (or *hammada*) and *reg* are often used to classify the rock and gravel deserts. An *hamada* is a desert pavement from which wind has removed most of the fine sediment leaving bare rock surfaces scattered with large rocks, and a *reg* is a desert plain from which the fine sediment has been removed leaving a surface strewn with gravel and pebbles. The term *hamada* is an Arabic

S. A. Ghazanfar and M. Fisher (eds.), Vegetation of the Arabian Peninsula, 175-190.

Figure 7.1 (A) Zalawt plains, traversed by wadis, at the foot of Jebel Semhan in Dhofar in southern Oman. (B) Gravel plains of the limestone plateau of the Jiddat al Harasees in central Oman.

word which is mostly used with reference to North African rock deserts and the term *reg* comes from the Berber language, adopted in French, to define gravel deserts. The term 'plains' is used here to describe both the rocky and gravel desert plains of the Arabian Peninsula.

The various plains of the Peninsula fall under differing climatic, edaphic and topographical regimes, and this is expressed as regional differences in plant species composition. In general the plains lie within areas of low precipitation, often with less than 100 mm per year (see Figure 2.7), and water supply (as rain, dew or fog) is the most important abiotic factor influencing the presence, composition and density of vegetation in these areas (Deil & Müller-Hohenstein 1985, Ghazanfar 1997, Mirreh & Al Diran 1995, Omar 1995). Natural springs are rare on the plains, although several underground aquifers exist, especially in the pediplains. Bore holes and artesian wells are drilled in these areas and water used for domestic purposes and, in some areas where the water is suitable and plenty, for irrigation. The extensive use of water from aquifers on the coastal plains has led in many places to the incursion of sea water, and water salinities of up to 7,000 μS (see Table 10.1 for an explanation of salinity units) have been recorded from some bore holes on the Batinah Coast of northern Oman. Depletion of aquifers, lowering of the water table and irregular irrigation practices have led to an increase in soil salinity in many areas of the plains, which in turn has led to the increased establishment of salt-tolerant species and the decline of salt-sensitive species.

Where there are no natural sources of water, the vegetation of the plains is very sparse. In general this vegetation consists of hardy, halophytic, shrubby perennials. Both species richness and cover tends to increase in depressions, shallow runnels and other low-lying areas in which water accumulates after rain. Since water penetration in fine silty soils is slow and evaporation rapid, rain in many areas of the plains produces ephemeral pools and a subsequent flush of germination of annual plant species, most of which barely reach the seedling stage before withering (Ghazanfar 1997). If sufficient rain falls, short-lived annuals may prolong their flowering or even complete two life-cycles in the same season.

Two main phytogeographical regions influence the species composition of the vegetation of the plains: the Somalia-Masai phytochorion in the southeastern and western areas of the Peninsula and the Saharo-Sindian phytochorion in the rest of the Peninsula (see Figure 4.2). The *Acacia-Commiphora* deciduous bushland with emergent scattered trees and succulent shrubs characteristic of the Tihamah and Dhofar plains is an extension of the northeastern African flora. The *Acacia* communities and halogypsophilous vegetation described by White (1983) is characteristic of the Saharo-Arabian subzone of the Saharo-Sindian region, and the characteristic *Cornulaca monacantha-Sphaerocoma aucheri* community of the Batinah Coast of Oman, the United Arab Emirates and the Bahraini coastal plains is an extension of the Nubo-Sindian flora (Kürschner 1986a).

In this chapter the vegetation of the plains is described by geographical regions. The description of the vegetation of the coastal plains is restricted to the inland communities, though there is inevitably small areas of overlap with the description of the coastal and sabkha vegetation in Chapter 9.

7.2 Western and Southern Coastal Plains

The coastal plains of western and southern Arabia consist of the Tihamah coastal lowlands of Yemen and Saudi Arabia and the Salalah and Zalawt plains (the latter two collectively referred to here as the Dhofar plains) of southern Oman. These plains are dissected by shallow wadis which arise from the mountains, depositing alluvium along their course. Most of the wadis have seasonal water flow, and where permanent streams are present the plains are extensively cultivated.

Habitation, cultivation and grazing have vastly altered the natural vegetation of the coastal plains. At one time a large part of the Tihamah and Dhofar plains were covered with open woodlands of *Acacia* species (Hepper 1977), but today only a few woodland relicts remain. The impact of agriculture, tree cutting for timber and firewood and continually increasing livestock numbers has degraded the natural woodlands and other vegetation and caused the increasing dominance of unpalatable species. The abundance of species such as *Aizoon canariense*, *Acacia oerfota*, *Calatropis procera*, *Tephrosia purpurea*, *Jatropha villosa*, *Anisotes trisulcus* and the succulent *Zygophyllum simplex* are indicators of mans's disturbance of the natural vegetation. Extensive damage to both soil cover and natural vegetation has also been caused by indiscriminate off-road driving. Damage is particularly extensive in the semi-desert tall grassland areas of Dhofar where total areas lost to tracks in the grassland of the Salalah plains was estimated to be more than 1% in 1980 (Sale 1980), and is considerably greater today.

Only a few plant species are endemic to the plains of southern Arabia. These are *Odyssea mucronata*, *Heliotropium pterocarpum* and *Crotolaria microphylla* on the Tihamah plains, and *Heliotropium cardiosepalum* and *Dicanthium micranthum* on the Dhofar plains.

7.2.1 The Tihamah and Southern Yemen

The coastal plains of the Tihamah extend from Jeddah in Saudi Arabia to Bab al Mandab. Geographers and historians have variously defined the actual geographical boundaries of the Tihamah (Stone 1985), but the region situated below the Asir highlands is traditionally known as the Tihamat al Asir, and politically, the region from al Lith to al Birk in Saudi Arabia is called Tihamat al Sham and that up to the Yemen border, Tihamat al Janub. That part in Yemen is known as Tihamat al Yemen.

The southern coastal plains of Yemen are relatively narrower and more ill-defined than those of the Tihamah. Sandy bays are present in the region of the Gulf of Aden, whilst the mountain ranges of the Hadhramaut in Yemen and Jebel Qamar in Dhofar rise close to the coast leaving only narrow plains.

The vegetation can best be described along a transect from the sea to the foothills of the mountains. Throughout these plains the first zone above the high tide mark, consisting of soft, sandy, highly saline soil, is usually bare of vegetation and may extend up to 5 km in some areas. However, where wadi fans occur or where the plain is cut by water courses and the soil is better drained, trailing plants such as *Ipomoea pes-caprae*, halophytic shrubs such as *Suaeda fruticosa*, *Cressa cretica*, *Limonium axiillare* and *L. cylindrifolia*, and salt tolerant grasses and sedges such as *Aeluropus lagopoides, A. massauensis, Cyperus conglomeratus, Sporobolus arabicus* and *Urochondra setulosa* occur. Further inland is a sand dune belt consisting of compact hillocks formed by marine deposits overlain by wind-blown sand, supporting a sub-xerophytic shrubland with an abundance of halophytic shrubs. The dominant perennial species are *Suaeda monoica* and *Salsola spinescens,* and after rain *Zygophyllum simplex* is the dominant annual species. Trees are absent, but palms are cultivated along wadis (Al-Hubaishi & Müller-Hohenstein 1984, Wood 1985). Further inland, halophytic shrubs are replaced by *Jatropha pelargoniifolia, Leptadenia pyrotechnica* and *Odyssea mucronata*. These form communities of varying densities with associated species such as *Aerva javanica*, *Dipterygium glaucum*, *Heliotropium pterocarpum*, *Panicum turgidum* and *Tephrosia purpurea*.

Further towards the mountains, gravel plains with an *Acacia-Commiphora* bushland form the dominant vegetation. This community is relatively dense in places with trees 2-4 m apart (Wood 1985). Common species are *Commiphora gileadensis*, *C. myrrha*, *Acacia ehrenbergiana*, *A. hamulosa* and *A. tortilis*. These are associated with other shrubs such as *Rhigozum somaliense*, *Saltia papposa*, *Cymbopogon schoenanthus*, succulents such as *Aloe niebuhriana*, *Euphorbia triculeata,* and the climbers *Cissus quadrangularis* and *C. rotundifolia*. Where sand dunes extend into this zone they are vegetated with an open, sparse woodland of *Acacia ehrenbergiana*. Hardy, unpalatable species such as *Cassia italica* and *Blepharis ciliaris* are also present.

The last zone, proximal to the mountains, is covered with silty alluvium soil and is densely populated and extensively cultivated. The alluvium plains are traversed by small gravel ridges which are vegetated with *Acacia oerfota*, *Aloe vera*, *Blepharis ciliaris* and *Lasiuris scindicus*. Trees of *Dobera glabra* with *Acacia ehrenbergiana* and *A. oerfota* occur on the uncultivated areas of the plains, often associated with species such as *Abutilon pannosum*, *Indigofera oblongifolia* and *Corchorus depressus*. In cultivated areas weedy species occur which include *Sesbania leptocarpa*, *Heliotropium longiflorum* and *Digera arvensis* (Wood 1985). This vegetation merges with the *Acacia-Commiphora* community of the foothills, with several common associated species such as *Adenium obesum*, *Euphorbia cactus*, *E. inarticulata*, *Indigofera spinosa* and *Chrysopogon plumosus*.

7.2.2 DHOFAR

The Dhofar plains form a 10-25 km wide coastal belt between the escarpment mountains and the Arabian Sea. They are composed of marine and aeolian sand and alluvial limestone gravels, and are traversed by a network of wadis which drain from the mountains (Figure 7.1: A). Where relatively stabilized sand bars have been formed between the sea and sea inlets, lagoons (*khawrs*) have formed. The vegetation of these water bodies is described in detail in Chapter 10.

The Dhofar plains come under the influence of the southwest monsoon (see section 2.2.2.3), and this is reflected in the abundance of the annual flora, and especially in the abundance of grass species (Cope 1985). The vegetation can be classified as a semi-desert grassland with scattered *Acacia* trees. The frequently heavy moisture from the fog and low cloud of the monsoon condenses in the canopy of trees and provides moisture by dripping to the ground (see section 2.3.5).

The inhabitants of the Dhofar coastal plains have local names for the zonation of soil type and vegetation from the sea to the foot of the escarpment mountains (Miller & Morris 1988). The *aj* is the area of soft sandy soil above the high tide mark and is devoid of vegetation. The soft alluvial plains, the *hazog*, is either bare or supports a community of halophytes with *Aeluropus lagopoides*, *Cressa cretica*, *Heliotropium fartakense*, *Limonium axillare*, *Cyperus conglomeratus*, *Urochondra setulsa* and the creeper *Ipomoea pes-caprae*. In the vicinity of *khawrs Paspalum vaginatum*, *Bacopa monnieri*, *Sporobolus* spp. and *Typha domingensis* subsp. *australis* are found. In the flat rocky areas, the *da'an*, the dominant vegetation is an open shrubland of *Acacia tortilis* and *A. ehrenbergiana*. *Boswellia sacra* occurs in the upper catchment areas. *A. tortilis* grows almost at ground level in several wind-swept areas. Species of *Commelina, Solanum* and *Withania*, with species of grasses and sedges such as *Dicanthium micranthum, Digitaria*, *Sporobolus*, *Ochtochloa* and *Cyperus,* form a dense cover under tree canopies.

The alluvial plains are interspersed with rocky outcrops, especially towards the foothills of the escarpment mountains. The soils in such areas are shallow and support a low shrubland of *Commiphora* and *Acacia* spp. Xerophytic shrubs such as *Cadaba baccarinii*, *C. farinosa*, *Caesalpinia erianthera*, *Commiphora habessinica*, *C. foliacea* and *C. gileadensis* grow in association with succulent species such as *Adenium obesum*, *Caralluma flava*, *Sansevieria ehrenbergii*, *Kleinia odora*, *Euphorbia cactus*, *Aloe dhufarense* and *A. inermis*. Towards the northern end of the Zalawt plains *Boscia arabica* grows, forming a very open scrub with the *Acacia-Commiphora* community.

Wadis flowing into the coastal plains have a considerable influence on the vegetation and around them there is a characteristic grouping of species. The areas around permanent water are extensively cultivated. Plant cover is normally dense, with *Tamarix nilotica*, often accompanied by the shrubs *Salvadora persica*, *Calotropis procera* and *Leptadenia pyrotechnica* and the climbers *Pentatropis nivalis* and *Cissus quadrangularis*. Wadi banks are mostly occupied by tall grasses and reeds such as *Desmostachya*, *Phragmites* and *Typha*.

7.3 Plains of Eastern Arabia

The plains of eastern Arabia include the sandy, coastal areas along the Arabian Gulf and the narrow gravel plains along the Gulf of Oman. These are the coastal lowlands of eastern Saudi Arabia, the plains of the United Arab Emirates, Qatar, Kuwait, the central plateau of Bahrain and the Batinah Coast of Oman.

7.3.1 LOWLANDS OF EASTERN SAUDI ARABIA

The coastal lowlands of Saudi Arabia form a belt along the Gulf with a maximum width of *c.* 100 km. To the north, around 27° 30' N, they border with the northern plains and south of 23° they merge with the Rub' al Khali. The northen plains rise gradually from sea level up to *c.* 400 m in the west and extend through Kuwait in the north. In the south the Summan plateau marks their limit. The soils of this area consist of late Pliocene or early Pleistocene alluvial deposits from the Wadi Rumah and al Batin drainage systems. They are silty with deposits of gravels and stones(Mandaville 1990).

The lowlands are generally sandy with some limestone outcrops. Large areas consist of sabkhas, which are highly saline, salt incrusted flat surfaces. Plant life is

Table 7.1 % of total shrub cover of the dominant perennial species (A) and density of the associated annual species (B) in the *rimth* saltbush shrubland in eastern Saudi Arabia. (Study area located 18 km southwest of Saffaniya, 27°53.4'N 48°37.8'E).Total shrub cover: 8.6%; total shrub density: 772 ha^{-1}. From Mandaville (1990).

(A)

Perennial species	% of total shrub cover
Haloxylon salicornicum	95
Panicum turgidum	3
Lycium shawii	2

(B)

Annual species	Density (m^{-2})	Annual species	Density (m^{-2})
Plantago boissieri	145	*Ogastemma pusillum*	4
Schismus barbatus	65	*Ononis serrata*	2
Astragalus spp.	14	*Picris babylonica*	2
Ifloga spicata	50	*Rostraria pumila*	1
Lotus halophilus	8	*Hippocrepis bicontorta*	1
Paronychia arabica	6	*Atractylis carduus*	0.4
Medicago laciniata	5	*Launaea mucronata*	0.4
Launaea capitata	4	*Euphorbia granulata*	0.4
Crucianella membranacea	4	*Stipa capensis*	0.4
Cutandia memphitica	2		

Table 7.2 Density of the associated annual species in a stand of *Rhanterium eppaposum* in eastern Saudi Arabia. (Study area located at 17 km NE of Qaryat al 'Ulya, 27°36.7'N 47°49.2'E). Total shrub cover: 16.3%; total shrub density: 3,145 ha^{-1}. From Mandaville (1990).

Species	Density (m^{-2})	Species	Density (m^{-2})
Plantago boissieri	22	*Launaea mucronata*	0.2
Schimus barbatus	12	*Polycarpaea repens*	0.2
Picris babylonica	7	*Horwoodia dicksoniae*	0.2
Neurada procumbens	4	*Gastrocotyle hispida*	0.2
Erodium laciniatus	1	*Emex spinosa*	0.1
Asphodelus tenuifolius	1	*Lotus halophilus*	0.1
Medicago laciniata	1	*Silene arabica*	0.1
Plantago ciliata	1	*Anisosciadium lanatum*	0.1
Rostraria pumila	2	*Ifloga spicata*	0.1
Astragalus spp.	1	*Reseda arabica*	0.1
Trigonella stellata	2	*Stipa capensis*	0.1
Paronychia arabica	0.5	*Loeflinggia hispanica*	0.1
Schimpera arabica	0.3	*Hypecoum pendulum*	0.04
Plantago ovata	1	*Leontodon laciniatus*	0.04

virtually absent on *sabkha*, although there is a characteristic fringing vegetation of halophytic shrubs. To the southeast, the coastal lowlands are covered with sand. Unstabilized and stabilized sand dunes occur in this area, with a sparse vegetation of naturalized date palms and *Tamarix* shrubs.

The vegetation communities of eastern and central parts of Saudi Arabia were described by Vesey-Fitzgerald (1957a) during anti-locust surveys, and that of the coastal and inland plains of Kuwait by Halwagy and Halwagy (1977) and Halwagy (1986). Here I follow Mandaville's (1990) classification and description of the plant communities of eastern Saudi Arabia.

The vegetation of the coastal lowlands and northern plains of eastern Arabia can be classified into four major plant communities. The first and most abundant is the *rimth* saltbush shrubland, dominated by *Haloxylon salicornicum* (Table 7.1), occurring from Iraq to the northern edge of the Rub' al Khali. In the central coastal lowland of Saudi Arabia this species is found on sandy substrates with high

Table 7.3 Species density and cover of the dominant and associated perennial species in the *thumam* grassland in eastern Saudi Arabia. (Study area located near al Ju'aymah, 26°47.1'N 49°54.5'E). Total shrub cover: 5%; total shrub density: 6,250 ha^{-1}. From Mandaville (1990).

Species	% Total Cover	Species	% Total Cover
Panicum turgidum	65	*Salsola baryosma*	9
Cyperus conglomeratus	4	*Pennisetum divisum*	1
Zygophyllum qatarense	12	*Leptadenia pyrotechnica*	4
Calligonum comosum	5		

underground water levels and can have a cover of 95% (See Table 7.1). On rocky surfaces other species, such as *Anabasis lacantha* and *Agathophora alopecuroides*, co-dominate.

The second plant community is the *'arfaj* shrubland (the *Rhanterium* steppe of Vesey-Fitzgerald 1957a). This community is dominated by *Rhanterium epapposum* and forms an open shrubland over much of the northern plains, extending into Iraq. The shrub grows up to 1 m in height and forms pure stands of varying densities ranging from stunted bushes several metres apart in sandy areas to robust shrubs *c.* 1 m apart on the limestone pavement of the northern plains. Annual plant species germinate after the winter rains (Table 7.2), with *Plantago boissieri* recorded as the most frequent and abundant (Mandaville 1990, Vesey-Fitzgerald 1957a).

The third community is the *thumam* grass-shrubland (Table 7.3). This community, dominated by *Panicum turgidum*, is widely distributed in the central coastal lowlands in relatively well-drained sandy soils. It is also found in restricted areas in Kuwait. Woody associates include *Lycium shawii*, *Leptadenia pyrotechnica* and *Calligonum comosum*. The latter species is of most frequent occurrence and is often found as a co-dominant.

To the west of the coastal lowlands lies a belt of limestone uplands known as the Summan plateau. This is bounded on the north and east by the northern plains and the coastal lowlands respectively, and merges in the south with the gravel desert at about the 23rd parallel. The Summan escarpment is rocky with poor soil cover, although there are some vegetated silty basins between the rocky outcrops. The southern part of the Summan plateau is more arid and consequently supports fewer species. On the rocky areas there is a sparse cover of stunted shrubs with *Haloxylon salicornicum* or *Anabasis setifera* being dominant, with scattered shrubs of *Ziziphus nummularia* growing in the basins. After favourable rains the basins support a good cover of annuals, although due to grazing pressures unpalatable species such as *Astragalus spinosus* tend to dominate.

7.3.2 The Arabian Gulf Coast

The plains of the Arabian Gulf region are mainly coastal. They are flat, saline and hypersaline areas extending 1-5 km from the coast, with numerous sea inlets, tidal lagoons (*khawrs*) and shallow bays. The vegetation of this coast is described in detail in Chapter 9 and is only mentioned briefly here for the sake of continuity.

In Bahrain, the coastal plains in the north of the island form the main inhabited area. The coastal plains lead to the central stony or gravel and sand desert. In this region urbanization, industrialization and agriculture have greatly changed the natural vegetation, which has been replaced by weeds, crops and date palm plantations. In addition, industrial activities, poor drainage and extensive irrigation have affected much of the soil surface, giving rise to the dominance of hardy, salt tolerant shrubs (Abbas & El-Oqlah 1992, Abbas *et al.* 1991a, 1991b, Doornkamp *et al.* 1980).

Coastal vegetation from northern Oman to Qatar is broadly characterized by the *Limonium-Zygophyllum* communities. (Babikir & Kürschner 1992, Batanouny 1981, Batanouny & Turki 1983, Ghazanfar 1991a, Ghazanfar in press-a) with *Limonium stocksii, Zygophyllum qatarense* and associated species. *Sphaerocoma aucheri* and *Cornulaca monacantha* colonize low dunes, and *Crotolaria persica*, *Atriplex leucoclada* and *Suaeda aegyptiaca* are the commonly associated species.

Inland, the sandy and low dune region is occupied by the dwarf shrub *Rhanterium epapposum*. This species forms the dominant vegetation and occupies vast areas of central and northeastern Arabia (Vesey-Fitzgerald 1957a, Zohary 1973). A variable group of species are associated with *R. epapposum*. In moderately loose soils *Cyperus conglomeratus* is co-dominant, and in shallow soils with rocky outcrops, *Haloxylon salicornicum* is co-dominant (Deil & Müller-Hohenstein 1996, Halwagy & Halwagy 1977).

The stony gravel plains of the Gulf support a vegetation similar to that of the Batinah Coast of northern Oman, described below, with an open woodland of *Acacia tortilis*, *A. ehrenbergiana* and *Lycium shawii*. Where the tree layer is destroyed or absent *Pulicaria glutinosa* is the dominant species. *Prosopis cineraria*, *Ziziphus* and *Ficus* occur along wadis. In regions with fine textured soil (*rodat*), *Z. nummularia* forms the dominant species, associated with common perennials such as *A. tortilis*, *L. shawii*, *Zygophyllum qatarense*, *Ochradenus* and *Ephedra foliata* (Batanouny 1981, Batanouny & Turki 1983).

Frankenia pulverulenta and *Zygophyllum simplex* characterize the ephemeral, salt tolerant, therophyte community with several grasses and other annuals. Common amongst these are species of *Aizoon*, *Anastatica*, *Arnebia, Astragalus*, *Asteriscus*, *Emex*, *Oligomerus*, *Spergula* and *Spergularia*.

7.3.3 The Batinah

The Batinah is a gravel plain, approximately 25 km wide, sloping gently from the eastern flanks of the northern mountains of Oman to the coast. The plain consists of broad alluvial terraces, limestone outcrops and interfluvial plains with wadis of coarse alluvial gravels and silt and very little soil development. The region is densely populated and extensively cultivated. Vegetable farms, date-groves and orchards form an almost continuous belt along the coast. Natural vegetation has largely been destroyed or modified, and where extant it is over-utilized by domestic livestock to the point where regeneration is minimal or absent.

The principal vegetation type of the Batinah is an open *Acacia-Prosopis-Ziziphus* woodland, consisting of *Acacia ehrenbergiana*, *Acacia tortilis*, *Prosopis cineraria* and *Ziziphus spina-christi,* and associated shrubs such as *Lycium shawii* and *Ochradenus arabicus.* Formerly, large areas of the coastal plain and alluvial lowlands were covered by this type of woodland (Kürschner 1986a), but anthropogenic influences have reduced it to degraded remnants (Figure 7.2).

Coastal limestone terraces along the Batinah are sparsely vegetated with an open *Commiphora-Euphorbia* community. Associated shrubs are *Convolvulus*

Figure 7.2 Degraded *Prosopis* woodlands in northern Oman.

virgatus, *Euphorbia larica*, *Fagonia indica*, *Lycium shawii*, *Rhazya stricta* and *Zygophyllum qatarense*. *Pteropyrum scoparium* and *Jaubertia aucheri* are the dominant species in wadis. Grasses include *Aristida* and *Stipagrostis*. The annual vegetation is characterised by several species of which *Zygophyllum simplex*, *Plantago ovata*, *Aizoon canariense* and *Cometes surrattensis* are the most common.

Stands of the mangrove *Avicennia marina* occur intermittently along the Batinah in tidal channels and permanently flooded lagoons and gullies. Due to the over-utilization of ground water and developmental projects along the beaches, the condition of mangrove stands has deteriorated in recent years (pers. obs). On sand dunes around tidal channels, mangroves are intermixed with a community of halophytic shrubs. The salt tolerant *Halopeplis perfoliata* and *Aeluropus lagopoides* are normally present at the borders of runnels, followed by *Indigofera oblongifolia* and *Suaeda aegyptiaca* on low sand dunes and elevated areas. Kürschner (1986c) has identified the concentration of Cl^- and $CaCO_3$ in the soil as determinants of the zonation of these two communities.

The composition of the annual vegetation and the growth and phenology of both the annual and perennial species on the Batinah is related to the timing and abundance of precipitation. Total precipitation affects the onset of flowering in the annuals and geophytes, but does not affect the trees and shrubs, and late rain delays the onset of all phenological phases (growth, flowering, fruit and seed set) in all life-forms (Figures 7.3, 7.4). In normal years, flowering in shrubs and trees occurs from 4 to 6 weeks after rain and in the annuals from 2 to 8 weeks after

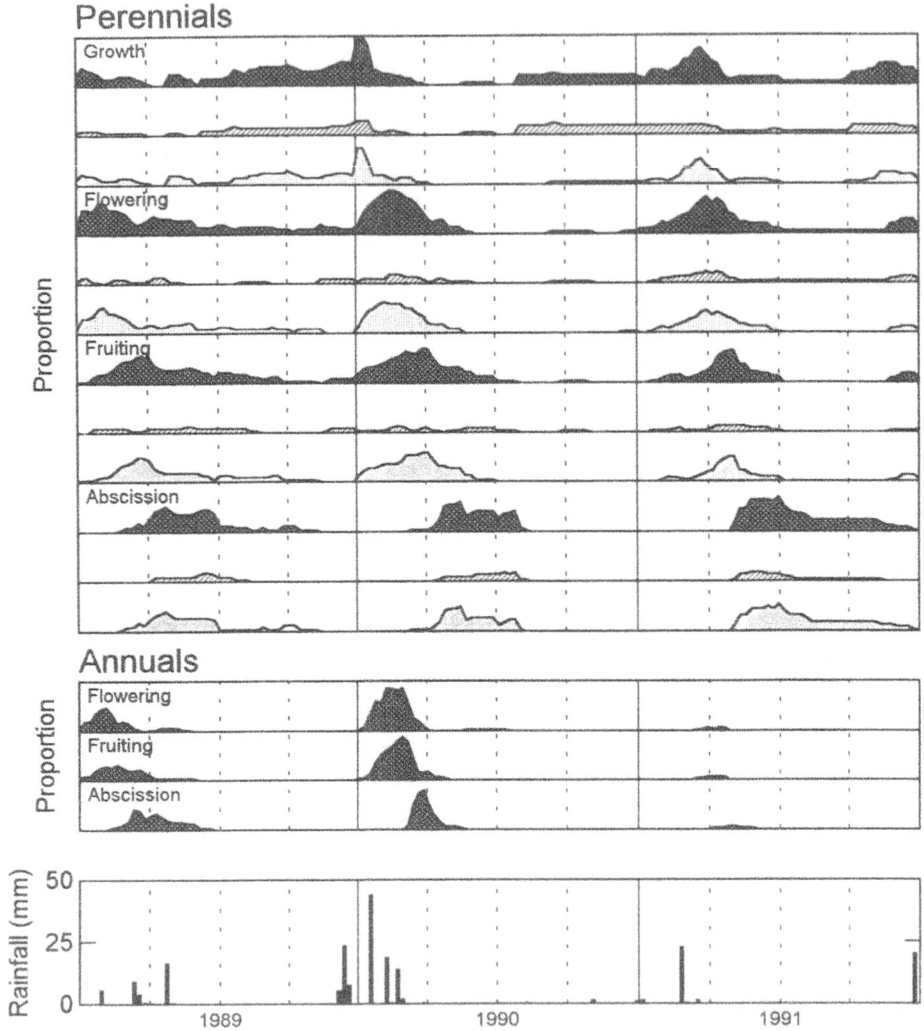

Figure 7.3 Phenological events (growth, flowering, fruiting and abscission) of the proportion of perennial and annual plant species in a wadi on the Batinah coastal plain of northern Oman (all scales are 0-1) and total weekly rainfall for 1989-1991. Total perennial and annual species (filled), phanerophytes (hatched) and chamaephytes (speckled). After Ghazanfar (1997).

rain. The perennial vegetation of the plains, especially the trees, are least affected by drought years, whilst annual species show a marked decline in richness. A decrease of 10% in species richness has been reported in drought years in Kuwait (Mirreh & Al Diran 1995, Omar 1995) and northern Oman (Ghazanfar 1997).

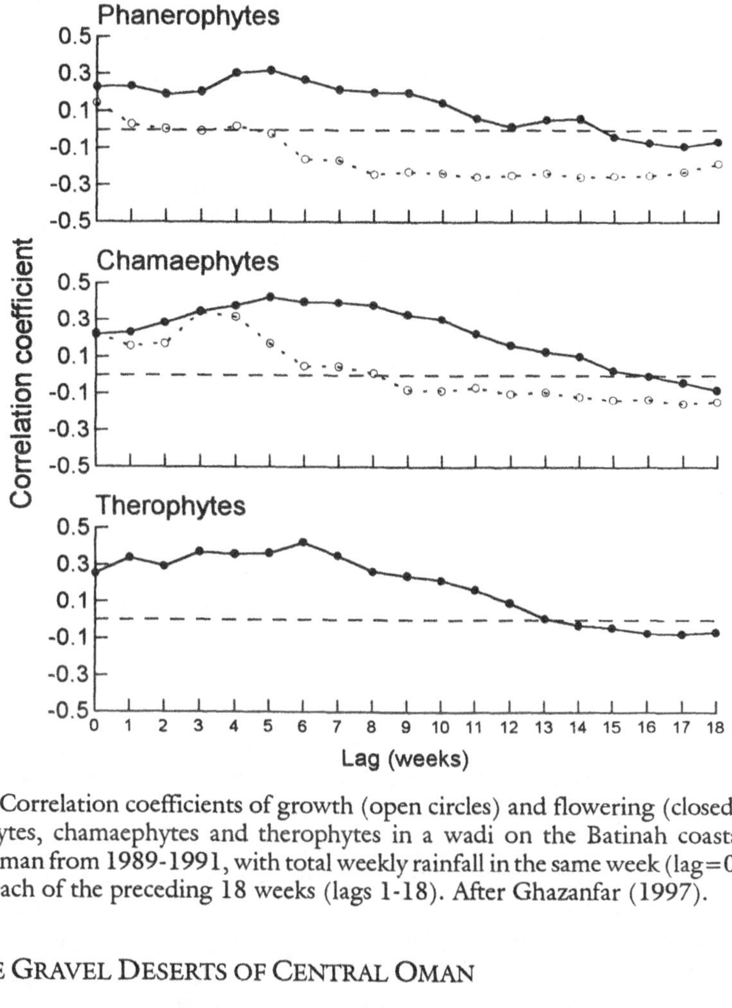

Figure 7.4 Correlation coefficients of growth (open circles) and flowering (closed circles) of phanerophytes, chamaephytes and therophytes in a wadi on the Batinah coastal plain of northern Oman from 1989-1991, with total weekly rainfall in the same week (lag=0) and total rainfall in each of the preceding 18 weeks (lags 1-18). After Ghazanfar (1997).

7.3.4 The Gravel Deserts of Central Oman

The inland deserts of Oman are principally flat, sandy expanses overlain with rocks and gravel. The substrate is limestone, sandstone and shale covered with aeolian sands. There is little soil cover and vegetation is sparse. The distribution of species and the pattern of plant communities is largely dependent on geomorphology and the water availability. Since the area is hyperarid (rainfall <100 mm per year and frequently <50 mm) vegetation where present is largely restricted to depressions and runnels. Water accumulates in depressions after rain (Figure 7.5) with subsequent germination of annuals, few of which survive to maturity. Moisture from fog and dew also contributes to plant growth (Figure 2.13).

The vegetation consists of a sparse, open scrub of *Acacia tortilis*, *A. ehrenbergiana* and *Prosopis cineraria*. Trees are heavily utilized, and most have distinct browse lines. Where the substrate is sandy, low mounds (*nabkhas*), up to a metre in height and 3 m in diameter are formed from the accumulation of wind-

Figure 7.5 An ephemeral pool in the gravel desert plain of central Oman.

blown sand caught between roots and stems. *A. ehrenbergiana* grows on these mounds with *Heliotropium kotschyi*, *Fagonia ovalifolia* and *Arnebia hispidissima* as associated species. Large shrubs other than *Acacia ehrenbergiana* are few; *Lycium shawii* is not uncommon, associated with and often growing in the shade of the canopy of *A. tortilis* (Ghazanfar, 1997). Ground cover is dominated by members of the Chenopodiaceae and Zygophyllaceae. These include *Salsola rubescens*, *Cornulaca* spp., *Zygophyllum qatarense* and *Fagonia* spp. Grasses are sparse, ephemeral and heavily utilized when available. Unpalatable species such as *Tephrosia apollinea*, *T. purpurea* and *Cymbopogon schoenanthus* are also abundant.

In the southern region of the central plains a few species are added to the *Acacia* community. Representatives of the Zygophyllaceae and Chenopodiaceae are still the dominant species, but with *Cornulaca monacantha*, *Holothamnus bottae*, *Zygophyllum qatarense* and *Stipagrostis plumosa* being common. *Ziziphus leucodermis* and the dwarf palm, *Nannorrhops ritchieana*, occur in sandy depressions.

The largest geomorphological unit of the central gravel deserts is the limestone plateau of the Jiddat al Harasees, the climate, flora and fauna of which forms a discrete ecological unit (Figure 7.1: B). It is an area of *c*. 40,000 km^2, 100-250 m above sea level, of Tertiary limestone overlain with sand and gravel. It is delimited in the east by the Huqf escarpment and to the west by the sands of the Rub' al Khali. The Jiddah is relatively flat, interspersed with low lying areas (*haylah*) and ridges. The ridges are gravelly and the depressions sandy. Rainfall is low (39 mm per year, see climagram for Ja'aluuni in Figure 2.6) and irregular, but precipitation from fogs supports a relatively good cover of vegetation (Stanley-Price *et al.* 1988). Moisture collected from fogs ranges from 0.08 to 13.8 l m^{-2} day^{-1}, and although fogs can occur throughout the year, the greatest number of foggy days occur in the spring and autumn months (see Figure 2.13). The effects of fog on the Jiddat are considerable. Not only does the plain have a greater species richness than surrounding areas, but there is growth of epilithic cyanobacteria and lichens on stones and the bark of trees (see Chapter 5).

The principal tree species are the same as those found in the central desert plains, *viz. Acacia tortilis*, *A. ehrenbergiana* and *Prosopis cineraria*, but there is a marked increase of species in the shrubby vegetation. About 200 species are recorded from the Jiddah, including 12 endemic species, and the area is classified as a local centre of plant endemism (see section 12.2.2.2). The endemic *Ochradenus harsusiticus* is a dominant shrub distributed throughout the Jiddah. In sandy depressions *Pulicaria undulata*, *Fagonia ovalifolia*, *Dipterygium glaucum*, *Taverniera lappacea*, *Crotalaria aegytiaca, Cassia holosericea* and *Tephrosia apollinea* are common, with the grasses *Ochtochloa compressa*, *Cymbopogon schoenanthus* and *Panicum turgidum*. On the ridges and gravelly areas *Stipagrostis sokotrana*, *Limonium stocksii*, *Cyperus conglomeratus, Heliotropium kotschyi* and *Corchorus depressus* are common. Saline soils are dominated by species such as *Salsola rubescens*, *Zygophyllum qatarense, Z. hamiense* and *Suaeda* spp. In restricted locations, where there are sandy depressions and shallow wadis, the dwarf fan palm, *Nannorrhops ritchieana* occurs with *A. ehrenbergiana.* At the eastern edge of the Jiddah, at the Huqf escarpment, the composition of the species changes, and around a few permanent water pools that lie immediately beneath the escarpment species such as *Juncus rigidus*, *Scirpus maritimus*, *Cressa cretica*, *Rhazya stricta*, *Aerva javanica* and *Pluchea arabica* occur.

7.4 Plains of Central and Northern Arabia

The rock and gravel plains of central and northern Arabia consist largely of areas of crystalline basement rocks covered with volcanic lava flows, called *harrah*. The region of the western Nejd, containing Harrat Nawasif, Harrat Rahat, Harrat Khaybar and Harrat 'Uwayrid, consists of table-lands at about 800 m elevation, rocky plains and inselbergs. Soil cover is sparse, rainfall generally <100 mm and the vegetation cover low or absent. Where vegetation is present the dominant species are *Artemisia sieberi* and *Achillea fragrantissima*, the latter forming almost pure stands in certain localities. In the more sandy areas, *Haloxylon salicornicum* is abundant, associated with *Astragalus spinosus*, *Halothamnus bottae*, *Farsetia aegyptiaca* and *Stipagrostis* spp.

7.5 Summary

The flora of the rock and gravel plains is relatively poor with a diversity of only a few hundred species. Vast areas lack substantial soil development and have a relatively low vegetation cover. The dominant vegetation consists of thorny, halophytic and succulent shrubs. Annual vegetation is largely dependant on rain. The southern coastal plains have the highest species diversity, and the central plains the lowest. The low plateau of central Oman where the vegetation benefits from heavy dews and seasonal fogs is unusual both for its relatively high species richness and for its relatively high percentage of endemic species.

CHAPTER 8

Vegetation of the Sands

James P Mandaville

"... in el-Weshm they say, "The Nefud reaches in the north to Jauf el-'Amir, and southward to Sunn'a [San'a]."... but we have seen that they are not continuous."

C M Doughty (1888)

8.1 Introduction

The 19th Century citizens of al Washm, central Arabia, were nearly right of course. It is indeed possible to travel from the northern fringe of the Great Nafud to within 200 kilometres of the Yemeni capital without ever stepping outside Arabia's linked sand seas. Including the southernmost outlier, Ramlat as Sab'atayn, they extend over 15 degrees of latitude, have a total surface area of 795,000 km^2 (Wilson 1973) and account for nearly a third of the Peninsula's land area.

Arabia has thousands of additional square kilometers of shallow sand-covered surface, but the deep, wind-formed and often mobile sands of the *nafuds* comprise a unique biological habitat with special ecological constraints. Plant life in these dune bodies is characterized by low species diversity, uniformity of life-form and species composition, and adaptations to the unstable substrate. By far the greater part of the Arabian sand seas occupy inland basins. Three of them, however, actually or nearly touch the coasts. The vegetation of such near-littoral sands tends to be atypical of the main, inland dunes and, except incidentally here, are described separately in Chapter 9.

S. A. Ghazanfar and M. Fisher (eds.), Vegetation of the Arabian Peninsula, 191-208.

The term 'erg', from French usage in North Africa and derived from Arabic *'irq* (pl. *'uruq*), meaning literally 'root' or 'vein', has entered the geomorphological literature as a technical term for any large sand body or 'sand sea'. However, the people of the Arabian Peninsula use it (with a few minor exceptions) to refer to a single, elongated sand ridge, linear dune or dune chain. It would thus seem preferable, at least when discussing landforms of the Arabian Peninsula alone, to avoid use of the term 'erg' in the broader, North African sense. I therefore use 'sand sea', or *nafud* or *ramlah,* the standard Arabic topographical terms for 'sand body' preferred, respectively, in the northern and southern parts of the Peninsula.

8.2 Sand Characteristics and Sources

Sand textures in dunes throughout the areas described here tend to fall in the middle zone of the fine to medium range (0.125-0.5 mm, or 1-3 phi; Holm 1953, McClure 1984, Whitney *et al.* 1983). In most parts the sand tends toward a reddish colour resulting from grain coatings of iron oxides. These oxides appear to be derived from windborne clay minerals and tend to increase in colour intensity with age and distance of transport, as ferric hydrate (limonite) is converted to hematite (Walker 1979).

The Rub' al Khali sand sea originated through the reworking and transport of Pliocene sediments in the Quaternary Period (McClure 1978), and the northern sand bodies have also been interpreted as Quaternary products of increased wind activity (Schulz & Whitney 1986). During this time both the northern and southern sands experienced alternating arid and pluvial episodes (Sanlaville 1992).

Wind patterns today range from prevailing westerlies in the western portion of the Great Nafud to increasing northerly components in the sand seas to the east and south. At around the Tropic, an easterly influence becomes apparent in the central and western parts of the Rub' al Khali, while the eastern part of that body experiences more multi-directional winds. These trends are all apparent in the alignments and dune forms of the Great Nafud, the sweeping arcs of sand that become the Dahna and central inner *nafuds*, and the massive structures of the eastern Rub' al Khali (see also section 3.4.9 and Figures 3.6 and 3.7). Wind velocities are maximum in winter in the Great Nafud but tend to peak in early summer farther to the south and east under the influence of the *shamal* winds arising from the seasonal low pressure pattern over the Gulf.

8.3 Plant Ecology in Deep Sand Environments

8.3.1 Plant-soil Interactions

For plants, sand possesses a number of advantages over heavier, less pervious soils with respect to water absorption, root supply and water storage. It presents disadvantages also, particularly in active dune areas, with regard to mechanical stability and nutritive qualities. The vegetation of any particular sand area will thus represent the integrated effect of these factors with local climate.

Sand, compared to substrates with significant content of silt, clay and organic matter, has relatively high porosity, permeability and wettability. Virtually all impinging rainfall is absorbed; runoff is negligible, and moisture penetration is deep. Its relatively coarse texture and permeability results in high evaporation rates in the upper few centimetres at the surface, but these same factors break capillary channels that might otherwise wick deeper moisture to the surface. The result is the establishment of a poorly ventilated 'sealing' layer at the surface and, after rains, of a relatively persistent moist horizon below within the root zone. This relatively shallow sealed layer is exploited by the extensive horizontal root system of plants such as the important Arabian psammophyte *Calligonum comosum* (Warren 1988), enabling it to make optimum use of moisture.

Sand does not, per unit volume, hold as much water as finer textured soils, but unlike the case of finer silts and clays, the forces binding water molecules to the particles are relatively slight. This means that nearly all the soil water is readily available for uptake by roots at low suction pressures. Overall, it is an excellent example of the 'inverse texture' effect described by Noy-Meir (1973), whereby in arid environments the disadvantages of coarse soils with respect to rapid drainage and low holding capacity are more than offset by lesser evaporative loss. A typical rainfall event on a sand sea area may thus involve: (1) the full interception, without runoff, of, say, 25 mm precipitation, (2) the near-immediate drainage of this water to form a moist layer with a clear wetting front approximately 150-250 mm below the surface, (3) the rapid drying, within 2-3 days, of the surface sand to a depth of a few centimetres, and (4) the persistence of the lower moist horizon for several weeks or months and its availability for plant growth until it dissipates through diffusion and transpiration.

Environmental trauma for most sand-dwelling plants is perhaps greatest at time of seedling establishment, which is threatened by problems of both water supply and mechanical stability. Since the upper few centimetres of the soil surface dry out very quickly after rain, root growth must be rapid to ensure continued water supply as well as mechanical fixation. This problem may be overcome to some extent by the burial of seeds by sand movement to depths where moisture is better conserved.

Questions sometimes arise as to the possibility of the rise of moisture to the plant root zone by capillary conduction through dunes from an underlying water table. Bagnold (1941) has shown, however, that surface tension in sand grains is unable to cause any moisture rise beyond 40 cm at most. He also pointed out that intradune evaporation and re-condensation cannot account for moisture rise, as the requisite work would require considerably greater thermal changes in the dune body than in fact occur. Bagnold (1941) associated vegetated and moist areas on dunes with areas of poorly consolidated sand in steeply inclined slipface laminae, and his experiments indicated that such beds had better water infiltration rates than areas with horizontal bedding. In my experience, however, the only plants regularly associated with the soft sands of slipface deposits in our area are the tussock grasses, *Centropodia forsskalii* and *C. fragilis*. Other, more common, perennials occur in typical, firm accretion deposits (which may, however, overlay slipface strata).

The primary disadvantage of sand as an environment for plant life is its mechanical instability with respect to wind forces, and plant adaptations for psammic environments are probably concerned with meeting this challenge at least as much as that of moisture supply. The mechanical problem is first met at the time of a seedling's establishment, when it is particularly vulnerable to being excavated by deflation or overwhelmed by accumulation. For plants established in more active dune fields, the challenge is never-ending and is enjoined with adaptations of rapid shoot and adventitious root development.

Bendali *et al.* (1990) describe growth mechanisms of *Rhanterium suaevolens* and *Aristida pungens* that enable them to survive sand accumulation in Tunisian sand terrain. They conclude that the ability to develop adventitious roots as well as aerial parts at increasing surface levels were key factors. *Aristida*, in fact, was interpreted as an obligate sand accumulator which diminishes and dies without sufficient sand build-up. Danin (1991) reports similar obligate growth patterns in *Stipagrostis scoparia* in Sinai. He classes the important Arabian species, *S. drarii*, however, as an ecomorphological type tolerant of sand accumulation, deflation and stability (Danin 1996a). *Artemisia monosperma*, a common codominant of Arabia's inland deep sand habitats also has an ability to develop adventitious roots from sand-covered stems (Danin 1991). *Calligonum comosum*, one of the most important perennials in the more active Arabian sands, may thrive in conditions where sand burial acts as a growth stimulant (Warren 1988).

The survival of plants in areas of active sand deflation is concerned mainly with root exposure. Danin (1991) notes that *Convolvulus lanatus* and *Artemisia monosperma* have roots with protective corky bark and cites studies showing that *Cornulaca monacantha*, as well as other lignified Chenopodiaceae, have roots with internal active phloem and xylem elements protected from external desiccation. Such adaptations are probably significant in important Arabian sand perennials including *C. monacantha* and *Calligonum* spp.

Dune sands are generally characterized by low nutrient status and particularly so, as a result of their low organic content, with respect to nitrogen. Bowers (1982) cites measurement of total nitrogen in North American dune fields ranging between 4 and 415 ppm, or one to two orders of magnitude below those reported for non-dune desert soils. Endomycorrhizae increase absorption of nutrients, particularly phosphorus, and bacterial nitrogen fixation may be an important nutrient aid in the vascular plants of dune areas. Bowers (1982) lists several genera, including *Aristida* and *Artemisia* that have associated nitrogen fixing bacteria. Danin (1991) suggests that air-borne dust and cyanobacteria may provide nutrient inputs to dune plants. There is growing evidence that the rhizosheaths characteristic of the root systems of desert perennial grasses such as *Stipagrostis plumosa* may be important aids to nitrogen nutrition as well as water absorption. Nitrogen fixing bacteria have been found associated with the cylindrical root sheaths formed of matted root hairs and sand grains held by secreted mucilage (Danin 1996b).

8.3.2 Vegetation Dynamics

Short- to medium-term autogenic succession in communities of arid zone plants hardly exists, given the mechanical limitations to the accumulation of organic materials and other soil-developing factors. Allogenic succession however, depending on habitat changes unrelated to plant growth, may sometimes be significant (Kassas 1966). Warren (1988) has described such a process involving *Calligonum*, *Heliotropium* and *Euphorbia* in active dune fields of the Wahibah Sands, and this probably exemplifies a pattern common over much of the Arabian sand seas. Bowers (1982), from studies of inland dune vegetation in western North America, visualizes a potential for a succession toward adjacent non-sand communities but noted that continued sand supply and arid conditions would normally inhibit such a development. Both of these factors exist in the majority of Arabian dune environments and might well lead to community stasis.

8.4 Vegetation of the Sand Seas

8.4.1 The Great Nafud

The Great Nafud (known in Arabic simply as an Nafud, 'the Nafud') has an area of 72,000 km^2 (Wilson 1973) and is the northernmost of the Arabian sand seas (Figure 8.1). Transverse dunes and barchanoid ridges predominate in the southeastern and far northern sectors of this sand body, as determined by westerly winds. The north-central and northeastern parts are characterized by somewhat complex forms with an overall linear aspect. South-central and eastern areas display a complex mix of dune types including linear, transverse, star and domal forms. The increasingly northern wind components in the east lead to more pronounced linear forms merging downwind into the upper arms of the Dahna and inner sand arcs (Whitney *et al.* 1983). A striking and often described feature of the Great Nafud are the large, horseshoe-shaped stabilized hollows, sometimes 60 m or more deep and aligned in sublinear formations.

Whitney *et al.* (1983) argue persuasively that the larger dune forms of the Great Nafud, as well as other northern sand seas, are partially stabilized by vegetation and lag deposits. According to them active dunes of fine pale sand comprise less than 5 percent of the dune surface and are generally perched on the upper parts of the stable base forms characterized by coarser grain size and reddish colour. This view is in agreement with reports by early travellers such as the Blunts (Blunt 1968), as well as those of recent observers such as Chaudhary (1983). The vegetation of the Great Nafud is typical of northern Arabian sand seas in consisting of open communities of large and small psammophytic shrubs

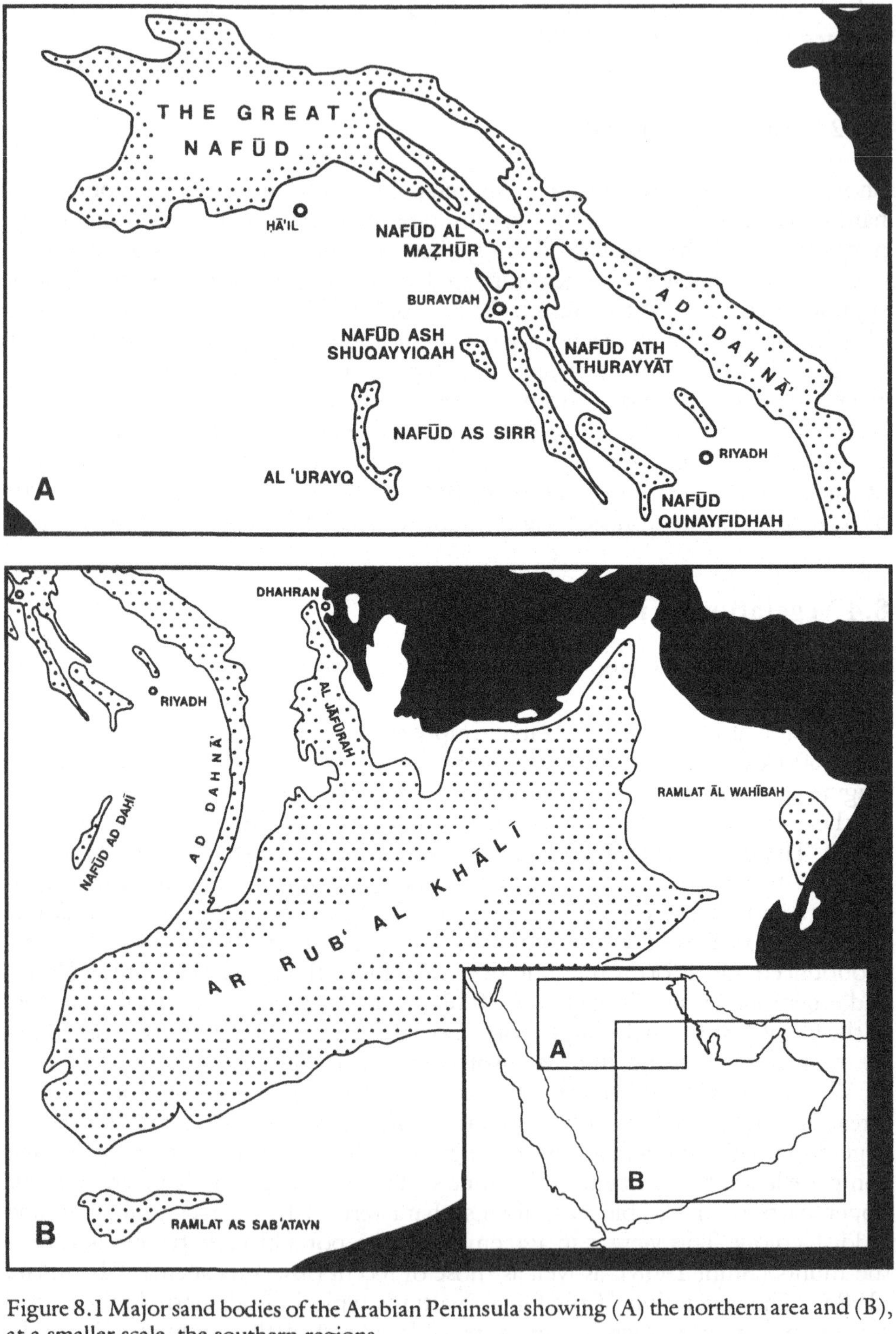

Figure 8.1 Major sand bodies of the Arabian Peninsula showing (A) the northern area and (B), at a smaller scale, the southern regions.

and in displaying a vernal aspect (depending on the quantity of each season's winter rainfall) with sometimes abundant annual herbs. Vesey-Fitzgerald (1957a) placed the Great Nafud in his "Central Arabian Red Sand Vegetation" type characterized by co-dominant *Calligonum comosum* and *Artemisia monosperma*.

I follow Chaudhary (1983) in the following description of the two core, deep sand communities of the Nafud, and all data, unless otherwise attributed, are derived from his valuable account.

8.4.1.1 *Calligonum-Artemisia-Scrophularia* Community

The *Calligonum comosum-Artemisia monosperma-Scrophularia hypericifolia* community is the most widespread community in the Nafud and corresponds directly with Vesey-Fitzgerald's (1957a) "Central Arabian Red Sand Vegetation" and the Calligono comosi-Artemisietum monospermae (*'abal-'adhir* sand shrubland) of Mandaville (1990)(Figure 8.2). It occurs on the mid-upper parts of the dunes, the sides of deep hollows and on undulating sheets of deep sands. Commonly associated perennials include *Stipagrostis drarii*, *Centropodia fragilis*, *Moltkiopsis ciliata*, *Monsonia heliotropoides* (or *M. nivea*) and locally, *Cyperus conglomeratus* (this much-cited macrospecies name is used here except with reference to the segregation of *C. eremicus,* Kukkonen 1995). *Calligonum* is well

Figure 8.2 The Dahna', at approximately 26°N, 47°E. Typical *Calligonum-Artemesia* community with *Artemisia monosperma* in foreground, right, and *Calligonum comosum* behind in right midground. Dry grass in foreground (and grazed by camel in background) is *Stipagrostis drarii*.

known as a source of long- and clean-burning firewood (Mandaville 1990), and its exploitation for this use in the Nafud sometimes leads to the damage or even local eradication of this dominant shrub.

8.4.1.2 *Haloxylon-Artemisia-Stipagrostis* Community

The *Haloxylon persicum-Artemisia monosperma-Stipagrostis drarii* community is analogous to the Haloxyletum persici (*ghada* shrubland) described by Mandaville (1990) but differs from it by the presence of *Artemisia*, which is largely absent in sand bodies of the farther south. In the Nafud it occupies very deep sand on the shoulders of dunes or in shallow hollows, sometimes extending to lower dune levels on leeward faces. Smaller perennial associates are *Moltkiopsis ciliata* and *Monsonia heliotropoides* (or *M. nivea*). Schultz and Whitney (1986) report *Haloxylon* growing also on sand-underlain calcareous crusts in some larger depressions.

The community led by this sub-arborescent shrub is important in at least some northern to northwestern parts of the Nafud. The Blunts observed during their 1878 crossing of the sands along the line from ash Shaqiq wells to al Jubbah that it was abundant north of al 'Ulaym (28°46'N, 40°27'E) but not seen south of that point (Blunt 1968). Schulz and Whitney (1986), however, reported it in a depression in the southern sands 70 km northwest of Ha'il. Musil (1927) described and photographed its striking presence while crossing the far northwestern arm of the Nafud. He also listed *Calligonum* on this route but it is not clear whether he was referring to mixed or alternating stands. In the more southern sand seas of the Peninsula this *Haloxylon* tends not to mix with *Calligonum* (Mandaville 1986, 1990). Perhaps even more than *Calligonum*, *Haloxylon* is prized as a source of firewood and sometimes may suffer considerably from damage near Bedouin encampments or from cutting and trucking to distant camps or settlements.

Chaudhary (1983) also describes two ecotonal or transitional communities of the Great Nafud. Both are generally characteristic of deep bottoms in the sands in areas adjoining non-deep sand communities. The first is basically a sand bordering community of *Haloxylon salicornicum*, with *Calligonum* and *Scrophularia* in microhabitats with deeper sand. The second is a modification of the *Rhanterium epapposum* community, again with *Calligonum* and *Scrophularia* in deeper, more mobile sand terrain. Associated perennials of smaller stature in both of these types include many found in the core communities described above.

8.4.1.3 Annual Vegetation

Chaudhary (1983) lists 149 annual and perennial non-woody plants collected from all habitats within the outer limits of the Nafud. Of these, somewhat over 100 are annuals of more or less importance in the springtime flushes of ephemeral herbs that may occur after favourable winter or spring rain. In general, these therophytes appear to be moisture-opportunists common to different community types. Table 8.1 lists, on the basis of my experience and reports of other workers,

Table 8.1 Annual plant species that are important contributors to biomass in dune and stabilized sand environments. *Plantago boissieri* is often the most abundant and the greatest contributor to biomass.

Anthemis deserti	*Lotononis platycarpa*
Arnebia decumbens	*Medicago laciniata*
Astragalus hauarensis	*Neurada procumbens*
Astragalus schimperi	*Paronychia arabica*
Cutandia memphitica	*Plantago boissieri*
Emex spinosa	*Rumex pictus*
Eremobium aegyptiacum	*Schimpera arabica*
Hippocrepis bicontorta	*Schismus barbatus*
Ifloga spicata	*Silene villosa*
Launaea capitata	

those species that are important contributors to biomass in dune and stabilized sand. The majority exhibit typical desert ephemeral life cycles with rapid germination, quick development and flowering, and considerable size plasticity.

8.4.2 The Dahna and Central Sand Seas

The Great Nafud leads to the southeast into major southerly extensions which, along with several separate sand bodies of central Arabia, have generally similar vegetation. They comprise the great arc of the Dahna in the east, Nafud al Mazhur southeast of the eastern end of the Great Nafud, leading into Nafud ath Thuwayrat, Nafud as Sirr and Nafud Qunayfidhah farther south (Figure 8.1). To these may be added Nafud ash Shuqayyiqah, al 'Urayq and 'Irq Subay' farther to the southwest (not shown in Figure 8.1), and Nafud ad Dahi standing below the front of the Tuwayq escarpment. Dune forms in these central Arabian sands often show the effects of multi-directional wind regimes and range from large linear structures in the Dahna and parts of Nafud al Mazhur to star and dome forms. Particularly striking are the massive dome-shaped dunes of Nafud ath Thuwayrat, which may be 1-1.5 km in diameter and 100-150 m in height. Similar structures are found in Nafud as Sirr, Nafud ash Shuqayyiqah and Nafud Qunayfidhah (Holm 1953).

The vegetation of these red sands is generally similar to that of the Great Nafud associations led by *Calligonum-Artemisia-Scrophularia* and *Haloxylon persicum*. The extent and present distribution of *H. persicum* is somewhat uncertain. Philby, during his 1917 traverse through the Qasim, saw stands of it on deep sands immediately southwest, northwest and west of Buraydah, noting that "Here for the first time in the Qasim I came across the graceful *Ghadha*, which is most plentiful to the eastward of Buraida and has gradually disappeared from central Qasim under the depredations of man and beast" (Philby 1928).

Figure 8.3 View southwest across the north-central Dahna', 26°36'N, 46°6'E. Typical aspect of the *Calligonum-Artemisia* community, with *Calligonum comosum* (in near right and middle foreground) and *Artemisia monosperma* with *Cyperus conglomeratus* in the background. Other perennials recorded here included *Moltkiopsis ciliata* and *Monsonia nivea*. *Eremobium aegyptiacum* was a common spring annual.

Draz (1978) reported the protection of *Haloxylon* in a *hima*, or traditional grazing preserve (see sections 12.3 and 12.5 for further discussion of the *hima* concept), to stabilize dunes near the city of 'Unayzah in the same general area.

The long, high sand ridges of the Dahna alternate in many parts with flat, interdune floors with shallow, stable sand or harder ground where the vegetation is not of the deep-sand type. The dunes themselves carry basically the same *Calligonum-Artemisia* community found in the Great Nafud with similar perennial and annual associates (Giacomini *et al.* 1979, Mandaville 1990, Vesey-Fitzgerald 1957a; Figure 8.3). Vegetation in the Nafud al Mazhur has been described as similar, with *Rumex pictus* particularly abundant among spring annuals (Vesey-Fitzgerald 1957a). The sand sea of al 'Urayq also carries *Calligonum-Artemisia* with *Cyperus conglomeratus* on higher dune parts, while *Anabasis articulata* appears in some places on the interdune floors. In some areas here *Haloxylon persicum* occurs in association with the *Calligonum-Artemisia* community, but at lower dune levels (Schulz & Whitney 1986).

The vegetation in parts of Nafud as Sirr and Nafud Qunayfidhah was studied by Baierle and Frey in their 100-km-wide transect in the vicinity of the Riyadh-Taif highway. *Calligonum comosum* was dominant on deeper sands with perennial vegetation, but *Artemisia* was not observed. *Cyperus conglomeratus, Moltkiopsis ciliata, Centropodia forsskalii,* and *Dipcadi erythraeum* were conspicuous perennial associates of the deeper sands, while the presence of *Panicum turgidum, Stipagrostis plumosa, Polycarpaea repens* and *Rhanterium epapposum* was also noted, presumably in the more stable habitats. The spring flush of annuals was seen concentrated at the bases of dunes and in sand hollows, where coverages of up to 35 percent were recorded. Frequent species in this annual herb layer were *Eremobium lineare* (*E. aegyptiacum*), *Astragalus schimperi, Astragalus gyzensis* (or *A. hauarensis*) *Plantago cylindrica* and *Neurada procumbens* (Baierle & Frey 1986). The vegetation of other, more remote central Arabian sand bodies, such as 'Irq Subay' with its remarkable star dunes studied by McKee (1966) is yet to be described.

8.4.3 The Jafurah

The Jafurah, extending south from about the 27th parallel to merge with the Rub' al Khali near the Tropic, is unusual by virtue of its very pale, non-reddish younger sands (Anton & Vincent 1986) and the fact that it touches or nearly touches the Gulf coast in some northern parts. In such coastal zones it is often bordered above the beach by coastal communities such as that led by *Haloxylon salicornicum*, or by sabkha halophytes (see Chapter 9). Dune forms consist largely of simple or compound barchans, transverse rounded ridges and in some areas parabolics (Anton & Vincent 1986). Relief, even in areas of the most active dunes, tends to be moderate, and active dune areas are in many parts interrupted or bordered by stretches of stabilized sand.

Calligonum comosum with *Cyperus conglomeratus*, but apparently without the co-dominant *Artemisia* typical of more inland sands, is characteristic of the higher, more active dune areas. Bordering sandy flats may carry communities led by *Haloxylon salicornicum, Zygophyllum mandavillei* or *Panicum turgidum*. I have observed *Calligonum* on higher dunes in an east-west transect at about latitude 24° 50'N, with *Haloxylon salicornicum* on neighbouring flats. During another crossing at latitude 25°50'N, *Calligonum* and *Cyperus* with *Stipagrostis drarii*, sometimes accompanied by *Centropodia fragilis* and *Moltkiopsis,* were observed in more active dune areas. The more stable, coastal sands east of the heavy dunes were occupied by *Zygophyllum* and *Panicum turgidum*. Rare examples of the subarborescent asclepiad, *Calotropis procera*, may be encountered in the northern Jafurah. Also in the north, *Tamarix* spp. occur in depressions where brackish groundwater is near the surface.

I have observed *Haloxylon persicum* along the eastern margins of the Jafurah (Figure 8.4). It may be accompanied by *H. salicornicum* (as seen at 24°55'N, 50°33'E) or form pure and sometimes dense stands (as found in 25°01'N, 50°32'E), sometimes forming massive sand hummocks in blowouts. Available data suggests that the community becomes more important to the south as the sands approach the Tropic. An important deep sand associate of the Rub' al Khali, *Limeum arabicum*, reaches its apparent northern limit in the Jafurah around the

Figure 8.4 A nearly pure stand of *Haloxylon persicum*, 1.3-2 m high, on sands fringing the eastern edge of the Jafurah near 25°N, 50°30'E.

25th parallel. *Dipterygium glaucum*, another southern sand component, extends north to around 26°.

The Jafurah lies within the area of winter and early spring rain (see Figure 2.11) but receives declining amounts of precipitation toward its southern parts. This is reflected in the relatively lesser proportion of therophytes in the total flora (Mandaville 1990). Spring flushes of annuals, when they occur after good rains, generally include the same species found on more northerly and inland sands, but with lesser diversity. ***Eremobium aegyptiacum*, *Plantago boissieri*, *Ifloga spicata*, *Neurada procumbens*, *Lotus halophilus*, *Astragalus hauarensis*, *Silene villosa*, *Cutandia memphitica*** and *Launaea* spp. are characteristic.

8.4.4 The Rub' al Khali

The Rub' al Khali, or 'Empty Quarter', has an area of over 500,000 km^2, approximately that of France. It occupies a sedimentary basin elongated on a SW-NE axis and declines in elevation from about 800 m in the far southwest to near sea-level in the northeast. Dunes, sometimes in massifs of great size, provide the only topographic relief and overlie alluvial deposits or, in the northeast, salt-flat (sabkha) terrain. In addition to sand sheets in northern parts, there is a great range of dune types (Breed *et al.* 1979). In the southwest they are predominantly linear, with parallel ridges up to 300 km long. The central parts exhibit less relief, with rolling sand sheets, barchans and complex transverse forms that merge in the east with great crescentic massifs up to 250 m high. These 'sand mountains', probably relicts of earlier and more energetic wind regimes, are now quite stable in outline and position and carry smaller and active dune forms on their backs.

There is a dearth of long-term climate data for this hyper-arid region but, extrapolating from peripheral climate stations, it is clear that the area lies within the contour of 50 mm mean total annual rainfall (see Figure 2.7). Estimates of annual precipitation range from 15 to 35 mm (Mandaville 1986). Any rain that falls in the northern Rub' al Khali does so mostly in the winter or spring months. Less is known about rainfall in the southern sands, but extrapolation indicates that any rain that falls is most likely to do so in the spring (Figure 2.11). Inter-annual rainfall variability is very high in this area (see Figure 2.12), and some parts may go for years without any precipitation.

Some sand-free interdune areas or fringing gravel flats are nearly devoid of higher plant life, and the saltflats in the eastern part of the basin are similarly sterile. On sand, however, only the rolling dunes of as Sanam in the north-central area are largely without plants, except for occasional tufts of *Cyperus eremicus*. Apart from this rather limited area, the vegetation may be described in terms of three important and widespread communities (Mandaville 1986), as follows.

8.4.4.1 *Cornulaca arabica* Community

In some areas, such as on the rolling sand sheets of southern Sanam around 22°N, 51°E, the endemic shrublet *Cornulaca arabica* (considered by Miller and Cope, 1996, to be a form of *C. monacantha*) may grow with spacings in the order of 20-100 m or more, accompanied by occasional *Cyperus eremicus*. Farther east, as at al Kidan, 22°10'N, 54°18'E, it is closer spaced and accompanied by *Cyperus, Limeum arabicum, Dipterygium glaucum,* by *Calligonum crinitum* on the upper dunes and by *Zygophyllum mandavillei* just above the saline floors. In Suhul al Kidan, farther west at 22°17'N, 53°22'E, *Cornulaca* was observed spaced at 3-4 m and flowering in November accompanied by much *Tribulus arabicus*, some *Cyperus*, and rarer *Calligonum* on the higher sands. *Cornulaca* is sometimes a host for the root parasites *Cistanche tubulosa* and *Cynomorium coccineum*.

8.4.4.2 *Calligonum crinitum* and *Dipterygium glaucum* Community

This is a rather poorly defined unit mapped over wide areas where *Cornulaca* and *Haloxylon persicum* are absent. It is led by an endemic woody shrub *Calligonum crinitum* subsp. *arabicum,* which is usually found spaced in the order of tens or scores of metres on higher sands. It is often accompanied by *Dipterygium* and *Cyperus*, and sometimes by *Limeum, Tribulus,* and *Zygophyllum* (Figure 8.5). In al Awarik, at 19°11'N 47°35'E, *Dipterygium* is locally dominant in this community

Figure 8.5 The northeastern Rub' al Khali, 22°37'N, 53°43'E. In foreground, *Tribulus arabicus*; immediately behind it, *Cornulaca arabica*. *Cyperus conglomeratus* joins *Cornulaca* in the middle background. Dune mass in background carries *Zygophyllum* immediately above the distant line of a salt flat, right, while *Calligonum crinitum* dots the crests and upper slopes.

and grows up to 1 m high, spaced at 3-5 m. *Centropodia fragilis* may be a locally important grass associate.

8.4.4.3 *Haloxylon persicum* Community

This is a clear-cut unit of rather restricted, patchy distribution in a belt across the northern and northwestern edges of the sands. It also may be found sporadically in the west, where Thesiger (1949) recorded it as far south as 19°30'N. *Haloxylon* is a large, ascending woody shrub to about 3 m high, often spaced in the order of 10-30 m on prominent hummocks. Associates may include *Dipterygium*, *Limeum*, *Cyperus*, *Stipagrostis drarii*, and in some northern fringes a few annual psammophytes such as *Eremobium aegyptiacum* and *Plantago boissieri*.

8.4.4.4 Other Constituents

Zygophyllum mandavillei (or *Z. hamiense*) may locally be a significant contributor to biomass, particularly in the lower, eastern parts of the basin where interdune floors are salt flats. It is typically found just above sabkha level in the lowest parts of dunes which carry *Cornulaca* and *Calligonum* above. On the western edge of the sabkha known as Umm as Samim, as at 21°45'N, 55°31'E, it ascends to higher dune levels and may appear locally dominant.

In some parts of the Rub' al Khali, non-psammophytes characteristic of specialized habitats may be found, such as *Seidlitzia rosmarinus*, sometimes with *Anabasis setifera*, around the saline flats of the east. Rare limestone outcrops may carry sparse *Salsola cyclophylla* and *Fagonia indica*, and Thesiger reported scattered *Tamarix* on open sabkha floors of some southeastern borderlands. The leguminous tree *Prosopis cineraria* may form conspicuous stands along the eastern and southeastern margins of the Rub' al Khali (Ghazanfar 1991a). It is characteristic of relatively shallow, stable sand, however, rather than the more mobile dunes where *Calligonum* is dominant.

Overall, the flora of the Rub' al Khali is notable for its very low species diversity; hardly ten species are of importance in the vegetation. Annuals are virtually absent except along the northern margins of the deeper sands. Even after rare excellent rains, *Eremobium, Neurada* and *Plantago boissieri* (these three being perhaps Arabia's best sand-adapted therophytes and with good populations immediately to the north) are hardly found south of the Tropic (Mandaville 1986). The meagre and infrequent rainfall in the core areas, coupled probably with substrate instability, does not appear to meet the cyclical moisture requirements for maintaining populations of annuals.

Popov (unpub.) recognized that the red dunes found not far inland from the southern and southeastern shores of the Gulf were distinct from the white coastal sands and were in effect extensions of the continental dunes of the Rub' al Khali. As pointed out by him, such areas well north of the tropic, as a consequence of greater winter rainfall (see Figure 2.11), carry a much more abundant perennial and annual flora, which has led to significant dune stabilization. Thus *Calligonum comosum* is accompanied locally by *Panicum turgidum, Leptadenia pyrotechnica, Pennisetum divisum* and *Prosopis cineraria*. Spring annuals include many of the species characteristic of inland sands in the central and northern parts of the Peninsula (Popov, unpub.).

8.4.5 Ramlat al Wahibah

Ramlat al Wahibah (the Wahibah Sands), with an area of about 9,400 km^2 (Dutton 1988a), lies in the far eastern part of Oman (Figure 8.1). Its northern and central parts are characterized by north-south oriented linear and relatively stable megadunes up to 100 m high, with broad interdune swales. In the south and southeast, relief is lower with more active and varied dune forms (Dutton 1988a). The sands reach the sea in the southeast, where there is considerable marine influence on the climate, particularly from the southwest monsoon (see section 2.2.2.3).

Munton (1988a) described two basic plant community types. These comprise a relatively well defined association of *Calligonum crinitum* (originally described by Munton 1988 and Cope 1988a as *C. comosum* but assigned to *C. crinitum* by Miller and Cope 1996) and *Cyperus* (listed as *C. aucheri* but which should be compared with the recently described *C. eremicus*) characteristic of higher, mobile dune sites, and an association of *Heliotropium kotschyi, Panicum turgidum, Euphorbia riebeckii* and *Indigofera* spp. on firmer, more stable sand. *Dipterygium glaucum* is common in small, active dunes of the southeastern parts. Dense stands of *Prosopis cineraria* are found locally along several fringing areas, and *Zygophyllum qatarense* may dominate on some parts of the periphery. An important grass of the Rub' al Khali and more northern Arabian sand bodies, *Stipagrostis drarii*, appears to be absent (see list by Cope 1988), but *Stipagrostis plumosa* and *S. sokotrana* are plentiful. *Panicum turgidum* is an important perennial in western parts of the sands (Munton 1988b). Available data indicate that the proportion of annual species is very low (see the list in Cope 1988), a characteristic shared with similar latitudes in the Rub' al Khali.

8.4.6 Ramlat as Sab'atayn

Ramlat as Sab'atayn is the southernmost major sand sea of Arabia (Figure 8.1) and one of the least described botanically. The predominant dune forms in its central parts are large-scale linear formations oriented generally southwest-northeast (Breed *et al.* 1979), with floor elevations ranging from 800-1,300 m. Maxwell-Darling (1937), who penetrated the southwestern edge of the sands for about 50 km via the Wadi Bayhan entrance, described the sand vegetation as sparse *Calligonum* (probably *C. crinitum*) with few if any perennial associates. Smaller plants noted by him in an area that received rain four months earlier included *Achyranthes aspera, Cerastium* sp., *Tribulus* sp., *Cyperus* sp., and *Neurada procumbens*. Al-Hubaishi and Müller-Hohenstein (1984) also list *Calligonum*, along with *Leptadenia pyrotechnica* and *Panicum turgidum*, as frequent in the dune fields. They indicate that annuals such as *Plantago ciliata* and *Anastatica hierochuntica* occur on dunes or dune valleys (my experience in more northern latitudes is that these two are more commonly found on silty bottoms). Shallow sands on floors between active barchanoid dunes near Ma'rib carry sparse *Panicum* and *Dipterygium glaucum* (ibid., pl. 27).

8.5 Summary

It becomes evident from this survey that plant life of deep sand habitats in the Peninsula holds much in common through 15° of latitude. There are, however, definite latitudinal shifts, related primarily to climatic factors, in the occurrences of important perennial associates (Figure 8.6). Overall, most striking is the prevalence of a community led by *Calligonum*, as one of the two closely related species (*C. comosum* and *C. crinitum* subsp. *arabicum*) from the Great Nafud in the north to the southern Rub' al Khali and Ramlat as Sab'atayn in the far south. Throughout this great area, the ubiquitous sedge *Cyperus conglomeratus* (replaced

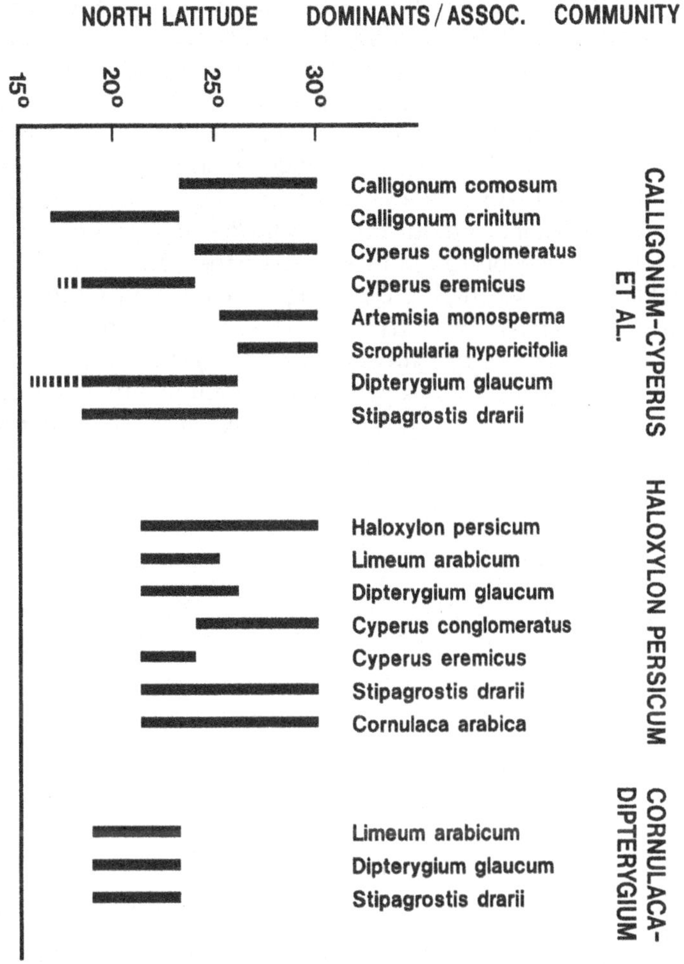

Figure 8.6 Latitudinal distribution of important community constituents in the deep sand vegetation of the Arabian Peninsula.

in the Rub' al Khali and possibly other parts south of about 24 degrees by *C. eremicus*) and, to nearly the same extent, the tussock grass *Stipagrostis drarii* accompany it. *Artemisia monosperma* and *Scrophularia hypericifolia* are strong associates in most northern parts of this range but appear not to be found south of roughly the 25th parallel. Approximately at this latitude *Dipterygium glaucum* begins to figure southward in the *Calligonum-Cyperus* community. South of this latitude winter and spring rain becomes more sparse(<100-150 mm yr^{-1}) and less dependable, with higher inter-annual variability (Figures 2.11 and 2.12), and we find a parallel decline in the numbers of annual species.

The distinct community of *Haloxylon persicum* occurs over smaller areas from the northern parts of the Great Nafud down into the Rub' al Khali to somewhat beyond the 21st parallel. These two major areas of its range are linked by infrequent discontinuous stands through the sands of central Arabia.

Continuing south beyond the Tropic on this imaginary transect we enter hyperarid territory where only the most sand-hardy annuals such as *Eremobium*, *Plantago boissieri* and *Neurada* can maintain populations. Even these have disappeared in the core sands another two degrees south. *Cornulaca monocantha* now becomes dominant over wide areas of the northeast and in the southern Dahna in the west, often leaving dune tops to *Calligonum* and sharing space with shrublets of southern distribution but lesser frequency. In the sand mountains of the east, *Tribulus* is seen on higher dune flanks, while *Zygophyllum* is on lower sands bordering saline flats. The outlying sand bodies of Ramlat al Wahibah and Ramlat as Sab'atayn also carry *Calligonum* and *Cyperus* in more mobile dune areas. However, they also include parts with dominant plants characteristic of less hyperarid and more stable sand conditions, such as *Panicum turgidum*.

Chapter 9

Coastal and Sabkha Vegetation

Ulrich Deil

9.1 Introduction

Coastal and inland saline habitats are highly stressed environments where salinity and the water holding capacity of the substrate are the dominant abiotic influences. However, a complex of environmental factors that includes the duration and degree of inundation by seawater and the input of both overground and underground fresh water also influence species richness and the composition of plant communities. Since degree of inundation by water, soil development and grain size of the substrate are largely governed by relief, coastal geomorphology and micro-topography have a strong influence on coastal vegetation.

The diverse geomorphology of the 8,000 km long coastline of the Arabian Peninsula exhibits three main geomorphological types: (1) low sandy dunes, (2) flat, silty-saline depressions (sabkha) and estuaries, and (3) cliffs and littoral mountains. Coasts with dunes and saline marshes are the most common landform along most of the western, eastern and parts of the southern coasts (the Batinah, Dhofar and Tihamah coastal plains), whilst rocky coasts dominate the northeast and northwest and parts of the southern coast.

There are three notable features of the coastal and sabkha vegetation of the Peninsula. Firstly, the number of species is often impoverished, with monospecific stands being common. Secondly, as a result of the diminishing influence of the sea landward, the vegetation forms zones, with each zone occupied by distinct communities (Zahran 1977). Thirdly, since each of the three main geomorphological types contains a variety of specialized habitats, the vegetation often forms distinct mosaics.

Coastal vegetation is one of the best studied vegetation types in Arabia, with the salt marsh vegetation along the western and the eastern coasts of the Peninsula

S. A. Ghazanfar and M. Fisher (eds.), Vegetation of the Arabian Peninsula, 209-228.

having received particular attention (Abdel-Razik & Ismail 1990, Aleem 1978, 1979, Al-Gifri & Al-Subai 1994, Al-Gifri & Hussein 1993, Babikir 1984, Babikir & Kürschner 1992, Batanouny & Turki 1983, Böer 1994, Böer & Warnken 1992, Danin 1983, De Clerck & Coppejans 1994, Deil & Müller-Hohenstein 1996, El-Demerdash & Zilay 1994, El-Sheikh *et al.* 1985, Frey & Kürschner 1986, Frey *et al.* 1984, Frey *et al.* 1985, Ghazanfar in press-a, Ghazanfar & Rappenhöner 1994, Halwagy 1986, Halwagy & Halwagy 1977, Halwagy *et al.* 1982, König 1987, Kürschner 1986c, Mahmoud *et al.* 1982, Mahmoud *et al.* 1985, Müller-Hohenstein 1992, Younes *et al.* 1983). Studies range from brief descriptions of vegetation types (Al-Gifri & Al-Subai 1994, Zahran 1993), to characterization of vegetation units by their complete floristic composition (Deil & Müller-Hohenstein 1996), measurement of environmental data (Abdel-Razik & Ismail 1990, Babikir & Kürschner 1992, Halwagy & Halwagy 1977), physiological investigations of photosynthetic pathways (e.g. Frey *et al.* 1984, Frey *et al.* 1985), and the mineral content of plants and soil (El-Shourbagy *et al.* 1987). Inland sabkha vegetation has been somewhat less studied (El-Sheikh *et al.* 1985).

The major part of this chapter (section 9.5) describes the vegetation communities of fourteen areas of coastal and sabkha vegetation that have been particularly well studied, the locations of which are given in Figure 9.1. Typical vegetation zonation and stratification of the communities are illustrated using three representative transects (Figures 9.4, 9.6 and 9.8), the locations of which are given in Figure 9.1 (III, I, and II respectively). Three representative maps are used to illustrate the typical mosaic-like patterns of the coastal vegetation: the Red Sea coast south of Luhayyah (Figure 9.5), the Indian Ocean coast near Muscat (Figure 9.7) and the Arabian Gulf coast near Qatar (Figure 9.9); for locations see Figure 9.1: C, A and B respectively).

9.2 Phytogeography

Key species in the saline habitats of Arabia are nearly always perennial (Figure 9.2). The predominant life-forms are succulent, semi-woody dwarf shrubs belonging to the families Chenopodiaceae, Zygophyllaceae and Plumbaginaceae, and hemicryptophytes with runners and spiny leaves belonging to the families Poaceae and Juncaceae. Annual succulents such as *Bienertia cycloptera* and *Zygophyllum simplex* are exceptions. Coastal species are either obligate halophytes like the representatives of the families Chenopodiaceae, Frankeniaceae and Plumbaginaceae, or salt tolerant genera from unspecialized families, such as *Sporobolus* and *Aeluropus* (Poaceae), or salt secreting species such as *Avicennia* (Verbenaceae) and *Limonium* (Plumbaginaceae). The most common coastal and salt tolerant species are *Arthrocnemum macrostachyum, Halocnemum strobilaceum, Halopeplis perfoliata, Salsola* spp., *Suaeda* spp., *Salicornia europaea, Seidlitzia rosmarinus* (Chenopodiaceae), *Aeluropus lagopoides, Odyssea mucronata, Sporobolus spicatus, S. consimilis* (Poaceae), *Juncus rigidus* (Juncaceae), *Zygophyllum* spp.

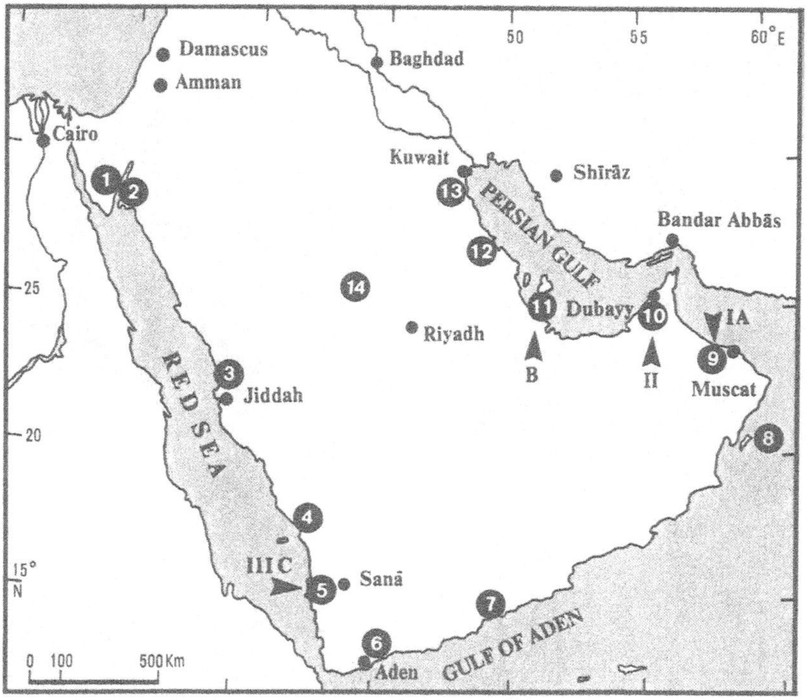

Figure 9.1. The location of coastal and inland sabkha studies (1-14), and the localities of coastal transects (I: see Figure 9.6; II: see Figure 9.8; III: see Figure 9.4) and vegetation maps (A: see Figure 9.7; B: see Figure 9.9; C: see Figure 9.5). 1, Western coast of the Gulf of Aqaba (Danin 1983, Frey *et al.* 1985). 2, Eastern coast of the Gulf of Aqaba (Mahmoud *et al.* 1985). 3, Red Sea coast north of Jeddah (Frey *et al.* 1984, Mahmoud *et al.* 1982, Younes *et al.* 1983). 4, Tihamah coast at Jizan (El-Demerdash *et al.* 1995, El-Demerdash & Zilay 1994, König 1987). 5, Tihamah coast north of wadi Siham (Wood 1983b, Deil & Müller-Hohenstein unpubl.). 6-7, Gulf of Aden coast (Al-Gifri & Al-Subai 1994, Al-Gifri & Hussein 1993, Kürschner *et al.* in press). 8, Masirah Island (Ghazanfar & Rappenhöner 1994). 9, Qurm Nature Reserve, Muscat (Frey & Kürschner 1986, Kürschner 1986c). 10, Dubai (Deil & Müller-Hohenstein 1996). 11, Qatar (Abdel-Razik 1991, Abdel-Razik & Ismail 1990, Babikir 1984, Babikir & Kürschner 1992, Batanouny 1981, Batanouny & Turki 1983). 12, Al Jubail, Saudi Arabia (Böer 1994, 1996d, Böer & Warnken 1992). 13, Kuwait (Halwagy 1986, Halwagy & Halwagy 1977, Halwagy *et al.* 1982). 14, Inland sabkha at al Qassim, Saudi Arabia (El-Sheikh *et al.* 1985).

(Zygophyllaceae), *Limonium* spp. (Plumbaginaceae) and *Avicennia marina* (Verbenaceae).

The transition from the extra-tropical to the tropical region in Arabia (see Figure 4.2) is also seen in the characteristic species of the coastal plant communities. Whilst comparing the Red Sea coast north and south of Jeddah, Vesey-Fitzgerald (1955, 1957b) was the first to recognize the difference between the salt marsh flora on either side of the Tropic of Cancer, and Freitag (1991) showed that the tropical and extra-tropical distribution of the halophytic coastal

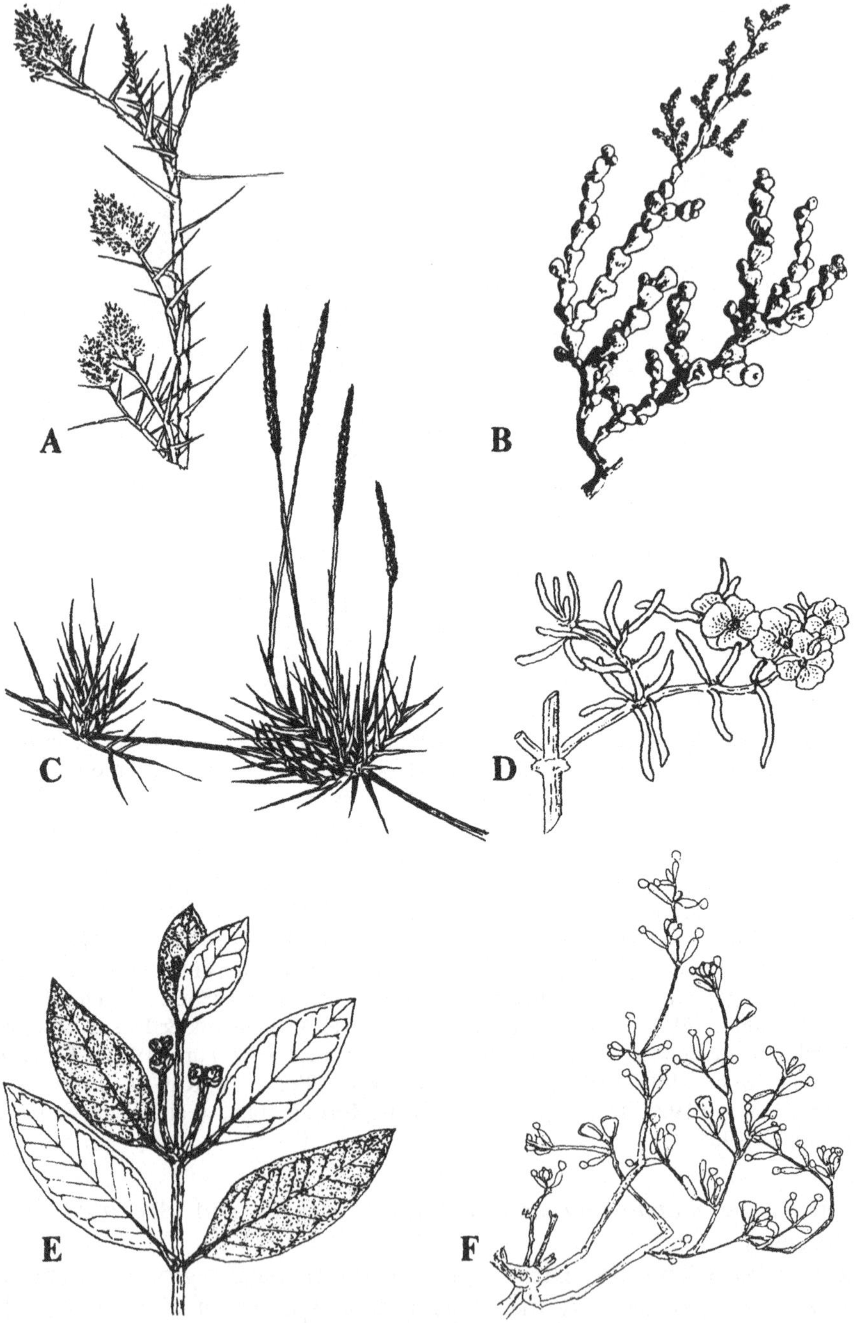

Figure 9.2. Habit of some characteristic species of the coastal vegetation. (A) *Odyssea mucronata* (Wood 1983b). (B) *Halopeplis perfoliata* (Täckholm 1974). (C) *Sporobolus spicatus* (Täckholm 1974). (D) *Seidlitzia rosmarinus* (Zohary 1966). (E) *Avicennia marina* (Täckholm 1974). (F) *Zygophyllum qatarense* (Boulos 1978).

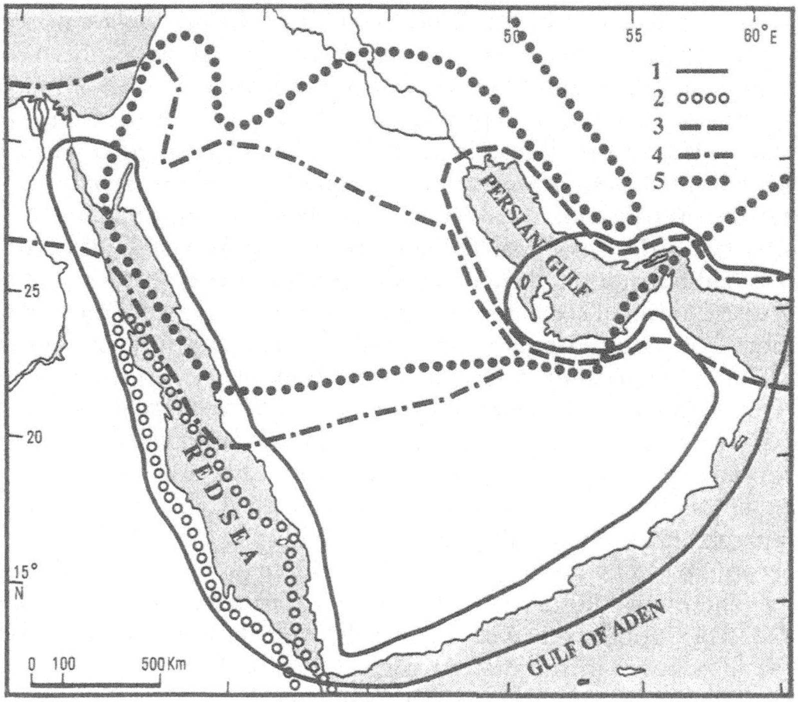

Figure 9.3 Distribution of selected coastal halophytes. 1, *Halopeplis perfoliata* (Freitag 1991). 2, *Cornulaca ehrenbergii* (Boulos 1992). 3, *Salsola drummondii* (Freitag 1991). 4, *Salsola schweinfurthii* (Freitag 1991). 5, *Seidlitzia rosmarinus* (Freitag 1991).

species of the Chenopodiaceae is similar to that of the non-halophytic species.

Halopeplis perfoliata is an example of a species with a typical circum-Arabian distribution in the Nubo-Sindian zone of the Sahara-Sindian phytochorion (Figure 9.3: 1). *Arthrocnemum macrostachyum* is a bi-regional species with a Sahara-Sindian/Mediterranean distribution. *Halocnemum strobilaceum* is a pluri-regional species occurring in the Mediterranean/Saharo-Sindian/Irano-Turanian phytochoria with a distinct southern distributional boundary. *Salsola schweinfurthii* is a Saharo-Arabian species (Figure 9.3: 4), and *Seidlitzia rosmarinus* has a Saharo-Sindian and Irano-Turanian distribution, not occurring south of Jeddah or Musandam (Figure 9.3: 5). *Suaeda monoica* is a tropical Saharo-Sindian species commonly distributed in Sudan and Eritrea and with its northernmost limit on the Diimaniyat Islands off the coast of Muscat (Ghazanfar 1992b); it is replaced by *Nitraria retusa* further north (Freitag 1991, Kassas & Zahran 1967).

The distributional limits of *Seidlitzia rosmarinus* delimits to a large extent the extra-tropical from tropical coastal vegetation complexes (Figure 9.3: 5). It occurs in seasonally wet inland saline habitats, often replacing the *Halocnemum* community on drier habitats, and usually forming a community of its own, which in the Irano-Turanian region includes several halophytic annuals. *Arthrocnemum macrostachyum* is replaced by the truly tropical *Halosarcia indica* in southeast Pakistan and western India, and by *Halopeplis perfoliata* in the southern coasts of the Arabian Peninsula. *Odyssea mucronata* is not distributed north of Jeddah, and similarly *Limonium axillare* is replaced by *L. pruinosum* north of the Tropic of Cancer. Other extra-tropical taxa include *Cornulaca ehrenbergii* (Figure 9.3: 2), *Gymnocarpos decander*, *Anabasis setifera* and *Halopyrum mucronatum*.

The north-south distribution of coastal species is more distinct on the Red Sea coast, with the border lying near Jeddah, than on the Arabian Gulf Coast where there is a broad transitional zone lying between Qatar and northern Oman. The east-west distribution of coastal species is not as distinct as that of the north-south distribution. The eastern elements are either restricted to the coasts around the Arabian Gulf (e.g. *Salsola drummondii*) or are Irano-Turanian species extending into the Gulf region (e.g. *Bienertia cycloptera* and *Seidlitzia rosmarinus*). Some east-west species are closely related vicariants, such as *Salsola drummondii* restricted to eastern Arabia (Figure 9.3: 3) and extending eastwards to India, and *S. schweinfurthii* distributed mainly from eastern Saudi Arabia to Jordan, with an outlier recorded from Oman (Figure 9.3: 4) (Miller & Cope 1996).

There are several vicariant groups of halophytic species in the Arabian Peninsula. These include species in the genera *Cornulaca*, with *C. monacantha* distributed from southwest Asia eastwards to Pakistan (Boulos 1992) and *C. aucheri* distributed in the eastern regions of the Peninsula, Iraq, Iran and southwest Pakistan, *Salsola*, with *Salsola chaudharyi* (treated as *S. villosa* in Miller & Cope 1996) in central Saudi Arabia (Botschantzev 1984) and *S. omanensis* in the coastal plains of Dhofar (Boulos 1991), and *Suaeda* (Freitag 1991) with *Suaeda moschata* restricted to the Barr al Hikman peninsula and Hallaniyat Islands in Oman (Scott 1981). Other examples include the *Cyperus conglomeratus*-complex (*C. aucheri*, *C. conglomeratus*, Kukkonen 1991) and the *Limonium axillare*-group (*L. axillare*, *L. stocksii*, *L. carnosum* and *L.* cf. *stocksii*). East-west- and littoral-inland vicariance is well illustrated in the genus *Zygophyllum* section Mediterranea, with *Z. coccineum* mainly distributed in the northern coasts of the Red Sea, *Z. qatarense* in the Arabian Gulf and Gulf of Oman (Boulos 1987), *Z. hamiense* in the southwestern corner of the Arabian Peninsula, *Z. mandavillei* in the southern Rub' al Khali and Hadhramaut, and *Z. migahidii* in the Nafud (El-Hadidi 1977, El-Hadidi 1980).

Sevada schimperi, a monotypic genus within Chenopodiaceae is endemic to the coastal habitats around Bab al Mandab (Freitag 1989a).

9.3 Physiological Adaptations

Of the few physiological studies on Arabian coastal species, most have been on germination biology and photosynthetic pathways. Seed germination studies on *Zygophyllum qatarense*, a species with a wide distribution, has shown that germination rate is reduced at high temperatures and salinities, and that germination is delayed until the cooler and wetter winter season, when rain partly leaches the salts below the rooting zone of the seedlings (Ismail & El-Ghazaly 1990, Ismail 1983). *Aeluropus massauensis* (Mahmoud 1984), *Halopeplis perfoliata* and *Limonium axillare* (Mahmoud *et al.* 1983) have a similar germination strategy.

The correlation between environmental stress and the photosynthetic pathway used by plants is well established. Hygrohalophytic and mesohalophytic species such as *Avicennia* and *Halopeplis* use the C_3 pathway, whilst xerohalophytes such as *Atriplex* and *Suaeda* and species within the spray zone such as *Aeluropus* and *Halopyrum* use the C_4 pathway (Babikir & Kürschner 1992, Frey *et al.* 1984, Frey *et al.* 1985). C_3 halophytes use salt secretion or salt accumulation for survival in saline habitats, whilst C_4 halophytes are able to survive both salinity and drought stress without further adaptations.

9.4 Intertidal Vegetation

Several areas around the coasts of the Arabian Peninsula are dominated by reef building corals, macroalgae and seagrasses. In areas with shallow coral reefs where the substrate is hard, macro- and turf-algae tend to dominate (Sheppard *et al.* 1992), whilst in soft, sandy substrates seagrasses are often dominant. Mangroves occupy rather small areas of the coast, and are heavily utilized by man and his livestock (see also Chapter 10). Intertidal communities, both plant and animal, and the mangal ecosystems of Arabia are dealt with in detail by Sheppard *et al.* (1992).

9.4.1 Seagrasses

Seagrasses are the only group of angiosperms which are able to live permanently submerged in the sea. The occurrence and distribution of the eleven species recorded from the Arabian Peninsula is controlled by a complex of environmental factors which includes substrate quality, depth, temperature, salinity and light penetration (Sheppard *et al.* 1992). There is a significant correlation between total seagrass cover and latitude, with cover increasing towards the lower, tropical latitudes (Price 1990). Shallow coastal bays (<10 m deep) often have well developed seagrass beds, such as along the shallow southeast coasts of Bahrain, where the species are restricted to shallow waters with good light penetration. Relatively dense seagrass beds occur in central and southern Oman (Jupp *et al.* 1996) and the Gulf of Aden.

The seagrass flora of the Red Sea is relatively diverse, with 10 of the 11 recorded species in Arabia occurring there. Four species are recorded from southeast Arabia and the Gulf (Jupp *et al.* 1996, Sheppard *et al.* 1992), with most communities dominated by the smaller-bodied species *Halodule uninervis*, *Halophila ovalis* and *H. stipulacea*. The larger *Syringodium isoetifolium* occurs in the Gulf, but is relatively rare. In contrast, several larger-bodied and wide-leaved seagrasses such as *Thalassadedron ciliatum*, *Thalassia hemprichii*, *Cymodocea rotunda* and *C. serrulata* occur in the Red Sea (Aleem 1978, Aleem 1979, Jupp *et al.*

1996). It has been suggested that the effects of seasonal upwelling along the southeastern coasts of the Arabian Peninsula, which causes large fluctuations in sea temperature, are responsible for the impoverished seagrass beds (Basson *et al.* 1977, De Clerck & Coppejans 1994) and the occurrence of only small-bodied hardy species (Jupp *et al.* 1996).

9.4.2 MANGROVES

Mangroves occur throughout the coasts of the Arabian Peninsula, bordering bays and creeks, some offshore islands and several sea lagoons. Of the three recorded species, *Avicennia marina* (Figure 9.2: E) is by far the commonest and most abundant (Frey & Kürschner 1989, Sheppard *et al.* 1992), being tolerant of low temperatures (12-35°C) and high salinities (40-50%) (Böer 1996c, Sheppard *et al.* 1992). The distribution of mangroves indicates that cold winter temperatures rather than salinity limit their northernmost extent, and mangroves may formerly have been more common in the Gulf and Red Sea than they are at present (Sheppard *et al.* 1992, and references therein).

Avicennia marina, originally described from al Luhayyah on the Red Sea coast of Yemen, occurs southwards from latitude 26° N along the Red Sea coast and in the Gulfs of Aden and Oman. The northernmost populations of *A. marina* are recorded from *c.* 27°N in the Wildlife Sanctuary near Jubail on the Arabian Gulf coast of Saudi Arabia (Böer & Warnken 1992, see also section 12.4.5.3) and the Gulf of Suez and Sinai coast in the Gulf of Eilat (Danin 1983). Dense stands of this species occur on Mahout Island in Gubbat al Hashish in central Oman, where the trees are up to 4 m in height, and where the mangroves sustain shrimp, crab and other fisheries of commercial importance (Fouda & Al-Muharrami 1996). *Rhizophora mucronata* is known from Gizan (south of Jeddah) and the Farasan Islands (El-Demerdash 1996) and from the Gulf of Aqaba and Bahrain. *Bruguiera gymnorhiza* has been recorded from the offshore islands near Hodeida (Zahran 1975), though its presence there is unconfirmed (Sheppard *et al.* 1992). Little is known about the productivity of mangroves in Arabia, which show a gross productivity in poorly developed stands of <1 kg m^{-2} yr^{-1} (see references in Sheppard *et al.* 1992).

9.5 Terrestrial Vegetation

Much of the Arabian shore line is low lying and therefore has wide supra-tidal and intertidal areas. The supra-tidal environment receives intense insolation and has highly saline and desiccated soils, and many of these low-lying, flat saline areas form sabkha. The Arabic term 'sabkha' literally means mud, but it is now generally used to define a flat, salt incrusted area which may occasionally have standing water. Sabkhas are characterized by soil crusts of salt and gypsum, and associations of cyanobacteria and algae over a black reducing layer (see also section 3.4.9.2).

Coastal sabkha is one of the commonest landforms of the Arabian coastline, and two of the largest coastal sabkhas include Sabkhat Matti in Abu Dhabi and the Barr al Hikman Peninsula in central Oman.

9.5.1 COASTAL VEGETATION

In this section the typical communities of coastal and sabkha vegetation are described from the Gulf of Aqaba in the northwest around the coast to Kuwait in the northeast. Figure 9.1 gives the locations (1-13) of the cited studies.

9.5.1.1 Gulf of Aqaba

The vegetation of the western coast of the Gulf of Aqaba in the vicinity of Nabg has been described by Danin (1983) and Frey *et al.* (1985) and that of the eastern coast at al Magawah near Ras Sheikh Humeid by Mahmoud *et al.* (1985) (Figure 9.1: 1-2). On the western coast near Eilat the first zone consists of *Avicennia marina* on mud deposits protected from strong waves by coral reefs. This is followed by a *Limonium axillare* zone and then by a *Nitraria retusa-Zygophyllum album* zone. *Salvadora persica*, forming sand mounds, occurs on alluvial fans over fresh water. In other places a *Suaeda monoica-S. vermiculata* zone can be seen near the shoreline. *Hyphaene thebaica* has its northernmost distribution near Eilat. On the eastern coast the first zone, in which the coast is frequently inundated by the sea and the top soil is high in salt and the water table shallow (30-70 cm), is occupied by *Arthrocnemum glaucum*. A sterile sabkha is followed by a *Suaeda pruinosa* zone which is not inundated during high tide. The water table is at a depth of 85-150 cm. A *Nitraria retusa* zone follows with the water table at 100-140 cm, with the associated species *Zygophyllum album, Z. coccineum* and *Tamarix* spp. The last zone, a *Zygophyllum coccineum* zone, occurs on coarse textured sand with *Cyperus conglomeratus* and *Fagonia bruguieri* as associates. The water table in this zone is at a depth >200 cm.

The vegetation pattern of the southern part of the Gulf of Aqaba has mangroves and littoral salt marshes in the mid-littoral zones, and also colonizing offshore coral reefs, followed by an unvegetated, frequently inundated region along the shore. A perennial halophytic dwarf shrub community with *Limonium pruinosum* and *Zygophyllum album* is present in the supra-littoral zone. This is the southern distributional limit of *Limonium pruinosum,* replaced further south by *L. axillare*. A *Salvadora persica* open shrubland occurs on sand mounds and in alluvial fans where fresh water is close to the surface, and an open shrubland with *Nitraria retusa* and *Zygophyllum album* is present on aeolian sands where the ground-water is salty.

9.5.1.2 Red Sea Coast North of Jeddah

The littoral salt marsh zonation of the Red Sea coast north of Jeddah has been described by El-Shourbagy *et al.* (1986), Frey *et al.* (1984), Mahmoud *et al.* (1982) and Younes *et al.* (1983) (Figure 9.1: 3). The littoral salt marsh communities consist of the mangrove *Avicennia marina* in the first zone followed by a *Halopeplis perfoliata* zone in the moist but not waterlogged soil fringing the shoreline. On soft aeolian deposits overlaying mudflats *Aeluropus massauensis* occurs in the third zone and on coarse soils where the water table is below 1.5 m, a *Zygophyllum coccineum* or a *Limonium axillare-Suaeda pruinosa* zone is present.

9.5.1.3 Tihamah Coast

The vegetation of the Tihamah coast at Jizan has been described by El-Demerdash *et al.* (1995), El-Demerdash and Zilay (1994) and König (1987), and that of the Yemeni Tihamah north of wadi Siham by Deil and Müller-Hohenstein (ined.) and Wood (1983b) (Figures 9.1: 4-5, 9.4, 9.5).

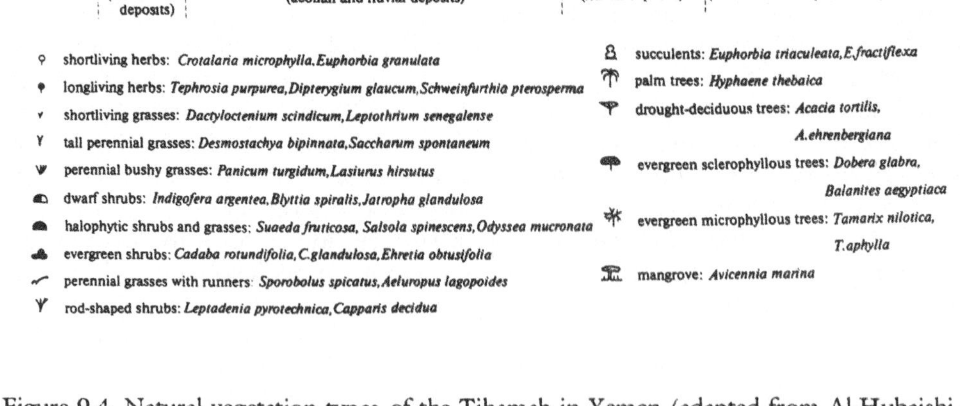

Figure 9.4. Natural vegetation types of the Tihamah in Yemen (adapted from Al-Hubaishi & Müller-Hohenstein 1984); length of the transect is *c*. 25 km. For location see III in Figure 9.1.

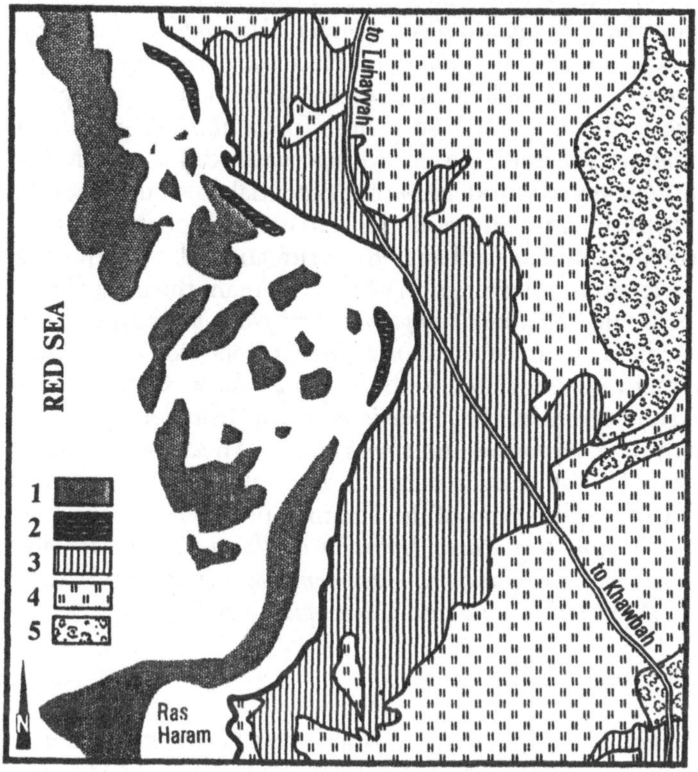

Figure 9.5. Vegetation map of coast between Luhayyah and al Hudayda in Yemen, typical of the Tihamah coast. The vegetation complex belongs floristically to the southern Red Sea type, and there is a clear zonation pattern. For location see C in Figure 9.1. An offshore coral reef, 1, protects the coast from strong wave action. The vegetation types are: 2, A few small areas of *Avicennia* mangrove; 3, a partly tidal sabkha, largely without vegetation, but with a mosaic of *Sueda fruticosa* in runnels, an *Aeluropus-Sporobolus* grass community on microdunes, and a *Zygophyllum hamiense-Suaeda fruticosa* community in sandy depressions; 4, A *Panicum turgidum-Odyssea mucronata* community on mobile dunes; 5, A *Salsola spinescens-Schweinfurthia pterosperma* community on consolidated dunes, and depressions which are cultivated in rainy years, with a *Shouwia purpurea* weed community developing on abandoned fields.

Along the Yemeni Red Sea coast northwards from Wadi Siham the following zonation occurs: An *Avicennia marina* zone followed by a *Limonium cylindrifolium-Suaeda fruticosa-Limonium axillare* community, forming hummocks. A sterile sabkha is present after which raised beaches above the high tide level are covered by *Atriplex farinosa, Zygophyllum hamiense, Aeluropus lagopoides* and *Halopyrum mucronatum*. Sand dunes towards the seaward side are colonized by *Suaeda monoica* and *Salsola spinescens*, and the inland dunes by *Odyssea mucronata*, *Jatropha pelargoniifolia* and *Leptadenia pyrotechnica*. Further inland, away from the sea, species such as *Indigofera argentea*, *Blyttia spiralis* and *Jatropha glandulosa*, perennial grasses such as *Panicum turgidum* and *Lasiurus scindicus* and tall shrubs such as *Cadaba rotundifolia, C. glandulosa*, *Ehretia obtusifolia* and succulents such as *Euphorbia triaculeata* and *E. fractiflexa* replace the halophytic coastal species. The dum palm *Hyphaene thebaica* characterizes silty and sandy depressions. In the wadis a *Tamarix nilotica-T. aphylla* gallery forest and tall grass communities of *Saccharum spontaneum* and *Desmostachya bipinnata* occur along the banks. An open *Acacia-Balanites-Dobera* woodland is the potential natural vegetation type of the foothills along the escarpment mountains. In general zonation of the species of the Yemeni Tihamah is similar to that of the Eritrean coasts (Hemming 1961).

Further north in the Jizan region of the Tihamah, the infra-tidal zone consists of seagrass beds with *Thalassia hemprichii*, *Halophila ovalis* and *Halodule uninervis*. The first zone is made up of *Avicennia marina*, followed by a zone of *Halopeplis perfoliata*, sometimes associated with *Suaeda pruinosa*. Coastal sand marshes are characterized and dominated by *Suaeda monoica, Aeluropus lagopoides, A. littoralis, Sporobolus spicatus, S. virginicus* and *Cressa cretica*. The *Suaeda monoica* community can also be found in saline depressions further inland. The non-saline sand dunes are covered by the typical flora of the Tihamah plain which consists of *Leptadenia pyrotechnica*, *Panicum turgidum*, *Tephrosia apollinea*, *T. quartiniana*, *Crotalaria microphylla*, *Dipterygium glaucum*, *Glossonema boveanum,* and scattered trees of *Acacia tortilis*, *A. ehrenbergiana* and *Calotropis procera*.

9.5.1.4 Gulf of Aden Coast

The coastal vegetation of the Gulf of Aden coast between Aden and Mukalla has been studied by Al-Gifri and Al-Subai (1994), Al-Gifri and Hussein (1993) and Kürschner *et al* (in press) (Figure 9.1: 6-7). The southwestern corner of the Arabian Peninsula is characterized by the occurrence of a new coastal species, *Odyssea mucronata,* endemic to this part of Arabia. *O. mucronata* is a clump-forming, spiny, rhizomatous perennial which colonizes semi-mobile dunes and flat sandy areas. Depending on the depth of sand, an *Odyssea mucronata-Suaeda monoica* community can be distinguished on flat sandy layers overlying saline silts, and an *Odyssea mucronata-Panicum turgidum* community on deeper sand. Other species in the first community are *Sporobolus spicatus*, *S. consimilis* and *Halopyrum mucronatum*, and in the second, *Salsola spinescens* and *S. vermiculata*. Lagoons on this coast are fringed with *Typha domingensis*, *Cyperus laevigatus* and *Bacopa monnieri*, and along wadis *Salvadora persica* and *Tamarix aphylla* are the dominant species. Towards the east, coastal dunes are covered with *Ipomoea pes-caprae* and the endemic, *Conyza cylindrica*. *I. pes-caprae* is found up to the central part of the eastern coast of Oman, which delimits its northernmost distribution.

9.5.1.5 Hadhramaut Coast

The Hadhramaut coast has been studied by Kürschner *et al.* (in press) at Felek, east of Mukalla. This area is situated in the transition zone from the southeastern to the southwestern vegetation type. This is seen from the *Cyperus conglomeratus* associations, where the Omano-Makranian element, *Coelachyrum piercei* and the Eritreo-Arabian element *Odyssea mucronata* are common members. The coastal vegetation is also rich in endemics and shows a strong phytogeographical relationship with the coasts of northeast Africa. The species zones are: (1) Coastal dunes colonized by sedges and grasses (*Cyperus conglomeratus*, *Halopyrum mucronatum*, *Odyssea mucronata*, *Coelachyrum piercei* and *Panicum turgidum*); (2) Sandy-salty depressions colonized by the endemic *Urochondra setulosa* association, with the co-dominant *Arthrophytum macrostachyum*, *Limonium cylindrifolium* and *Crotalaria saltiana*; (3) Clayey-salty, relatively wet areas colonized by monospecific stands of *Arthrophytum macrostachyum*; (4) Sandy coastal plains colonized by the endemic *Anabasis ehrenbergii-Pulicaria hadramautica-Zygophyllum hamiense* association; (5) The karstic limestone plateau colonized by *Stipagrostis paradisea*, *Commiphora gileadensis* and *Euphorbia rubriseminalis*.

9.5.1.6 Oman

Ghazanfar (in press-a) has described the coastal vegetation of Oman, Ghazanfar and Rappenhöner (1994) the coastal vegetation of the islands of Masirah and Shagaf, and Frey and Kürschner (1986) and Kürschner (1986c) the vegetation of Qurm Nature Reserve near Muscat (Figure 9.1: 8-9, Figures 9.6 and 9.7). There are four main coastal vegetation communities in Oman: (1) A *Limonium stocksii-Zygophyllum qatarense* community in northern Oman where the coasts are mainly sandy and interspersed with rocky limestone headlands. (2) A *Limonium* cf. *stocksii-Suaeda aegyptiaca* community characteristic of rocky shores with narrow beach areas and a wide spray zone. (3) An *Atriplex-Suaeda* community characteristic of the vegetation of offshore islands, flat sandy beaches and coastal sabkhas (dominant and associated species are *Atriplex coriacea*, *A. farinosum*, *A. leucoclada*, *Arthrocnemum macrostachyum*, *Suaeda aegyptiaca*, *S. vermiculata*, *S. monoica*, *S. moschata* and *Halocnemum strobilaceum*), and a *Limonium axillare-*

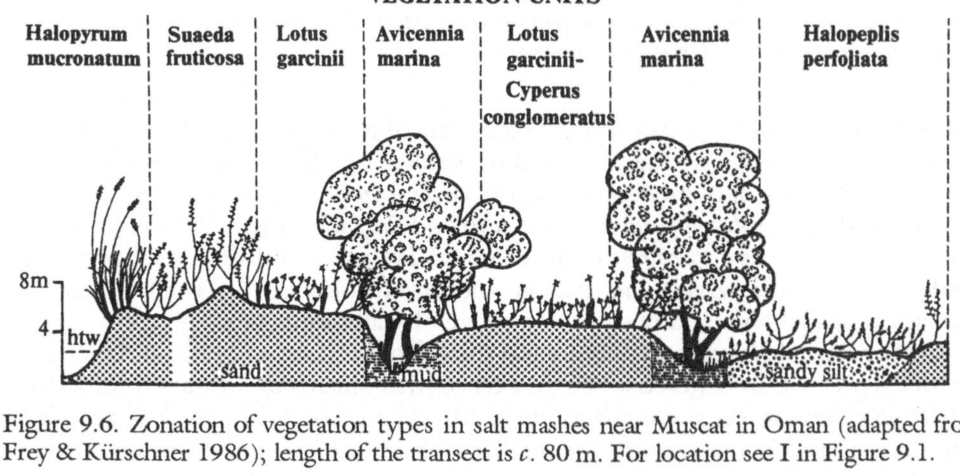

Figure 9.6. Zonation of vegetation types in salt mashes near Muscat in Oman (adapted from Frey & Kürschner 1986); length of the transect is *c.* 80 m. For location see I in Figure 9.1.

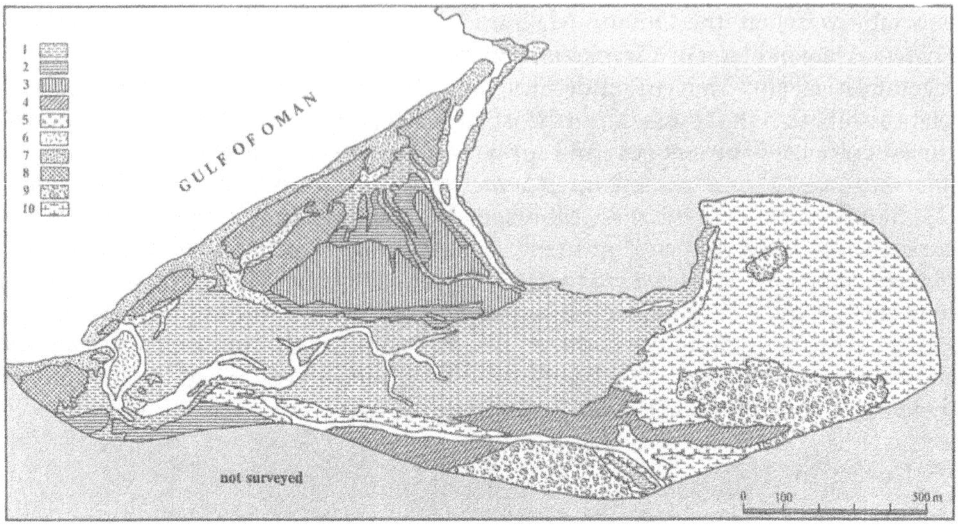

Figure 9.7. Vegetation map of Qurm Nature Reserve near Muscat in Oman (adapted from Kürschner 1986c). For location see Figure 9.1: A. The area is the estuary of Wadi Adai, sheltered from the open sea by a dune belt parallel to the coast. There are two tidal channels, and the large variety of habitats is created by the effects of a superposition of subterranean freshwater on tidal saltwater. 1, A network of channels, gullies and runnels that are permanently or periodically flooded, with mangrove forest and mixed formations of *Avicennia* shrubland and halophytic low shrub communities. On the central sabkha adjacent to the Mangrove runnels is, 2, a *Halopeplis perfoliata* community on moist sandy clay with high top soil salinity, and, 3, a *Sueda fruticosa-S. vermiculata* community on dries sites without salt crusts. 4, Mixed formation of halophytic dwarf shrub communities and ephemeral forb communities. 5, *Aeluropus lagopoides* salt meadow. 6, Salt desert. 7, Beach zone with marine deposits, flooded at high tide and without vegetation. 8, A *Halopyrum mucronatum* community on mobile coastal dunes and a *Sphaerocoma aucheri-Cornulaca monacantha* community on more or less stabilized dunes. 9, A sand and gravel alluvial plain with open *Acacia tortilis-Prosopis cineraria* woodlands and *Commiphora myrrha-Euphorbia larica* shrubland. 10, Cultivated areas, ruderal and fallowland communities. Recent development of part of the eastern area of the reserve as a park has modified or removed much of the natural vegetation (see section 12.4.3.3).

Sporobolus-Urochondra community characteristic of the vegetation of the southern coasts, with *Limonium axillare, Urochondra setulosa* and *Sporobolus* spp. associated with several other species depending on coastal geomorphology. (4) Coastal lagoons with *Sporobolus virginicus*, *S. iocladus* and *Paspalum vaginatum* as the main bordering species, and *Phragmites australis* and *Typha* spp. forming the bordering reeds. In addition, *Avicennia marina* occurs throughout coastal Oman in discontinuous patches and over a wide range of water salinities.

On the Barr al Hikman Peninsula and the offshore island of Masirah *Avicennia marina* is present in sheltered lagoons, a halophytic shrub community dominated by *Atriplex farinosa* and *Suaeda moschata* occurs on low coastal dunes which receive salt spray, and a *Halopyrum mucronatum-Urochondra setulosa* community occurs on more or less stabilized dunes. An *Arthrocnemum macrostachyum-Suaeda vermiculata* community occurs on the saline, silt plains and a *Limonium stocksii-Cyperus conglomeratus-Sphaerocoma aucheri* community on shallow sands. The non-salty coastal foot-plains are colonized by a *Pulicaria glutinosa-Stipagrostis masirahensis* community. The reeds *Phragmites australis* and *Juncus rigidus,* with *Tamarix mascatensis* and *Cressa cretica* form a mosaic integrated with the halophytic vegetation. *Zygophyllum qatarense, Limonium stocksii, Sphaerocoma aucheri* and *Echiochilon jugatum* reach their southern limit at Masirah and the coastal area of central Oman.

9.5.1.7 United Arab Emirates

Deil and Műller-Hohenstein (1996) described the zonation of coastal vegetation near Dubai (Figure 9.1: 10). A transect through the coastal dunes and sabkha that is typical of the dry haloseries within the Omano-Makranian sector of the Arabian Gulf is illustrated in Figure 9.8. Four plant communities are associated in the *Limonium stocksii-Zygophyllum qatarense* vegetation complex: (1) The seaward dunes colonized by the *Cornulaca monacantha-Sphaerocoma aucheri* community (the Salsolo-Suaedetalia of Knapp 1968); (2) The landward dunes colonized by *Halopyrum mucronatum* (stabilizing the sand), *Atriplex leucoclada* and *Suaeda aegyptiaca*; (3) Salty depressions which may be temporarily inundated with sea-

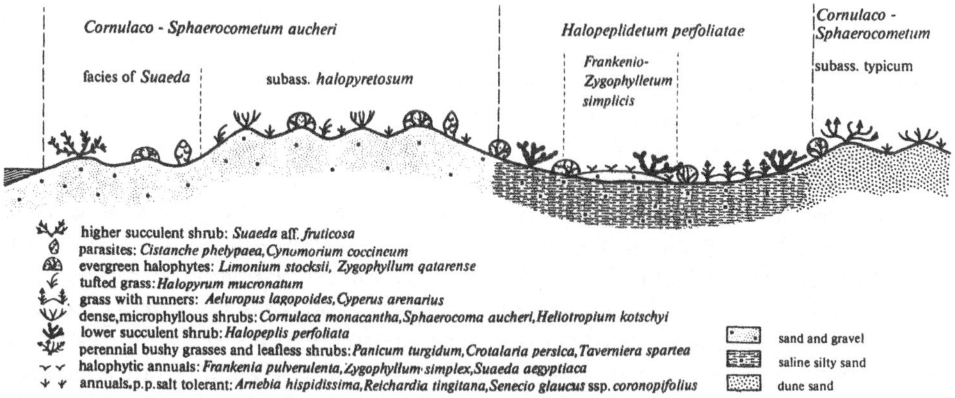

Figure 9.8. Transect through the coastal dunes and sabkhas near Dubai in the United Arab Emirates (adapted from Deil & Müller-Hohenstein 1996); length of the transect is 300 m. For location see Figure 9.1:II. There are four plant communities in this vegetation complex: 1, Seaward dunes colonized by a *Cornulaca monacantha-Sphaerocoma aucheri* community (the Salsolo-Suaedetalia of Knapp 1968) with the associated sand-stabilizing *Halopyrum mucronatum,* and *Atriplex leucoclada* and *Suaeda aegyptiaca.* 2, Salty, temporarily inundated depressions characterised by *Halopeplis perfoliata.* 3, An ephemeral, salt-tolerant *Frankenia pulverulenta-Zygophyllum simplex* community growing on depressions overlain with sand. 4, Landward dunes away from the influence of slat spray dominated by *Cornulaca monacantha* and *Sphaerocoma aucheri*, associated with glycophytic dune species such as *Panicum turgidum, Crotalaria persica, Lotus garcinii, Taverniera spartea* and *Indigofera intricata.*

water are colonized by *Halopeplis perfoliata*; (4) An ephemeral, salt tolerant *Frankenia pulverulenta-Zygophyllum simplex* plant community growing in depressions with sandy overlays. The landward dunes, away from the influence of salt spray, are also dominated by *Cornulaca monacantha* and *Sphaerocoma aucheri*. They are associated here with glycophytic (i.e. non-halophytic) dune species such as *Panicum turgidum, Crotalaria persica, Lotus garcinii, Taverniera spartea* and *Indigofera intricata*. The wet haloseries of Abu Dhabi, Dubai, Fujairah and the offshore islands is similar to that of Qatar described below and has been briefly outlined by Western (1982, 1983, 1987).

9.5.1.8 Qatar

The halophytic vegetation of Qatar (Figure 9.1: 11, Figure 9.9) has been well studied (Abdel-Razik 1991, Abdel-Razik & Ismail 1990, Babikir 1984, 1986, Babikir & Kürschner 1992, Batanouny 1981, Batanouny & Turki 1983). Along a transect from the mangrove zone to the sabkha plain there is a distinct floristic and edaphic gradient with the following zonation: (1) *Avicennia marina*, (2) *Arthrocnemum glaucum*, (3) *Halocnemum strobilaceum*, (4) *Juncus rigidus-Aeluropus lagopoides*. Associated species are *Zygophyllum qatarense*, *Halopeplis perfoliata* and *Anabasis setifera*.

Some geographical variability is present in the zonation series in Qatar. In a coastal littoral plain in southwestern Qatar around the Gulf of Salwa, seven interconnected halophytic plant communities form a mosaic. These are : (1) the *Halopeplis perfoliata* community on sandy beaches along the Gulf shore and surrounding depressions, not inundated by the sea; (2) the *Halocnemum strobilaceum* community, which colonizes the depressions; (3) the *Halopyrum mucronatum-Sporobolus "arabicus"* (= *S. consimilis*) community on calcareous sands; (4) *Limonium axillare*, *Suaeda vermiculata* and *Cistanche tubulosa* forming sandy mounds; (5) the *Zygophyllum qatarense* community growing in shallow depressions and runnels on coarse textured soils, associated with *Cornulaca monacantha, Robbairea delileana, Stipagrostis plumosa* and with *Zygophyllum simplex* making the transition to the glycophytic vegetation; (6) inland, *Panicum turgidum* and *Pennisetum divisum* tussocks are present on fine sand and *Anabasis setifera* on coarse sand; (7) the *Suaeda vermiculata* community on fine textured soils, but restricted to the southwestern area. In northwestern Qatar (ad Dakhira), an *Avicennia marina* association is present in the supra-littoral border followed by an ephemeral halophytic forb community, the *Salicornia europaea-Suaeda maritima* association, in the intertidal zone. An *Arthrocnemum macrostachyum* association is present in the supra-tidal area with *Halopeplis perfoliata* sometimes associated with it. The *Aeluropus lagopoides-Tamarix passerinoides* association is present on dunes and the *Salsola cyclophylla-Panicum turgidum-Anabasis setifera* association on windblown, sandy accumulations at the foot of limestone cliffs. The limestone plateau itself is colonized by a xeromorphic, very open dwarf shrubland of *Zygophyllum qatarense, Helianthemum lippii* and *Lycium shawii*.

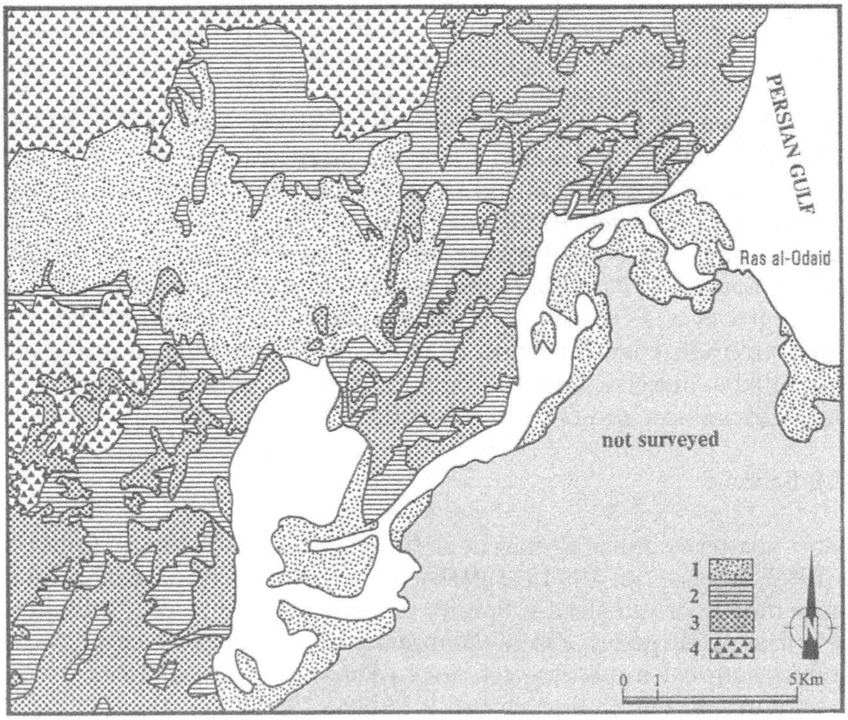

Figure 9.9. Vegetation map of the Khor al Odaid area in southeastern Qatar, a transition zone from calcareous *hamada* plains to a tidal delta (adapted from Babikir 1984). For location see Figure 9.1: B. The mapped area is representative of the numerous inlets and sabkhas of the Qatar coast, formed by the uplift of the Qatar plain during the Quaternary and the emergence of ancient estuaries. Phytochorologically, the mapped area belongs to the northern Gulf type. 1, Sandy beaches and tidal deltas without vegetation. 2, Sabkhas with a *Limonium axillare* community nearest to the shore and a *Halopeplis perfoliata* community further inland; species associated with *Limonium* are *Cistanche tubulosa*, *Salsola schweinfurthii*, *Aeluropus lagopoides* and *Sporobolus spicatus*, and species associated with *Halopeplis* are *Salsola vermiculata*, *Anabasis setifera* and *Halocnemum strobilaceum*. 3, Sand dunes with a *Cyperus conglomeratus-Panicum turgidum* community. 4, Rocky *hamada* plains with an overlay of windblown sand, vegetated with a *Zygophyllum qatarense* community; important associated species are *Haloxylon salicornicum* and *Seidlitzia rosmarinus*. Soil types in the three habitats are similar in texture and grain size, but differ in salt content. Electrical conductivity decreases from 8,250 μS in sabhkas to 4,430 μS in sand dunes and 2,400 μS in the *hamadas*.

9.5.1.9 Al Jubail (Saudi Arabia)

The vegetation of the Arabian Gulf coast in the vicinity of Jubail (Figure 9.1: 12) has been described by Böer (1994, 1996d) and Böer and Warnken (1992). The zonation of species within the intertidal zone from the sea landward is given here with tide range and water salinity (as electrical conductivity, EC, see Table 10.1): (1) *Avicennia marina* (1.3-1.7 m; EC = 40,000 μS); (2) *Salicornia europaea* (1.45-1.75 m; EC = 30,000 μS); (3) *Arthrocnemum macrostachyum* (1.6-2.1 m; EC = 10,000 μS); (4) *Halocnemum strobilaceum* (1.8-2.35; EC = 15,000 μS); (5) *Halopeplis perfoliata* (>2 m and above the intertidal zone; EC = 4,000 μS); (6) *Limonium axillare* (>2 m and above the intertidal zone; EC = 2,400 μS); (7) *Zygophyllum qatarense* (>2 m and above the intertidal zone; EC = 17,660 μS); (8) sterile sabhka (sea-level; 50,000 μS). The outer fringe consists of the *Seidlitzia rosmarinus* community on embryonal dunes followed by *Rhanterium epapposum*, *Haloxylon salicornicum*, *Panicum turgidum* and *Calligonum comosum* on non-saline sands. Mangroves and frequently inundated marsh vegetation occur in regions with high chloride concentrations in the soil. *Halocnemum* and *Zygophyllum* habitats, sabkha margins and the upper intertidal zone have high gypsum concentration and low sodium chloride content.

9.5.1.10 Bahrain

A zonation pattern similar to that at al Jubail occurs in Bahrain (Abbas & El-Oqlah 1992, Abbas *et al.* 1991a, 1991b). Total cover on the coastal plain is 8%, consisting mostly of sub-shrubs. Species richness is low in supra-littoral habitats with only four plant species. *Zygophyllum qatarense*, a dominant coastal species, has a wide distributional and ecological range (Abbas & El-Oqlah 1992, 1996) and occurs both in saline and non-saline soils (Abbas 1995).

9.5.1.11 Kuwait

Halwagy and Halwagy (1977), Halwagy *et al.* (1982) and Halwagy (1986) distinguished 13 communities on the coastal salt marshes of Kuwait (Figure 9.1: 13). Vegetation zonation is related to the microtopography of the marsh. *Salicornia europaea* grows on low, frequently inundated mud banks or along creeks, sometimes associated with *Aeluropus lagopoides* and *Bienertia cycloptera,* or with *Juncus rigidus* on the fringes of creeks. A *Halocnemum strobilaceum* community occupies the lower marshes along the shoreline with the seaward edge inundated *c.* 440 times per year by tides and the inland edge *c.* 10 times per year. A *Seidlitzia rosmarinus* community occurs further inland, followed by *Nitraria retusa* above the high tide mark dominating the middle marshes, and finally the *Zygophyllum qatarense* community on elevated, coarse sandy sites on the landward edge of the marsh. The salt marshes are fringed by non-halophytic communities such as the *Cyperus conglomeratus* community, the *Rhanterium epapposum-Convolvulus oxyphyllus-Stipagrostis plumosa* community and the *Haloxylon salicornicum* community, the latter covering most of the territory of Kuwait.

9.5.2 Inland Sabkhas

Inland sabkhas are highly saline, flat areas of the desert plains formed by the drying out of lakes or pools and continued evaporation from the soil. The largest inland sabkha is Umm as Samim in northwest Oman bordering the sand desert of the Rub' al Khali and covering an area of *c.* 5,000 km^2. Sabkhas are usually devoid of vegetation, although some halophytic species occur on the fringes. The fringing vegetation of Umm as Samim is very sparse since rainfall is scanty (<50 mm per year) and temperatures high (see the climagram for Fahud, Figure 2.6). The few species present are *Aeluropus lagopoides, Cornulaca monacantha*, *Haloxylon salicornicum*, *Salsola* cf. *drummondii*, *Suaeda aegyptiaca* and *Zygophyllum qatarense.* Therophytes consist of *Zygophyllum simplex* and *Tribulus longipetalus* (Ghazanfar 1992a). The vegetation of sabkhas around al Khari springs southeast of Riyadh and that of al Qassim in the Nefud of Saudi Arabia (Figure 9.1: 14), consist of *Seidlitzia rosmarinus* where water is present at depths of 35-75 cm and a salinity of 50,000 μS, and a *Zygophyllum decumbens-Suaeda pruinosa-Salsola baryosma* (=*S. aegyptiaca*) community where the water is at a depth of 60-120 cm and salinity 500 μS (El-Sheikh & Youssef 1981, El-Sheikh *et al.* 1985).

9.6 Summary

The coastal and sabkha vegetation of the Arabian Peninsula is influenced by the diverse geomorphology of the coastline and by the distribution patterns of tropical and extra-tropical species. The dominant features of this vegetation are the paucity of species, marked zonation patterns determined by the declining influence of the sea landwards on both plants and substrates, especially evident in coastal sabkhas, and the occurrence of vegetation mosaics. Inland sabkhas are virtually unvegetated, with a few fringing halophytic species. Seagrass beds consist mostly of large-bodied species in the western and smaller-bodied species in the eastern sea.

Although three species of mangroves have been recorded, the dominant and most abundant is *Avicennia marina*, which has a patchy distribution all along the coasts of the Peninsula. Several coastal halophytes exhibit a north-south distribution gradient, with the northern limit of the tropical species extending to the Tropic of Cancer and the northern Mediterranean and Irano-Turanian species extending down to Jeddah in the west and to northern Oman in the east. There are several north-south and coastal-inland vicariant species groups in the coastal and halophytic species.

Transects from the sea inland exhibit a marked zonation of species. Where mangroves are present they form the first zone, followed by various halophytic communities, the composition of which depends on both biotic and abiotic conditions. Sandy-salty depressions are colonised by *Arthrocnemum macrostachyum*. *Halopyrum mucronatum* occurs above the high tide level, often with *Atriplex*,

Zygophyllum, *Suaeda* and *Aeluropus* spp. These are usually followed by *Suaeda* and *Limonium* species, often forming hummocks.

Chapter 10

Water Vegetation

Shahina A Ghazanfar

10.1 Introduction

In a region in which the scarcity of water determines what survives and what perishes, inland and coastal water bodies and their associated vegetation are a unique landscape feature. There are freshwater springs, albeit not abundant, in both the mountains and plains of the Arabian Peninsula, and these locations form the oases of the desert. The sources of most of the water flows have been highly modified by man, usually covered and with the water regulated so that it rarely forms standing pools. However, some natural pools do occur in the highlands, especially in the higher rainfall areas of the western and southern mountains, where water collects to form pools or small lakes supplemented with runoff from rain. These pools often have a good growth of submerged and emergent aquatic vegetation. Besides freshwater springs, sulphurous and thermal springs occur in several locations, though they do not generally support much vegetation.

Brackish water pools, inlets and lagoons (generally known as *khawrs*) occur intermittently along the coasts of the Peninsula. These generally support an abundance of vegetation and are used extensively for fishing, livestock grazing and drinking, and when the water is not too saline, for irrigation.

In general the composition and ecology of the vegetation of the fresh and brackish water bodies of Arabia has received little attention, and the major part of this chapter is based on work that has been carried out in Oman.

S. A. Ghazanfar and M. Fisher (eds.), Vegetation of the Arabian Peninsula, 229-240.

10.2 Water Bodies and their Sources

10.2.1 Thermal and Sulphurous Water Springs

Some of the better known thermal springs in the western range of the northern mountains of Oman are at Rustaq, Hazm, Fanja and Nakhl, and there are also springs at al Ain in the United Arab Emirates, and some in the Gulf of Suez. Water temperature in these springs ranges from 35 to 70°C, and vegetation when present consists mostly of blue-green algae (cyanobacteria). Sulphurous springs occur in a few locations in the deserts of southern Oman and southern Yemen. A sulphurous spring at Mughshin in the inland desert plain of southern Oman is covered and tapped, but water seepage has created a small marsh. The main plants there are *Zygophyllum*, *Tamarix* and the reed *Phragmites*.

10.2.2 Freshwater Pools

The freshwater springs in the mountainous areas of the Peninsula mostly flow all year, though some have a reduced water flow during dry months. In many places freshwater is directed from the source by systems of open irrigation channels, known as *falaj* in Oman, or as *qanat* in Bahrain. *Aflaj* (plural) may carry water for distances of up to 5 km from the source. Generally, the freshwater springs are covered at the source, though some have large cemented holding tanks. In only a few cases does water from springs collect or seep out to form permanent pools or small lakes (Figure 10.1).

10.2.3 Brackish and Saline Water Pools

Brackish water pools occur mostly along the coastline, though a few pools are found inland where they are usually formed by seepages from coastal hills. In Dhofar in southern Oman brackish water pools occur at the mouth of wadis flowing out of the escarpment mountains to the sea (Figure 10.1). The ***khawrs*** are unique coastal water bodies, separated from the sea by sandbars and fed by overground and underground freshwater from the land, and underground seawater from the sea. The salt level in ***khawrs*** varies from being brackish at the landward side to saline towards the seaward side. The salinity also varies with season, local rainfall and the amount and quality of ground-water seepage. Salinity (measured as electrical conductivity, see Table 10.1 for an explanation) can range from 5,500 μS to 13,000 μS (Figure 10.3). During the southwest monsoon in southern Arabia (June-September) rough and heavy seas wash away the sand bars so that at high tide seawater enters the ***khawrs***, resulting in higher salinities. During the winter months, beach sand is built up again and freshwater seeps in from the land, thus lowering the water salinity. Occasional heavy rains can cause flooding, and raised water levels in the ***khawr*** may open it to the sea, or flood-

A

B

Figure 10.1 Fresh and brackish water pools in southern Oman. (A) a permanent fresh water pool at Wadi Darbat in the Dhofar mountains. (B) Khawr Rawri, a brackish water pool on the Dhofar coastal plain.

Table 10.1 Water types defined by their salinity. Salinity levels are usually expressed in terms of the electrical conductivity (EC) of water; the higher the salt level, the greater the conductivity. The unit of conductivity is Siemens (S) m^{-1} in the International Systems of Units (1 mho = 1 S). 1dS m^{-1} = 1 mmho cm^{-1} ≈ 0.06% NaCl ≈ 0.01 mole l^{-1} NaCl. 10,000 ppm = 10 o/oo (parts per thousand) = 10 g l^{-1} = 1.0%. To convert salinity from o/oo to μS: Salinity(o/oo) x 1.56 = conductivity (m S cm^{-1}); conductivity (μS) x 640 = salinity (o/oo).

Water type	Dissolved Salts (0/00)	EC (μS)
Fresh	<1.25	<195
Slightly saline	1.25-2.5	195-390
Moderately saline	2.5-5	390-780
Brackish	5-32x10^6	780-50,000
Sea water	32x10^6-36x10^6	50,000-56,000

water may wash away the sand bar, allowing sea water to enter the ***khawr*** again (Anon. 1993, Figure 10.2).

Tidal saline pools and sea lagoons also occur along the coast, and are also often referred to as *khawrs*. In a few instances, such as the highly saline pools along the Dhofar coast in southern Oman, the pools and lagoons may be separated from the sea by wide sand bars. In such cases freshwater seepage from land is minimal and water salinity can reach up to 60,000 μS during the dry months (Figure 10.3).

10.2.4 Seasonal Salinity Fluctuations

Freshwater pools show the least seasonal fluctuation in salinity, whilst brackish and saline water pools which receive freshwater runoff may be subject to large seasonal fluctuations in electrical conductivity(Figure 10.3). In Dhofar salinity has been recorded to be relatively higher in the months immediately prior to the monsoon (May to June) and relatively lower during and immediately after the monsoon (June to September). Salinity levels also decrease during April if there is rain, but are highest from November to March, which are generally dry months (Ghazanfar 1993b).

10.3 Vegetation

The most important abiotic factor affecting aquatic vegetation in arid environments is the salinity of the water. Species richness and the composition of plant communities is largely related to the level of salt. Fresh and brackish water bodies are usually the most species rich, whilst saline water bodies are relatively poor in species (Figure 10.3).

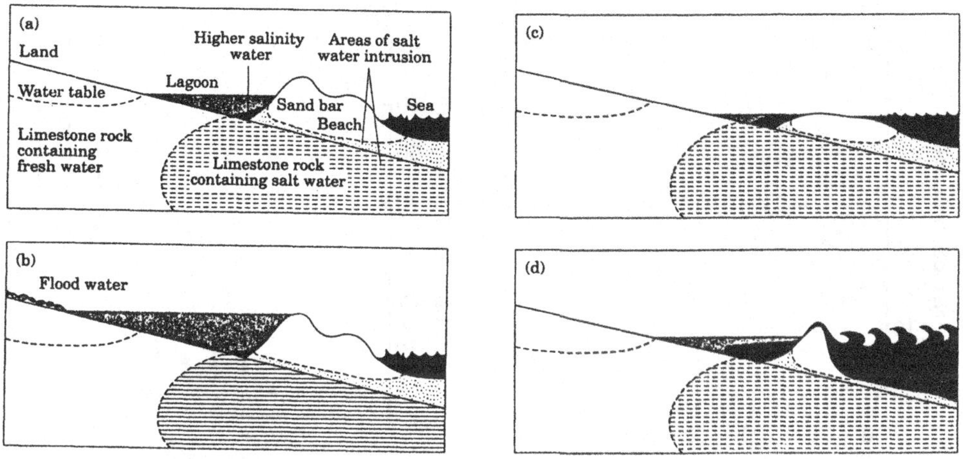

Figure 10.2 Salinity variations in *khawrs*. (a) In the winter months sand is built up by the sea, and there is freshwater seepage from the land and consequently higher salinities at the seaward end. (b) An irregular flood event from localised rains, with the water level rising in a few hours. (c) A flood event in which the water washes away the sand bar, resulting in a low water level which allows sea water to enter at high tide, giving higher salinities; the sand bar is rebuilt within a few days; (d) The high energy seas of the monsoon removes beach sand, and high tides overtop the sand bar, resulting in higher salinities. After Anon. (undated).

10.3.1 Classification of Life-forms

In the Arabian Peninsula three main life-forms of water plants occur. The classification is based on that of Hejny (1960):

Hydatophytes: These are submerged or floating, free or anchored plants. In order to reproduce they require their vegetative parts to be submerged or supported by water. This category includes species in which the reproductive structures are submerged such as in *Najas* and *Chara,* where the reproductive parts are above the water surface such as in the fern *Ceratopteris thalictroides,* where the vegetative parts are aerial and free floating such as in *Lemna*, or where the vegetative parts are rooted such as in *Schoenoplectus litoralis*.

Tenagophytes: These are plants which may or may not require to be submerged or supported by water, but must have some vegetative parts above water for sexual reproduction. This category includes those plants which occur in wet, seasonally flooded areas. Species such as *Polygonum senegalense, Halocnemum strobilaceum,*

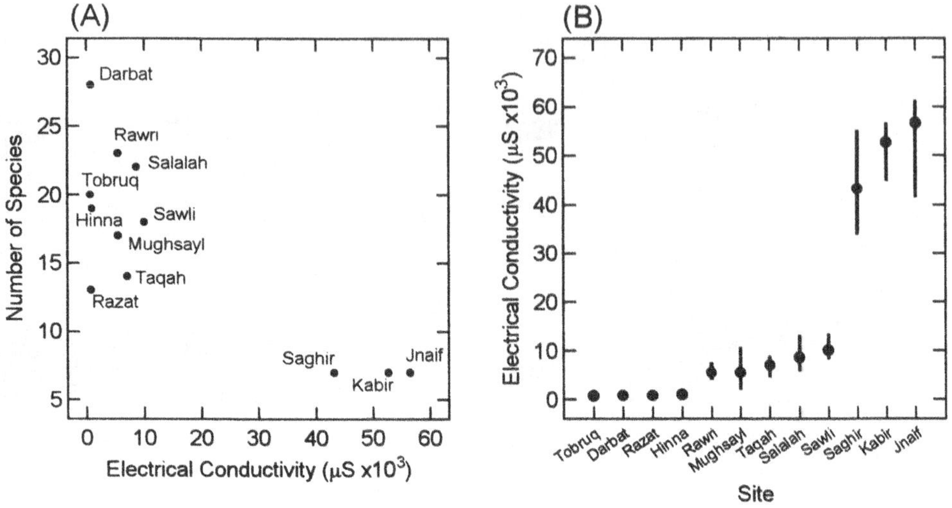

Figure 10.3 Variations in plant species richness with mean electrical conductivity (A) and seasonal fluctuations (bars) of mean electrical conductivity (B) in the fresh, brackish and saline pools of Dhofar. (Fresh water pools: Tobruq, Darbat, Razat, Hinna. Brackish water pools: Rawri, Mughsayl, Taqah, Salalah, Sawli. Saline water pools: Saghir, Kabir, Jnaif). After Ghazanfar (1993b) and Anon (1993).

Eclipta alba, Ipomoea aquatica, Lippia nodiflora and *Cyperus* spp. are included in this category.

Trichophytes: These are terrestrial plants which can tolerate seasonal flooding. This category includes species on the margins of pools such as *Bacopa monnieri, Vernonia arabica, Sesbania sesban* and *Saccharum* spp.

10.3.2 Ecological Categories

Water vegetation consists of an assemblage of algae, mosses, ferns and flowering plants. These are classified as macrophytes and are divided into two distinct ecological categories: (1) those attached to rocks and other substrates or rooted or anchored in the substratum, and (2) those which are free-floating.

(1) Macrophytes which occur always attached to rocks or other solid objects in springs are the mosses and algae. These occur in both fast- and slow-flowing water. Mosses are unable to photosynthesize bicarbonate ions and are therefore

confined to water regions where there is an adequate supply of dissolved carbon dioxide, such as where the water emerges from the ground. Mosses are also found in turbulent flow, such as at water falls. Downstream, at lower carbon dioxide concentrations, mosses decrease in abundance and algae tend to increase. Partly submerged rocks usually have a good cover of mosses. Rooted plants include ferns, flowering plants and the macrophytic green alga, *Chara*. These do not normally occur in deep water pools or rough and rocky areas of pools where rooting and establishment is difficult. The number of species in fast-flowing water is also considerably less than in slow-moving water. *Schoenoplectus litoralis* occurs in running water in some pools in the Dhofar mountains of Oman (Ghazanfar 1993b). In deep water pools (>1 m) with a low current speed or still water, species with tough, resilient stems such as *Ipomoea aquatica, Polygonum senegalense* and *Potamogeton nodosus* form fringing vegetation. Where the water is turbid and more than 2 m in depth, angiosperms are usually absent.

2) Floating plants are not as abundant as those which are rooted. *Lemna gibba*, which floats with its leaves on the water surface (Figure 10.4: E), occurs in Saudi Arabia, and *Utricularia* spp., with submerged leaves, is recorded from both Oman and Saudi Arabia (Collenette 1985, Ghazanfar 1993b). In a freshwater pool on the escarpment mountains of Dhofar *L. gibba* is recorded in shallow water near the source, often collecting and growing amongst the roots of other plants, but generally absent some distance (<50 m) away from the source. *Utricularia* spp. are recorded from deeper water with a depth of *c.* 1.5 m. Where the water is flowing, the grass *Isachne globosa* forms floating mats on debris.

10.4 Plant Communities

A number of plant communities, consisting of both submerged and emergent species, can be distinguished in fresh, brackish and saline waters. Overgrazing and trampling of the fringing vegetation of water pools and extensive usage of water for domestic purposes has led to severe degradation of several water pools in Oman. Soil erosion around the pools is evident and the vegetation is often heavily grazed or dead.

10.4.1 Plant Communities of Freshwater

10.4.1.1 *Isachne globosa* Community

In this community *Isachne globosa* is the dominant species and occurs at the edges of pools. It often forms floating mats over debris, and although the leaves are often submerged the inflorescence is always above the water. Common associates are *Ceratopteris thalictroides* and *Echinochloa colona*. Where the water is deep (up to 1 m) *Chara* spp. and *Ceratophyllum demersum* (Figure 10.4: G) form the chief

Figure 10.4 Common fresh and brackish water plants. (A) *Polygonum senegalense*. (B) *Potamogeton nodosus*. (C) *Potamogeton lucens*. (D) *Najas marina*. (E) *Lemna gibba*. (F) *Potamogeton pectinatus*. (G) *Ceratophyllum demersum*. (H) *Ruppia martima*.

associates. *Isachne* is chiefly a tropical Asiatic genus, and *I. globosa* is distributed in east and southeast Asia and Australia. In Arabia is occurs only in Dhofar in southern Oman, in pools on the escarpment mountains. *Ceratophyllum demersum* is a rootless, submerged perennial which is cosmopolitan in distribution. It is found in the deeper parts of pools, where it can attain a biomass of up to 172 g dry wt m^{-2} (Jupp 1993).

10.4.1.2 *Echinochloa colona* Community

This community occurs in damp and moist areas as fringing vegetation at the edges of pools, but is never submerged. It often occurs on moist ground away from the water edge on the landward side of *Isachne globosa*. Associates are *Bacopa monnieri* and *Lippia nodiflora*. The plants in this community are largely cosmopolitan, being present throughout the warm tropical and subtropical regions of the world.

10.4.1.3 *Potamogeton nodosus* Community

Potamogeton nodosus (Figure 10.4: B) is the dominant species occurring in waters deeper than 0.5 m, and is usually present in monospecific stands at the margins of pools. Associates are *Ipomoea aquatica*, *Najas* spp. and *Polygonum senegalense*, but these are generally not intermixed with *P. nodosus*.

The genus *Potamogeton*, of which three or four species are present in the Arabian Peninsula, is chiefly a temperate genus. It occurs in fresh and slightly brackish water bodies. *P. nodosus* (Figure 10.4: B) is a rhizomatous freshwater species, with submerged and floating leaves. In Arabia it is one of the most common aquatic plant species in permanent mountain pools, and often the only hydrophyte present. In shallow muddy pools in wadis it is present with *Ipomoea aquatica* and *Najas* spp. *Potamogeton pusillus*, a submerged aquatic recorded in water down to 0.5 m deep in running water channels, is apparently rare in the Peninsula.

10.4.1.4 *Polygonum* Community

This community consists of *Polygonum senegalense* or *P. amphibium* which grow at the margins of shallow pools ($<$0.5 m) on wet ground. Common associates are *Bacopa monnieri*, *Eclipta alba* and *Chara* spp.

Polygonum senegalense (Figure 10.4: A) occurs throughout tropical Africa and Egypt. It is a rooted hydrophyte, located at the edges of pools in wet mud and often covers vast areas. It can withstand seasonal drying, but will not tolerate long periods without moisture. In western Saudi Arabia *P. amphibium*, a Mediterranean/European species, replaces *P. senegalense* (Collenette 1985). *P. setulosum* is also associated with *P. senegalense*, but this species is recorded from only a few locations in western Arabia.

10.4.1.5 *Chara-Potamogeton* Community

This community is found in clear, still or slow-flowing water of 0.5-1.5 m depth. Characteristic species are *Chara* spp., *Potamogeton pectinatus,* and the common associate is *Najas graminea*. Species of *Urticularia* are also present but are not common.

Chara is one of the most common macrophytes, occurring in deep or shallow pools. It can tolerate moderately saline and calcareous water. A biomass of 264 g dry wt m^{-2} (Jupp 1993) has been recorded from a freshwater pool in Dhofar. *Potamogeton pectinatus* (Figure 10.4: F) is a cosmopolitan species, common throughout the Arabian Peninsula in pools in the hills and plains. It is a profusely branched, rhizomatous perennial with submerged filiform leaves. It can tolerate salinities of up to 5,700 μS and is common in the *khawrs* of Dhofar.

10.4.1.6 *Pteris vittata-Adiantum capillus-veneris* Community

Composed characteristically of the ferns *Pteris vittata* and *Adiantum capillus-veneris*, this community occurs on the banks of pools, especially where they are rocky or stony. Both species grow on wet soil and in shade. *Lindenbergia fruticosa* is often an associated species. This fern community is common throughout the Arabian Peninsula in the mountains and foothills. Both ferns grow commonly in moist ground by *aflaj* (irrigating water channels) and at the source of springs.

10.4.1.7 *Schoenoplectus litoralis* Community

This community occurs in shallow to deep, fast- to slow-moving water. *Schoenoplectus litoralis* is the dominant species, often forming monospecific stands. The leaves are submerged but the inflorescence axis is aerial. Associated species are *Chara* spp., *Cyperus* spp., *Fimbristylus* sp. and *Eleocharis geminata*. Where the water is deep (up to 1 m), *Potamogeton pectinatus* forms the chief associate.

Schoenoplectus litoralis occurs from Africa to Australia and is widespread in the Middle East. The plant is tufted, up to 2 m tall and is tolerant of salinities of up to 10,000 μS.

10.4.2 Plant Communities of Brackish and Saline Water Pools

10.4.2.1 *Avicennia marina* Community

Avicennia marina occurs in saline water bodies which do not receive a large inflow of freshwater and which have salinity levels of 20,000-30,000 μS. Associated with the mangrove, and usually occurring at the edges, is the macrophytic green alga, *Enteromorpha*. *Phragmites australis* is associated with *A. marina* where there is some seasonal freshwater inflow and the water salinity is relatively low. On low ground on the banks around the pools *Aeluropus lagopoides* and *Cressa cretica* are common with *Sporobolus* spp. (*S. virginicus* and *S. iocladus*) and *Urochondra setulosa* as the co-dominant species.

10.4.2.2 *Paspalum vaginatum* Community

In this community, the grass *Paspalum vaginatum* is the dominant species, growing on low wet ground at the margins of pools and submerged at high tides or during floods. *Bacopa monnieri* is often an associate. *P. vaginatum* is a stoloniferous, creeping grass distributed on shores and tidal mud-flats in the tropics and subtropics. On the Arabian Peninsula it is present only in southern Arabia.

10.4.2.3 *Najas marina-Potamogeton pectinatus* Community

This community consists of the submerged species *Najas marina, Potamogeton pectinatus, Ruppia maritima* (Figure 10.4: D, F, H) and *Chara* spp. In brackish water pools where salinity ranges from 1,080 μS to 14,500 μS, *N. marina* and *P. pectinatus* are the dominant species. Where salinity is low ($<$1,200 μS), *Ruppia maritima* forms the dominant species. *P. lucens* (Figure 10.4: C) is reported from some brackish water pools in the United Arab Emirates but is apparently not common.

Najas marina is a cosmopolitan species of fresh and brackish water. It consists of long, somewhat spiny stems, rooting at the lower internodes. The leaves are linear-oblong, in pseudo-whorls and have dentate-spinose margins. It is one of the most common submerged species in the brackish pools in Dhofar, with a biomass of up to 561 g dry wt m^{-2} in some *khawrs* (Jupp 1993). *Ruppia maritima* occurs in brackish waters throughout the temperate and tropical regions of the world. It is a branched herb, with linear, minutely denticulate leaves. *Potamogeton pectinatus* is also a cosmopolitan species common in fresh and brackish water. The stems are profusely branched, rhizomatous, with tuberous winter-buds. The leaves are narrowly linear. It is most abundant in freshwater pools but also occurs in brackish water.

10.5 Summary

The scarcity of fresh and brackish water bodies in the Arabian Peninsula is reflected in the paucity of aquatic and brackish water plant species, with less than 20 species recorded. Some water bodies have only monospecific stands of aquatic vegetation, and most aquatic communities are composed of only two or three species. Species diversity is higher in freshwater than in brackish or saline waters. Brackish water pools, the *khawrs*, are unique aquatic systems in which salinity levels are variable, being controlled by the season, local rainfall and the seepage of ground-water. Most of the aquatic and brackish water species which occur throughout the Arabian Peninsula have a wide distribution range and occur in the warm tropical and subtropical regions of the world.

Seven fresh water communities are distinguished, consisting of both submerged and emergent species. Amongst the common species which form the fringing vegetation on moist soil are *Echinochloa colona*, *Isachne globosa*, *Polygonum* spp., *Schoenoplectus littoralis* and the ferns *Adiantum* and *Pteris*. Amongst the submerged species, *Ceratophyllum demersum*, *Potamogeton* and *Chara* spp. are common. Plant diversity in brackish and saline waters is poor, consisting mainly of the mangrove *Avicennia marina*, the grass *Paspalum vaginatum* and the submerged *Najas*, *Ruppia* and *Potamogeton* spp.

Chapter 11

Plants of Economic Importance

Shahina A Ghazanfar

"Economic plants may be defined as those that are utilised either directly or indirectly for the benefit of Man. Indirect usage includes the needs of man's livestock and the maintenance and improvement of the environment; the benefits may be domestic, commercial or aesthetic."

SEPASAT Newsletter No. 1 (Anon. 1983)

11.1 Introduction

Prior to the relatively recent exploitation of oil resources, rural communities in the Arabian Peninsula relied largely on local raw materials in their everyday lives. Houses were made from stone, date-palm and other timber, mud and grass was used for plaster, palm-fronds and rush for making baskets, mats and other utilitarian objects, and native and exotic plants were used for medicines, dyes and fragrances. With increasing dependence on imported and locally manufactured goods for the majority of household and other needs, the use of plants and the practice of native crafts are becoming a rarity. However, plants are still utilized, albeit rarely, to make utilitarian objects and as medicines, perfumes and dyes.

In this chapter I describe the utilization of indigenous plants for traditional herbal medicines, dyes, perfumes, as building material for houses and for the production of traditional household implements. I also describe the favoured

S. A. Ghazanfar and M. Fisher (eds.), Vegetation of the Arabian Peninsula, 241-264.

rangeland and fodder plants. Cultivated crops of economic importance, such as coconut, *kat*, lime and others are not described here as information on these is fairly well-documented in published literature and databases (Anon. 1990 and references therein, Wickens *et al.* 1989 and references therein). Since the date-palm forms a significantly important crop both historically and contemporarily over much of the Arabian Peninsula and is used for more than just its fruit, a brief account of its uses is included.

11.2 The Date-Palm

The date-palm (*Phoenix dactylifera*) was the first fruit tree to be taken into cultivation in the Old World (Zohary & Hopf 1994), and charred date stones have been excavated in eastern Oman which date back to 2,500 years BC (Berthoud & Cleuziou 1983, Cleuzion & Constantini 1980). Dates have formed a staple diet for the Arabs since early times and "*... [the] fruit almost takes the place of bread to the hardy Arab...*"(Miles 1901).

In Arabia, date-palms are cultivated wherever water is sufficient for irrigation and the temperature is suitable. The plant requires a warm, dry climate for growth, and a high temperature with low humidity for fruit-set and ripening. Date-palms can grow in relatively saline soils and can withstand irrigation with brackish water. Young plants produce basal suckers used for vegetative propagation. Trees mature in five years and can yield up to 200 kg of fruit per tree. Pollination is by hand and in a date-palm grove one or two male trees are planted for approximately 40 female trees. Good quality date plants are highly valued and are sold for as much as $175 per plant (1997 prices).

It is believed that the date-palm is blessed by God with gifts not found in other trees. All parts of the tree are useful and it is undoubtedly a versatile plant of great economic importance to the people of the Peninsula. Over 40 varieties of dates are recorded from Bahrain, Oman and Saudi Arabia, each variety with its distinct shape, colour, taste and fruiting time. At one time, date cultivation was so extensive in Bahrain that it was known as the land of a million date-palms. Now processed and packed in factories, Arabian dates are also exported outside Arabia. Several products are made from the fruit, including date-syrup (*dibs*), often called date-honey, a product unique to Arabia, the use of which dates back several thousand years. The trunk and leaves provide material for making thatch, screens, baskets, mats, brooms, ropes and hives for traditional bee-keeping (Figure 11.1).

11.3 Medicinal Plants

The use of native Arabian plants for medicinal purposes goes back to pre-Islamic civilizations, when colocynth, (*Citrullus colocynthis*, Figure 11.3: B), Christ's thorn (*Ziziphus spina-christi*, Figure 11.3: A) and capers (*Capparis spinosa*), amongst others, are known to have been used. Archaeo-botanical evidence from the Near

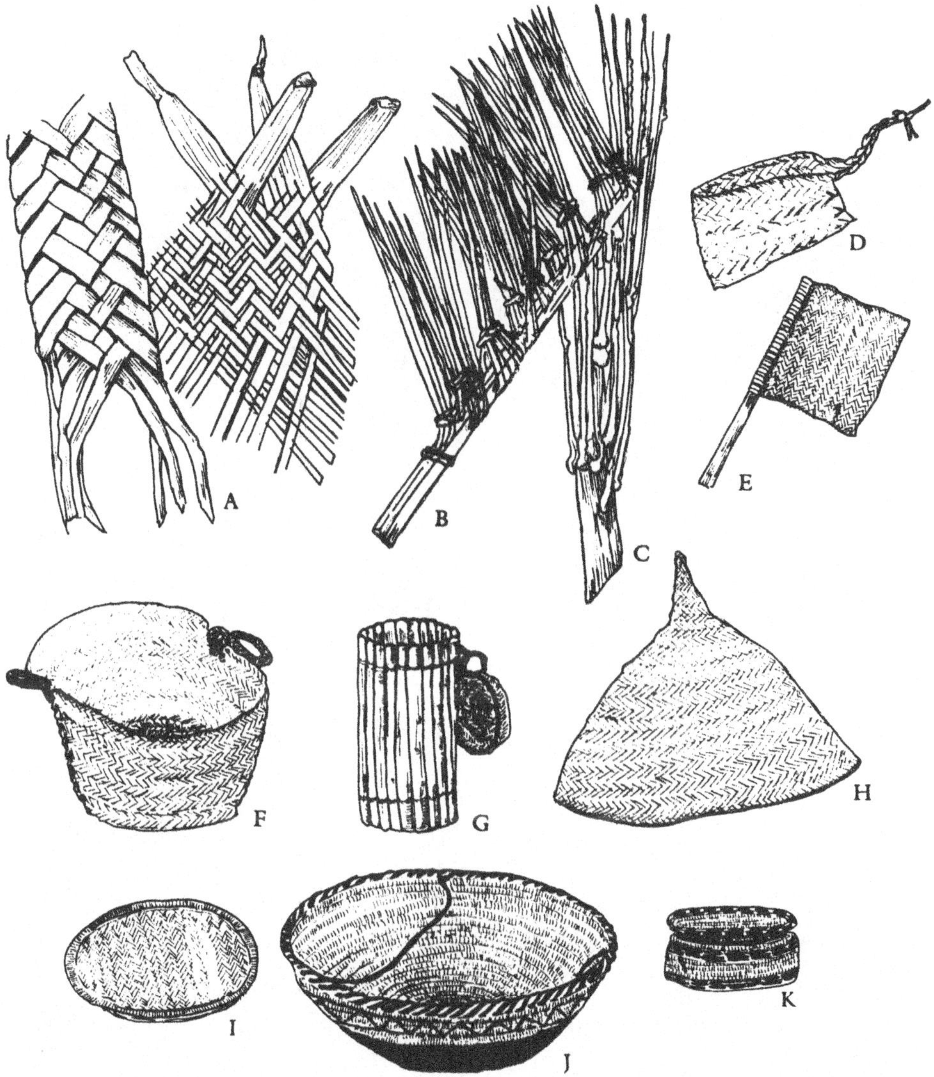

Figure 11.1 Products made from the fronds of the date-palm (*Phoenix dactylifera*) and dwarf-palm (*Nannorrhops ritchieana*). (A) The basic woven plait used for making mats and baskets. (B) Broom (*mekasha*) made from date-palm frond. (C) Broom (*asu*) made from the female inflorescence, normally used for sweeping the place where animals are kept. (D) Small bag for storing *halwa*, a traditional Omani sweet. (E) hand-held fan (*mahafa*). (F) basket (*qafir*) with a round bottom and straight sides. (G) Long cylindrical container (*meshan*), *c*. 50 x 15 cm, made from date-palm leaf stalks, used for storing dates. (H) Conical cover for food (*mekaba al shut*). (I) Shallow tray (*saf*), used for cleaning grains. (J) Bowl with a goat leather base, made from the leaves of the dwarf palm (*kurma*, *mabrad*), used for keeping milk and yogurt in the southern region of Oman. (K) Small basket with a tight fitting lid (*daraj*, *fatih*) made from leaves of the dwarf palm, used for holding items of jewellery and cosmetics.

Figure 11.2 (A) Frankincense tree (*Boswellia sacra*) in its natural habitat in Jebel Qara, Oman. (B) Flowers. (C) Resin (frankincense) exuding from a slash made in the tree trunk. (D) Traditional frankincense burners.

East indicates that domesticated crops such as garlic and fig, mentioned in old medical texts, were also cultivated (Dawood 1967, Zohary & Hopf 1994).

Traditional medicine in the Arabian Peninsula is based primarily on the Greco-Arab system of medicine (*unani tib*), in which a holistic approach is taken to treatment. This method is based on the classical Greek medical system of the four bodily humours of blood, phlegm, yellow and black bile, and their relation to the elements, earth, water, air and fire. In addition, the concept of the four qualities of wet, dry, hot and cold are integrated with that of the bodily humours such that each of the four humours are assigned a quality (Foster & Anderson 1978). Maintenance of a balance between the humours is vital for perfect health, and any imbalance in the four humours leads to ill health (Chadwick & Mann 1950). Medicinal herbs and certain foods are also assigned the qualities of being hot, cold, wet or dry. If diagnosis of a disease results in the identification of an imbalance in a particular bodily humour, medicinal herbs are prescribed in order to restore that imbalance. In addition, foods, especially vegetables and fruits, are prescribed to help restore the imbalance.

The concept of bodily humours comes from the ancient Greek medicine of Hippocrates. The bodily humours and the 'qualities' of these humours and treatments were acquired by the Arabs through the works of Dioscorides (1st century AD) and Galen (AD 131-201). During the 5th century Galen's works were translated by the Nestorians and taken with them to Persia. They also translated several Indian medical books from Sanskrit. During the 9th to the 13th centuries, with the spread of Islam, medical works were translated into Arabic and adapted by Arab scholars into the Arabian system of medicine. Arab physicians such as Ibn Sina (Avicenna, AD 980-1037), Al Razi (Rhazes, AD 865-925) and Al Zahrawi (Albucassis b. AD 936) and Ibn Rushd (Averroes, b. AD 1200) contributed greatly to medical knowledge. Ibn Sina's epic book, *Al Qanoon al Tibb*, *The Cannon of Medicine* (in 14 volumes), was translated into Latin and formed the basis of European medicine well into the 18th century.

In the Arabian Peninsula more than 260 indigenous plants are used medicinally (Ghazanfar 1994a). In addition, several plant preparations are imported from Africa, India, Iran and Pakistan which are commonly sold and used for treatment. In the last decade Ayurvedic clinics have opened in the Arabian Gulf countries, prescribing plant-based medicines that originate primarily from India.

Plants for medicines are rarely used as fresh material. Normally they are dried and stored in plastic, metal or glass containers. From time to time they are spread out in the sun and checked for insect pests. Roots, stems and leaves are the more commonly used plant parts; flowers are usually used for fragrances or dyes, and seeds normally for oil extraction. Herbal medicines are rarely made from single plants, instead a number of plants are prescribed for treating a particular disease or condition. Herbal medicines are used as extracts, infusions and decoctions. Measurements are usually a *qahwa* (coffee) cup or a handful. The concept of 'hot' and 'cold' foods and the balance of bodily humours is considered when herbal medicines are prescribed, and the same plant may be used for the treatment of several diseases.

In addition to plants, minerals, such as black salt, mercurials (Hardy *et al.* 1995), sal ammoniac, potash, lead (Woolf 1990, Worthing & Sutherland 1996) and pearl ash, and animals products such as ambergris, the gall bladder of sheep, and shells of various species of molluscs, are also used for treatment.

Herbal medicines are very commonly used for digestive problems and the treatment of colds and coughs. Fevers and headaches are commonly treated by keeping the body cool and applying herbal pastes to the forehead (Ghazanfar & Al-Sabahi 1993). One of the popular uses of herbal medicines is in childbirth and for pre- and post-natal care (Colfer 1990). Plants are also used as aphrodisiacs, for fertility and for the regulation of menstruation. They are used for various dermal infections, bites and stings, for alleviating muscular pain (Ghazanfar 1994a), and as health tonics, sedatives and during convalescence (Ghazanfar 1996b).

Treatment by herbal medicines or other traditional methods are especially popular for diseases or problems in which modern treatment is prolonged or proves unsatisfactory. Jaundice is often treated by cauterization, with dietary supplements of lime, sweet orange, mango, grape or pomegranate (Ghazanfar 1995). Diabetes, relatively common in Oman, is often treated with herbs. Herbal healers consider diabetes as a 'hot' and 'dry' disease, and consequently herbal medicines prescribed for treatment are 'cold' and 'wet' in order to redress the imbalance of the humours. Usually the medicine consists of several plants mixed together and boiled in water and the infusion is taken in regular doses for a number of days. *Rhazya stricta* (*hermel*) is a popular plant regarded highly for its anti-diabetic properties. Other plants include *Acacia nilotica* (in which pods and seeds are used), mango leaves, garlic, bitter gourd and *Teucrium* spp.

Tables 11.1-11.7 list the medicinal plants commonly used for the treatment of different classes of disease and problem. The information has been collated from Ghazanfar (1994a and references therein), Ghazanfar & Al-Sabahi (1993), Miller & Morris (1988) and Schopen (1983). Several species listed in these tables have been identified by WHO for use at the primary health care level for the Arabian Peninsula (Anon. 1985c).

11.3.1 Carminatives, Laxatives and Anti-diarrhoeals

Amongst the most commonly used plant-based treatments are those for the digestive system (Table 11.1). The value of plants as carminatives, laxatives and digestives has been exploited since ancient times. For instance, the use of dill, coriander, cumin and nutmeg as carminatives, senna and colocynth as laxatives, and the multiple uses of summak from a tanning agent to providing relief from earache, toothache, dysentery and gangrene, were known in Arabia and Egypt as early as the 2nd millennium BC (Simpson & Conner-Ogorzaly 1986). Volatile oils in the family Umbelliferae, to which coriander, fennel, cumin and dill belong, are responsible for the carminative properties and are now widely used in commercial pharmaceutical preparations.

Table 11.1 Plants used as carminatives, laxatives and as anti-diarrhoeals.

Plant	Part used	Preparation and treatment
Allium cepa, *A. sativum*	bulb, cloves	juice of fresh bulb and cloves used for colic and abdominal pain
Cadaba farinosa	leaves	infusion of leaves taken for colic
Carissa edulis	berries	eaten for abdominal colic and constipation
Citrullus colocynthis	leaves, roots and seeds	infusion of crushed leaves or roots with goat's milk, or seeds taken with food, as a laxative
Croton confertus	leaves	slightly crushed in water, and solution taken orally for constipation and stomachache
Cynomorium coccineum	whole plant	eaten as a laxative
Dorstenia foetida	seeds	eaten for flatulence and indigestion
Euclea schimperi	leaves	eaten for colic, diarrhoea and indigestion
Euphorbia hadramautica	seeds	eaten to cure flatulence, colic and indigestion
Ficus carica	fruit	eaten as a laxative, and as a general tonic
Heliotropium fartakense	leaves	soaked in water, taken for indigestion and colic
Ipomoea pes-caprae	seeds	eaten as a purgative
Jasminum grandiflorum	flowers	infusion of dried flowers taken for colic, dysentery and abdominal pain
Lavandula spp.	leaves and stems	boiled in water, and solution taken for colic and stomachache
Maerua crassifolia	leaves	infusion of leaves taken for constipation and colic
Matricaria aurea	flowers	infusion of flowers taken as tea for colic and stomach cramps
Moringa peregrina	seed oil	taken orally for constipation and stomach cramps
Nepeta deflersiana	leaves	infusion of leaves taken as tea for indigestion
Ocimum basilicum	seeds	infusion of leaves or crushed seeds taken as tea for diarrhoea
Plantago coronopus, *P. major*	whole plant, seeds	infusion of plant taken as a laxative, and seeds of *P. major* crushed and swallowed with water for diarrhoea.
Reichardia tingitana	leaves	dried powdered leaves with water, to treat colic and constipation
Rhazya sticta	leaves	mixed with senna and milk, taken for abdominal colic and constipation
Ricinus communis	seed oil (castor oil)	used as a purgative
Senna alexandrina, *S. holosericea*, *S. italica*	leaves and seeds	decoction of leaves, often mixed with other herbs, used as a laxative and for stomach cramps
Tamarindus indica	fruit	soaked in water, solution taken as a laxative, blood cleanser and as a general tonic
Teucrium polium, *T. mascatensis*	leaves and stems	fresh or dried, boiled in water for colic and stomach pains
Thymus vulgaris	leaves	infusion of dried, powdered leaves to which salt and cumin is added, taken for colic, coughs and colds

Table 11.2 Anthelmintic plants.

Plant	Part used	Preparation and administration
Artemisia sieberi	leaves	crushed and an infusion made, taken orally (the plant contains essential oils, which are reported to be toxic against *Ascaris*)
Capparis decidua	fruit	eaten for expelling worms, also to treat constipation, hysteria and other psychological problems
Capparis spinosa	leaves, root, bark	soaked in water, extract used as an anthelmintic, expectorant and as a tonic
Carica papaya (cultivated)	leaves and seeds	pounded and eaten as an anthelmintic; also used treat diarrhoea; raw fruit is digestive and carminative
Cyperus rotundus	tubers	dried or eaten fresh to expel worms; also eaten to regulate menstruation; powdered and used as an insecticide
Fumaria parviflora	whole plant	extract as an anthelmintic, laxative and for treating dyspepsia
Peganum harmala	leaves and seeds	tea made from leaves or powdered seeds mixed with water, taken orally as a vermifuge against tapeworms and for removing kidney stones; seeds mixed with senna and honey used for stomachache
Punica granatum (cultivated)	fruit rind	dried, mixed with thyme and flour, baked as bread and eaten as an anthelmintic and to cure diarrhoea
Rhazya stricta	seeds	powdered, eaten to remove stomach worms; in central Oman dried leaves powdered with halwa (sweets) and leaves of *Cassia senna* are mixed with milk and used for abdominal pain, colic and constipation; the mixture is drunk on alternate days; also given to infants from about three months to make their joints strong
Vernonia cinerea	roots and seeds	seeds are eaten and decoction of roots used as an anthelmintic and diuretic

Table 11.3 Plants used for treating muscular pain and swollen joints.

Plant	Part used	Treatment
Calotropis procera	leaves	oil is rubbed on the site of pain, then leaves are heated and placed on it
Cissus rotundifolia	leaves	heated and used as a poultice for backache
Conyza incana	leaves	heated and used as a poultice for muscular pain
Dracaena serrulata	gum resin	mixed with water and made into a paste, applied on legs and feet to relieve pain
Euphorbia amak	latex	mixed with oil and massaged on painful joints
Heliotropium fartakense	leaves	pounded with salt, turmeric, ginger and made into a paste, applied to sprains and swellings
Juniperus excelsa subsp. *polycarpos*	leaves	soaked in oil, massaged over painful joints, muscles and paralysed limbs
Kleinia odora	stems	juice from the stems applied over painful muscles
Psiadia punctulata	stems	heated and placed on site of pain
Tephrosia apollinea	leaves	powdered with salt, heated and placed on swellings or swollen joints

Table 11.4 Plants used for skin disorders, burns, wounds, bruises, stings and bites.

Plant	Part used	Treatment and application
Acacia gerardii, *A. nilotica*	gum resin, leaves	applied to soothe burns; leaves are pounded into a paste and used as a poultice on boils and swellings or applied around boils to draw out the pus
Achyranthes aspera	roots	crushed and applied on scorpion stings to reduce itching and swelling
Adenium obesum	sap, crushed bark	pounded and applied on wounds and skin infections
Allium cepa, *A. sativum*	bulb	juice rubbed to remove spots from face or treat skin rash
Allophylus rubifolius	leaves	pounded and placed on skin boils to bring out the pus
Aloe dhufarensis, *A. tomentosa*, *A. vera*	leaves	juice extracted from the leaves is either used by itself or mixed with indigo and applied on skin rash, burns and wounds; also used as disinfectant for burns and wounds; also applied around the eyes to relieve itching and pain
Amaranthus graecizans, *A. viridis*	leaves	crushed and applied on scorpion stings, snake bites and itchy skin rash
Anagallis arvensis	whole plant	crushed and applied to soothe skin rash, snakebite and skin ulcers
Aristolochia bracteolata	whole plant	rubbed on place of snake bite or scorpion sting
Aerva javanica	flowers	mixed with water, made into a paste and used as a wound dressing and to stop bleeding
Becium dhofarense	whole plant	boiled in water, infusion used to soothe skin sores, dry skin, itch and insect bites
Calotropis procera	leaves and latex	latex from stems is dropped around a pus filled wound to draw the pus; also applied on scorpion stings to relieve pain and swelling
Capparis cartilaginea	leaves and stems	crushed in water and heated, the solution is strained and applied on bruises, snakebites, swellings and itchy skin rash
Caralluma aucheriana	stems	extract from the succulent stems applied to soothe sunburnt or itchy skin
Citrullus colocynthis	leaves	poultice made from crushed leaves and garlic, applied on bites and stings
Citrus aurantifolia	fruit	dried and crushed, mixed with salt and water and heated to make a paste, applied on skin to extract thorns
Convolvulus arvensis	roots and leaves	crushed and applied on wounds and cuts to stop bleeding
Cyphostemma ternatum	stems and leaves	juice extracted, used to wash feet to remove fungal infections; heated with salt, is used as a poultice for skin infections
Eulophia petersii	pseudobulbs	juice extracted to treat eczema, ringworm, skin rash and sores
Euphorbia cactus, *E. larica*, *E. schimperiana*	whole plant	extract applied around boils to draw out the pus; also applied on sores, ringworm and skin ulcers; latex from *E. larica* is applied around pus filled boils and that from *E. schimperiana* is used to loosen thorns for extraction
Euryops arabicus	leaves and stems	heated leaves and stems applied to sore feet
Ficus carica, *F. cordata* subsp. *salicifolia*	latex, leaves	applied to burn warts; dried leaves are crushed, mixed with honey and applied to treat skin discolouration and freckles
Heliotropium fartakense, *H. kotschyi*	leaves and stems	pounded with salt, turmeric and ginger, applied on burns and ulcers; paste of leaves used for ringworm, eczema, wounds and skin sores
Impatiens balsamina	leaves and stems	extract used to soothe itchy and inflamed skin

Table 11.4 continued.

Plant	Part used	Treatment and application
Jatropha dhofarica	sap	applied over wounds and skin sores as an antiseptic and to stop bleeding
Medicago sativa	leaves	mixed with leaves of tamarind and salt, applied on bruises
Melilotus indicus	whole plant	crushed and used to soothe skin rash
Myrtus communis	leaves	paste applied on wounds and scorpion stings; ashes of burnt leaves applied on blisters and ulcer
Olea europaea	leaves and bark	pounded bark and leaves applied on blisters and skin ulcer
Pergularia tomemtosa	latex	latex applied on skin sores
Phoenix dactylifera	fruit	paste made from dates, mixed with salt, applied to bruises
Pluchea arabica	whole plant	pounded with water and boiled and solution applied to boils and swellings
Polycarpaea repens	whole plant	crushed and applied at site of snakebite
Psoralea corylifolia	seeds	mixed with thyme and made into a paste, for treating general skin disorders
Rhazya stricta	leaves	extract mixed with oil, *Nigella sativa* and ginger, or leaves crushed in water, applied on skin rash
Rumex vesicarius	leaves and seeds	eaten as an antidote for scorpion stings
Salvadora persica	leaves	fresh or dried, powdered, applied on skin blisters and scorpion sting
Sanserieria ehrenbergii	leaves	juice applied to skin sores and eruptions
Tamarix aphylla	leaves	dried leaves applied on wounds to stop bleeding
Teucrium polium	leaves	fresh or dried, boiled in water and poured on bites and skin ulcers
Trichilia emetica	seeds	crushed with sulphur, as a skin ointment

Table 11.5 Plants used for fertility, childbirth, and post- and pre-natal care.

Plant	Part used	Treatment
Anastatica hierochuntica	dried plant	soaked in water and the water taken orally at the time of childbirth
Delonix elata	leaves	infusion given for prolonged or difficult labour
Haplophyllum tuberculatum	leaves	an anal suppository is made with leaves of *Azadirachta indica*, resin of *Commiphora* spp. and thyme, given to the new mother after childbirth, to strengthen back muscles
Launaea nudicaulis	leaves	inserted in the vagina after childbirth to stop excessive bleeding
Moringa peregrina	seed oil	mixed with clove oil and cardamom oil, taken during labour
Morus nigra	fruit	infusion of fruits and berries of *Salvadora persica,* given as a general tonic to women and to regulate menstruation
Psoralea corylifolia	seeds	crushed with butter, applied to treat mastitis
Pulicaria jaubertii	seeds	infusion of seeds taken to improve lactation; a decoction of leaves taken to stimulate digestion
Rhazya stricta	seeds	crushed and added to milk, given to increase lactation
Trigonella foenum-graecum	seeds	boiled in water, mixed with egg and given to the new mother for one week after childbirth

Table 11.6 Plants used for colds, coughs, fever and headaches.

Plant	Part used	Treatment
Acridocarpus orientalis	seed oil	massaged on the forehead to relieve headache
Alkanna orientalis	leaves	infusion of fresh leaves, mixed with thyme, taken for sore throat
Aloe dhufarensis, *A. vera*	leaves	juice/gel applied to forehead for headache and to lower fever; applied around eyes to soothe pain and redness
Ambrosia maritima	whole plant	crushed, burnt and the smoke inhaled to relieve difficult breathing
Arnebia hispidissima	whole plant	infusion taken as tea to lower fever
Cichorium intybus	leaves	eaten or boiled in water to lower fever
Citrus aurantifolia	fruit juice	taken for cold and fevers; also used for digestive problems
Cocculus hirsutus	leaves and roots	used for lowering fevers
Commiphora myrrha	resin	for fever, smoke from burnt resin inhaled or resin soaked in water and solution taken orally
Croton confertus	shoots	dipped in butter, sucked to cure coughs
Fagonia indica	whole plant	boiled in water and water used to bathe body to reduce fevers
Glycyrrhiza glabra	roots	soaked in water, or powdered and mixed with water, taken for coughs and as an expectorant
Launaea nudicaulis	leaves	infusion or paste made with water is applied to cool the forehead and body
Mentha longifolia	leaves	infusion with honey, taken for coughs, headaches and as a carminative
Nigella sativa	seeds	eaten for chest congestion and difficult breathing
Phoenix dactylifera	fruit	paste made from dates, applied to forehead and eyes to relieve headache
Rhazya stricta	leaves and stem	burnt and smoke inhaled to ease chest pains; fresh ground leaves mixed with lemon or boiled in water, to bathe body to reduce fever
Senecio asirensis	leaves	infusion taken to lower fever
Sisymbrium irio	seeds	boiled in water, solution taken as a drink to reduce fever
Solanum nigrum	whole plant	boiled in water and water used to bathe forehead to reduce fever
Sonchus oleraceus	whole plant	as a coolant, diuretic, laxative and as a general tonic
Trigonella foenum-graecum	seeds	boiled with dates and figs in water, taken as a drink for bronchitis and coughs
Vernonia cinerea	leaves	decoction taken to reduce fever

Table 11.7 Plants used as health tonics.

Plant	Part used	Treatment
Citrus aurantifolia, *C. limetoides*	fruit	juice or fruit, eaten for general health
Ficus carica	fruit	juice or fruit, eaten for general health
Morus nigra	fruit	eaten for general health
Punica granatum	fruit	juice or fruit, eaten for general health and convalescence, especially when recovering from fever
Rhus somalensis	berries	eaten either raw or roasted as a general tonic
Caralluma aucheriana	stems	juice added to curdle milk, given to speed convalescence; also eaten raw to improve health
Sarcostemma viminale	whole plant	inner tissue of the stem eaten as a general tonic and to cleanse the digestive system

11.3.2 Anthelmintics

About 16 plant species are known to be used for the treatment of stomach worms (Table 11.2). Several of these are also used as laxatives and for stomachache.

11.3.3 Muscular Pain and Swollen Joints

Muscular pain, inflamation of joints and swelling are generally treated by massage with oils or by applying a poultice of medicinal leaves (Table 11.3). Gums and resins are also used for treatment.

11.3.4 Skin Disorders, Burns, Wounds, Bruises, Stings and Bites

Over 50 plant species are used for treating skin problems (Table 11.4). General skin disorders treated include eczema, skin rash and skin infections. Treatments are also available for removing freckles. Special 'cooling' pastes are applied topically for burns, sunburn and heat rash. Treatment for snake and scorpion stings includes both topical application at the site of the sting and an orally taken plant extract. Special plants are used for the treatment of pus-filled boils and for removing thorns.

11.3.5 Fertility, Childbirth, Post and Pre-natal Care

Several plant species are used to improve lactation and strengthen back muscles after childbirth (Table 11.5). Suppositories and tampons are prepared with a selection of plants to relieve backache after childbirth. A special diet including honey, fenugreek, ginger and garlic is given to mothers from 7 to 40 days after giving birth. Several herbal treatments are used for fertility, contraception and abortion (Colfer 1990, Musallam 1973).

11.3.6 Cold, Coughs, Fever and Headaches

Infusions of plants are commonly used for treating colds and coughs, and smoke from burning particular species is inhaled to reduce chest congestion and ease difficult breathing (Table 11.6). To lower a fever, the body is bathed in water to which plants have been added. The body is cooled down by applying herbal pastes and by imbibing 'cooling' drinks. Pastes and oils are applied or massaged on the forehead for relief from headaches.

11.3.7 Health Tonics

Approximately 15 plant species are used as general health tonics, of which five are commonly used (Table 11.7). Concoctions and infusions of plants are used for the frail and elderly, for children and during convalescence. Several species are used to 'give strength', or 'to make the body strong', especially after illness. Health tonics are also used to 'purify blood' and to cleanse the blood system.

11.4 Dye Plants

The use of certain plants as a source of dye has been established since ancient times. In the Old World the use of madder, indigo and saffron to dye fabrics, and henna (Figure 11.3: D) to dye hair, have been known since the 3rd millennium BC. The first use of indigo was recorded in China 6,000 years ago and it has been used as a fabric dye in India, Africa and Arabia for more than 2,000 years (Simpson & Conner-Ogorzaly 1986).

In Dhofar in southern Oman and in the Tihamah of Yemen, indigo (*Indigofera tinctoria, I. coerulea* or *I. arracta*) was cultivated in great quantities during medieval times and traded together with other commodities (Baldry 1982, Wilson 1985). It was also cultivated a great deal in northern Oman to dye homespun and imported cottons (Miles 1901). Today imported synthetic dyes have largely superceded the use of natural dyes, and traditional hand-woven indigo-dyed cloth is rarely seen. Pounded root bark of *Calligonum comosum* (*arta*), *Memecylon tinctorium,* (*wers*), *Crocus sativus* (saffron) and *Ziziphus spina-christi* (*sidr*) were also used to dye cotton cloth. Madder (*Rubia tinctoria*) is still used to dye locally produced goat wool in the mountains of northern Oman, but its use is not common and is restricted to a few weavers (Jones 1989).

The common plant species which are used or have been traditionally used as a source of dye are given below:

Acridocarpus orientalis (*qafas*) (Figure 11.4), and *Pulicaria glutinosa* (*muthedi, mohaddedi*) are used together as source of yellow dye in northern Oman. Leaves of *qafas* with stems of *muthedi* are boiled in water and ashes of *Haloxylon salicornicum* (*rimth*) are added as a mordant. Goat wool is immersed in the mixture and then left to cool. The yellow coloured wool thus obtained is used for the borders of camel rugs and camel trappings (Jones 1989). All three species are distributed in the foothills of the northern and southern mountains of Oman and are collected from the wild as required.

Aloe inermis (*sekel*) is distributed in southern Oman, Yemen and Somalia. It has been an important dye source in Dhofar. Dead and dried leaves are scorched lightly over fire and pounded into a coarse powder. This is boiled in water until a strong blue-black colour is obtained. Cloth and fibres for basketry and mats are

Figure 11.3 (A) *Ziziphus spina-christi* (*sidr*), leaves used as hair shampoo. (B) *Citrullus colocynthis* (*handhal*), used as a purgative. (C) *Salvadora persica* (*rak*), roots and stems used as tooth-cleaning sticks. (D) *Lawsonia inermis* (*henna*), paste of powdered leaves used as a cosmetic dye.

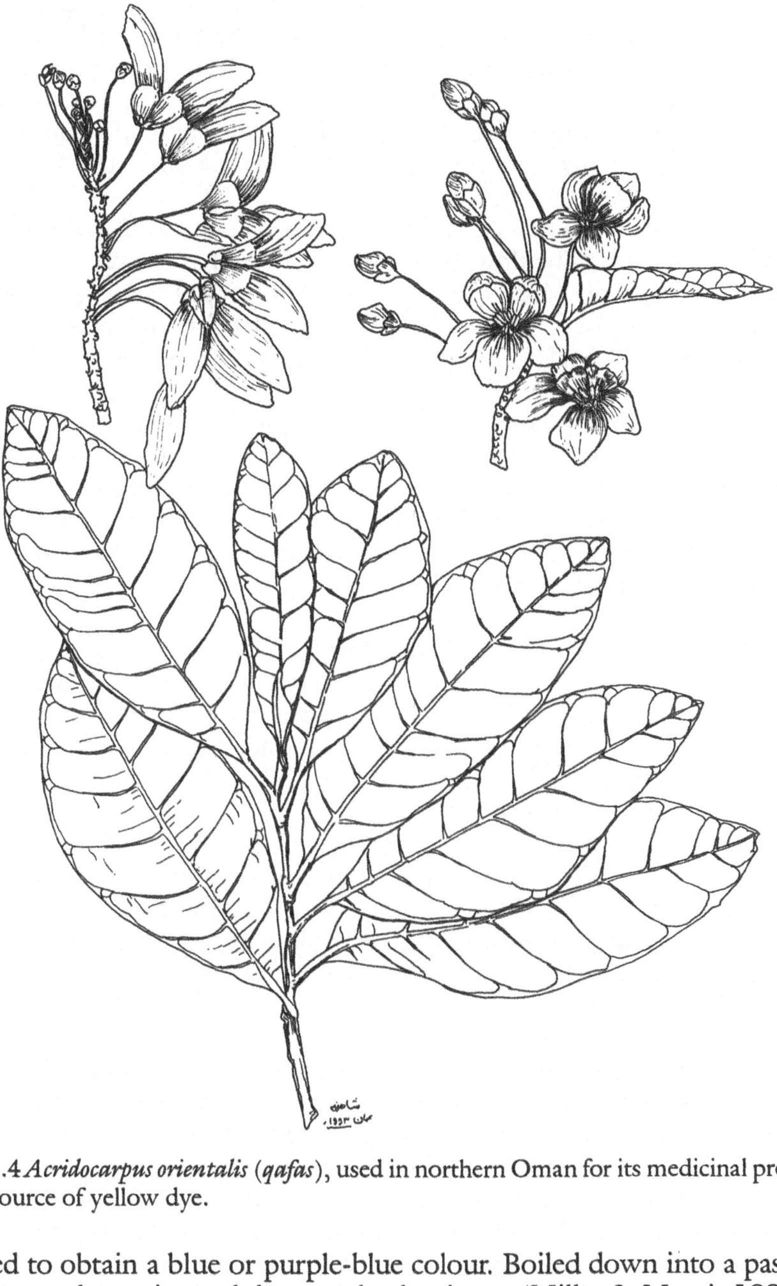

Figure 11.4 *Acridocarpus orientalis* (*qafas*), used in northern Oman for its medicinal properties and as a source of yellow dye.

immersed to obtain a blue or purple-blue colour. Boiled down into a paste, the dye is also used to paint and decorate leather items (Miller & Morris 1988).

Anogeissus dhofarica (*mishteh*) is endemic to Dhofar in southern Oman and to southeast Yemen, and is one of the dominant woodland trees on the southern escarpment mountains. The tree is of great economic importance and, especially in former times, all parts of the tree were used either as building material, for medicinal purposes, for dye or for fodder. As a dye the leaves are left to macerate in warm water until a dark colour is obtained. Material is placed in the dye, brought to boil and then kept simmering. The colour obtained is yellow or yellow-green. A yellow dye is also obtained from a paste made from the leaves, which is then used to decorate clay utensils (Miller & Morris 1988).

Carthamus tinctorius (*shawrina*), safflower, has been cultivated in several parts of the Arabian Peninsula. The flowers are dried and used as a source of red dye for dyeing cloth and wool. Made into a paste with oil it is used as a cosmetic and applied on the forehead or face. (See also Miles 1901 for a reference to its use in northern Oman as a facial cosmetic application).

Commiphora gileadensis (*shikuf*) grows throughout the mountains of southwest Arabia and has a variety of uses which also includes the use of its gum-resin as a fragrance. The dried wood and bark have been used as a yellow-green dye for dyeing cotton. The dried bark is boiled in water and left to develop a strong colour. Cloth is then immersed in the solution and left until the required colour is obtained (Miller & Morris 1988).

Cynomorium coccineum (*tarthuth*) is a parasitic perennial with short (20-30 cm), fleshy, usually unbranched stems. It is common throughout Arabia, often parasitic on *Zygophyllum* spp. The flowering spike bears many red flowers and has an unpleasant smell, attracting flies. The flowering stalk has been used as a source of crimson dye in Saudi Arabia. The roots are edible and the whole plant has also been used medicinally as a laxative.

Euclea schimperi (*kilit*), is a small, evergreen tree distributed in the mountains of southern and western Arabia. The underbark is cut into little pieces, boiled in water and left to macerate until a brownish colour is obtained. Cloth is dyed by re-boiling the water and leaving the cloth to soak until the colour is strong (Miller & Morris 1988).

Indigofera tinctoria (*nil*), indigo, is found as an escape and may be cultivated on a small scale in Yemen and Dhofar. Indigo was extensively cultivated in Arabia until about 50 years ago (Miles 1901, Stone 1985). In the Tihamah indigo dyeing is still used, though uncommonly. Indigo is prepared by immersing the leaves and branches of the plant in water in a container and allowing them to ferment. The sludge which collects at the bottom is collected and pressed into cakes, constituting the dye. Cotton cloth is immersed in large clay vats containing the dye and then hung out for the colour to develop. This is repeated several times in order to enhance the colour. The cloth is finally brushed with a flour paste, dried and beaten with mallets to make it shiny. Lengths of this cloth are used as a traditional dress for festivities in Dhofar and Yemen (Wilson 1985).

Lawsonia inermis (*henna*) (Figure 11.3: D) is a large shrub, commonly cultivated in date-palm groves and gardens. The leaves yield an orange-red dye which is used for dying cloth, but more commonly as a cosmetic for dying hair and staining the palms of the hands and soles of the feet.

Limonium axillare (*qataf*) is a common plant of coastal regions distributed throughout Arabia. The roots and stems are used in Dhofar for dying ropes and fishing nets. Roots and stems are ground with water and made up into a paste which is used directly on the material to be dyed, or ground material is boiled in water and the cloth immersed and allowed to simmer for some time. The dye apparently makes the cotton nets stronger and protects against the deleterious effects of sea water (Miller & Morris 1988).

11.5 Perfume And Cosmetic Plants

The use of incense and perfumes in the ancient Near East dates from over four thousand years ago, when they were used in temples for religious rituals. Incense was also employed in healing the sick, in exorcisms, and in festivities and other social events (Groom 1981, Morris 1984). Frankincense was very highly regarded by the ancient Romans and in great demand. By the first millennium BC, with the domestication of the camel, overland trade routes became established between southern Arabia and Palestine, and frankincense, along with other items of trade, was exported from Dhofar and the Hadhramaut as well as from Eritrea and Somalia.

The earliest recorded account of the use of frankincense and myrrh from Arabia by the ancient Greeks comes from Herodotus (Groom 1981), suggesting that by 500 BC a well established trade existed between southern Arabia and Greece. To ascertain the origin of frankincense, Alexander the Greek (356 to 323 BC), sent Anaxicrates to southern Arabia and his account was recorded by Theophrastus (*c*. 295 BC):

> *"The trees of frankincense and myrrh grow partly in the mountains, partly on private estates at the foot of the mountains; wherefore some are under cultivation, others not; the mountains, they say, are lofty, forest-covered and subject to snow, and rivers from them flow down into the plain. The frankincense tree, it is said, is not tall, about five cubits and is much branched ... The myrrh tree is said to be still much smaller in stature and more bushy."* (quoted in Groom 1981).

The incense trade expanded considerably in the first two centuries (AD) and Dhofar became the main region for the production of Arabian frankincense. By the fourth century, with the economic decline of the Roman Empire and the spread of Christianity, the frankincense and myrrh trade declined.

In the Arabian Peninsula today perfumes are widely used and incense (*bakhor*), made from various preparations of a mixture of frankincense, sandal-wood, aloes wood, ambergris, musk oil and molluscan shells, is burnt for most social events and festivities. Even though cheap imported perfumes and ready-to-use incense have largely taken over from traditional perfumes, frankincense, myrrh and rose water are still highly regarded for their fragrance and traditional value. Oman is still the chief producer of frankincense in Arabia, though the demand is now considerably less than in former times.

Frankincense is obtained from *Boswellia sacra* (Burseraceae, Figure 11.2), which grows in the upper plateau on the northern face of the escarpment mountains of Dhofar. Some trees also grow in the wadis which flow out onto the coastal side of the mountains. Classical and modern accounts of the harvesting and handling of frankincense in southern Arabia were given by Theophrastus (Hort 1916), Pliny (Groom 1981), Bent (Groom 1981) and Thomas (1932). Pliny's account suggests that by the first century the demand for frankincense had increased so much that two harvests were being extracted each year. In his book on frankincense and myrrh Groom (1981) gives detailed historical and modern accounts of the distribution, harvest and trade of frankincense and other incenses from southern Arabia. The account of the harvesting of frankincense given by the explorer Bertram Thomas (1932) more or less describes the method still in use today:

> *"The tree begins to bear in its third or fourth year. The collectors, women as well as men, come to make slight incisions here and there in the low and stout branches with a special knife. A gum exudes at these points and hardens into large lozenge-shaped tears of resinous substance which is known as frankincense (liban). After ten days the drops are large enough for collection, and the tree will continue to yield from these old incisions opened as necessary at intervals of ten days for a further period of five months. After this the tree dries up and is left to recover, the period varying from six months to two years according to its condition. Collection of the 'liban' is made chiefly during the monsoon months. It is stored in the mountain caves until the winter, when it is sent down to the ports for export, for no country craft is put to sea during the gales of the summer south-west monsoon. This delay enables the product to dry well, though normally it is ready for export in from ten to twenty days after collection."*

In Dhofar the best quality frankincense, called *laqat,* is collected from the ground after it has fallen and dried for sometime. This is pale yellow in colour and very clean. The lesser qualities are scraped from the trees and are reddish in colour (Mandaville 1977). Frankincense is burnt in special frankincense burners (Figure 11.2: D), and used widely in public and private places. Its use is particularly special in the Sultanate of Oman.

The fragrance of roses has always held a place of high esteem in Islamic culture. Rose water and rose oil is used in medicinal preparations, to sanctify mosques, on religious occasions and is one of the most favoured perfumes. The oldest historical evidence of rose cultivation comes from the Minoan civilization, from the remains of a fresco from the palace at Knossos, Crete (*c.* 1500 BC). This was identified as *Rosa damascena*, the damask rose, which is one of the most important roses for the rose oil distillation industry. The parentage of *R. damascena* is not fully known, but it is believed to be of hybrid origin, with *R. gallica* being one of its most probable parents (Widrlechner 1981).

During the 8th and 9th centuries, Greek texts on medicine, botany, and alchemy were translated and became available to Arab herbalists and physicians. Arab scholars, such as Al Jawbari (13th century) and Al Dimashqi (d. 1327) developed the techniques of distillation, and the production of perfumes and scented oils became a flourishing industry during the Islamic era (Al Hassan & Hill 1986 and references therein). With the establishment of caravan and pilgrimage routes between Arabia, north Africa and Asia, the knowledge of distillation techniques became more widely available, and by the 13th century Iran became one of the principal centres of rose oil production (Dawood 1967). The use of rose oil and rose water spread widely and was eventually introduced by a Turkish merchant into the Ottoman province of Eastern Roumelia (now part of Bulgaria) around the end of the 17th century. This area is now one of the world's largest rose oil producers (Widrlechner 1981). In addition to the production of rose water the Islamic perfume industry included the manufacture of preparations such as musk, *ban* (a perfume from *Moringa oleifera*), *ghalia* (a perfume from musk and ambergris) and others.

Today, rose water is produced in Oman and Saudi Arabia. Rose oil distilled from roses at Taif in Saudi Arabia is considered of a good quality and is highly priced (1 g of rose oil is priced at 6,500-7,500 US$ at 1997 prices). Omani rose water is not made in large quantities since the demand for it is not great and it is of poor quality. Rose water made from rose oil exported from India, Turkey or Iran is cheaply available throughout the Gulf countries and Saudi Arabia.

The origin of rose cultivation in the northern mountains of Oman and the southwestern mountains of Saudi Arabia is not fully known. Roses were presumably brought in from Syria to Taif to produce rose water for the use of pilgrims to Mecca and Medina (Zwemer 1986). In Oman roses have been cultivated for at least a couple of centuries and are believed to have been imported by the Persians. At present, small areas (*c.* 50 x 50 m) of rose bushes are cultivated in several villages in the central massif of the northern mountains. Locally the rose is called the 'Shiraz rose'; it is highly scented and similar to the damask rose in appearance. The method employed today for rose oil extraction is very simple and crude and gives the rose water a strong smoky smell. Fresh blooms are collected at dawn and stored in dark rooms. A handful of flowers are placed in a metal bowl, on top of which a smaller bowl is placed. Both are lowered into a narrow chimney over a wood fire. Another bowl, which is about the size of the opening of the chimney, is filled with water and placed on the top of the chimney, blocking its opening. As the vapours from the flowers rise, they condense and collect in the smaller inner bowl. The locally produced rose water is used as perfume, as a flavouring for tea and is added to a popular local sweet, 'halwa', to produce a dark coloured, characteristically smoky flavoured halwa.

In addition to frankincense and rose water, myrtle and jasmine are also commonly used. The leaves of myrtle (*Myrtus communis*) and flowers of jasmine

(*Jasminum* spp.) are often placed among clothes for their perfume. Other plants used in a similar way include leaves and flowers of *Lavandula* spp. and *Ocimum sanctum*.

Since imported cosmetics are cheap and freely available, the use of plants for adornment has declined. However, a few plants still retain their popularity. One of the most popular used throughout Arabia is *henna* (*Lawsonia inermis*) (Figure 11.3: D). The dried, powdered leaves are made into a paste with water to which oils, lime juice and other ingredients may be added to enhance the colour and perfume. The paste is then applied and left for some time to give an orange-red dye. Palms and backs of hands and feet are stained in intricate patterns for religious festivities, weddings and birthdays. *Henna* is also popularly used as a hair colour and for dyeing cloth. *Henna* plants are cultivated in most homes and villages, and *henna* powder imported from India, Iran and Pakistan is also available.

Oils of certain plants are used as face lotions. In Oman and Yemen the seed oil of *Moringa peregrina* (Figure 11.5) is used as a skin lotion (Ghazanfar & Rechinger 1996). The powdered leaves of *Ziziphus spina-christi* (Figure 11.3: A) are used as a shampoo and are believed to leave the hair clean and lustrous. Hair is often lacquered with frankincense. Roots and stems of *Salvadora persica* (Figure 11.3: C) are used widely throughout Arabia as teeth cleaning sticks.

Amongst cosmetics *kohl* is perhaps one of the best known Muslim adornments, widely used by both men and women in Arabia. *Kohl* is made of finely powdered antimony sulfide which is applied to the eyes with a thin applicator. *Kohl* is often made of soot collected from burning frankincense. Facial decoration with black or red dyes is also common in some parts of the Peninsula, especially in rural areas. In Yemen, a black paste, *khidab*, used for facial decoration, is made traditionally with oak galls and minerals such as copper oxide, potash and sal ammoniac (Schönig 1995). In central and northern Oman, paste made from sandal-wood oil and powdered *wers* (*Memecylon tinctorium*) is applied on the face to give a yellow complexion.

11.6 Plants used for Building and Utilitarian Objects

The use of native plant material for building and making household objects has declined during the last two decades in parallel with the decline in use of native plants for medicinal and other purposes. Houses are now popularly constructed with concrete bricks and tin roofs, and plastic and metal has replaced natural fibres. However, the dwarf-palm, *Nannorrhops ritchieana*, the dum-palm, *Hyphenae thebaica*, *Acacia tortilis* and an assortment of grasses and branches from various shrubs and trees are still used for building houses and for making utilitarian objects (Figure 11.1).

In the Yemeni Tihamah traditional houses are still built with the trunks and branches of *Tamarix aphylla*, *T. arabica* (*athl*), *Ziziphus spina-christi* (*sidr*) and *Cadaba rotundifolia* (*qadab*), using date-palm as scaffolding. Leaves of the dum

Figure 11.5 *Moringa peregrina* (*shu'*) seed oil is used for medicinal and cosmetic purposes. (A) Vegetative shoot. (B) Flowering shoot. (C) Flower. (D) Capsule.

palm and grasses such as *Lasiuris scindicus* (*thumam*) mixed with cow and donkey dung are used for the structure. Branches of *Indiofera oblongifolia* (*hasah*) are used for lathing and insulation and *Tephrosia purpurea* (*sanfah*) for protection against termites. Ropes are made from palm-fronds and the grass *Desmostachya bipinnata* (*halfa*) (Stone 1985).

Houses were commonly built in Oman from date-palm fronds and logs, especially in the northern regions. The houses were constructed with *c*. 4 m long panels (*da'an*) made from frond-stalks tied together with coir string. Palm-fronds were also used to make wind-catching towers which were mounted on the house at the beginning of the hot season and taken down in winter (Costa 1985). It is now rare to see a palm-frond house, though enclosures for keeping goats and storage rooms made in this way are still in use. In the higher areas of the central massif of the northern mountains of Oman trunks and branches of *Juniperus excelsa* subsp. *polycarpos* are used as the roof supports for small stone huts. Wood from *Anogeissus dhofarica*, *Olea europaea* and *Blepharispermum hirtum* has been used for house building in southern Oman. Wood and branches of several species of trees and shrubs, such as *O. europaea*, *Periploca aphylla*, *Maytenus dhofarensis*, *Euclea schimperi*, *Acacia senegal* and *A. tortilis* are used for making pens for livestock, and dried grass such as *Cymbopogon schoenanthus* is used as a floor covering for newly born lambs. Recently in Oman furniture and household objects made from date-palm fronds are being sold in a drive to improve and encourage the manufacture of traditional crafts.

Basketry is still a popular craft in rural Arabia. The technique is based on making plaits which are woven from the centre outwards to the rim, each round being joined to the next by a cord (Stone 1985). Young fronds which are still green and supple are used. They may be dyed and then woven to make colourful patterns in baskets and mats. Mats, rectangular and round, are made from lengths of plaited date-palm fronds stitched together. Ropes are made by twisting and rolling strips of the fronds. Flowering stalks and frond-stalks are used for making brushes, brooms, bird cages and bird traps (Figure 11.1).

In former times the dwarf-palm, *Nannorrhops ritchieana*, was of great economic importance in Dhofar (for uses, see Miller & Morris 1988). It is still used in some parts of Dhofar and central Oman, but its use is now mainly limited to the production of items for tourists. One of the most popular items made from this palm are bowls for keeping goat and camel milk (Figure 11.1: J). They are rarely used today for their originally intended purpose, but are produced for the tourist trade. Strands of leaves are plaited and worked round to make a bowl, and then the plaits are stitched together and a piece of tanned goat skin is wetted and stretched over the base to about half the height of the bowl. The outside wall of the bowl is decorated with leather embroidery. Plaited leather straps are attached on either side for hanging. *Dracaena serrulata* was mostly used for making ropes and fish nets, since the rope was stronger and withstood the corrosive effects of the sea water better than that made from the date-palm.

Many other household objects such as bedposts, ladles, bowls, dishes, pipes, *kohl* applicators, loofah scourers, spears, arrows and walking sticks were formerly made from the wood of native plants. Such items have however been largely replaced by imported, household wares and are now rarely seen.

11.7 Fodder And Rangeland Plants

One of the most important uses of plants in the Arabian Peninsula is for livestock fodder, and it has been estimated that about 80% of the total area of the Peninsula is used as rangelands for domestic camels, cattle, donkeys, goats and sheep (Anon. 1982, Chaudhary 1989b). Plant utilization was formerly regulated by herding in selected areas on a rotational basis in order to let grazing areas regenerate. Grazing was also controlled in selected areas by the traditional *hima* system, where plants were cut for fodder when grazing was poor (Lancaster & Lancaster 1990). Within the last fifty years there have been great increases in livestock holdings and most rangelands are now over-utilized, with the result that palatable species do not have a chance to regenerate and unpalatable species tend to dominate. The current *status quo* is largely a result of the availability of supplemental fodder and water tankers (for further details of the effects of grazing, see discussion in section 12.3).

The carrying capacity of the Arabian rangelands depends largely on the more permanent, perennial vegetation, with fodder from annual plant species only available after rain. Under severe grazing palatable annual species are eaten before they can flower and set seed and young shoots of perennials are continuously browsed, affecting both growth and flowering. Ultimately regeneration is affected, resulting in the long-term degradation of the rangelands.

Perennial grass species of *Bromus*, *Cenchrus*, *Cynodon*, *Digitaria*, *Lasiurus*, *Panicum*, *Pennisetum*, *Paspalum*, *Sporobolus*, *Stipagrostis* and other genera are widely available throughout the Peninsula and are both grazed and cut for fodder. Rhodes grass is cultivated in Oman and Saudi Arabia as supplementary fodder. Amongst the shrubby vegetation, favoured fodder shrubs are *Atriplex leucoclada*, *Haloxylon salicornicum*, *Lycium shawii*, *Salsola* spp. and *Rhanterium eppaposum*.

Lopping for fodder, which can induce new growth if carried out carefully, is often carried out when natural grazing is poor. Favoured species are those of *Acacia* and *Prosopis*, favoured in particular for their nutritious pods and leaves. In Oman, the leaves and young shoots of *Ceratonia oreothauma* subsp. *oreothauma*, *Acacia tortilis*, *A. gerardii* and *A. nilotica* are commonly lopped. *Prosopis cineraria* is favoured for camels. Fresh twigs and leaves of *Anogeissus dhofarica*, *Avicennia marina*, *Dobera glabra*, *Olea europaea*, *Ormocarpum dhofarensis*, *Ziziphus spina-christi* and most leguminous shrubs and trees are also cut and used for fodder where they occur.

Goats are indiscriminate eaters and will graze and browse most species. However, they ignore species containing bitter or poisonous alkaloids such as *Rhazya stricta*, *Nerium oleander*, *Calotropis procera*, *Solanum incanum*, *Tephrosia spp.* and *Adenium obesum* and the abundance of these species in any area is a good indicator of overgrazing. Whereas goats and camels browse on spiny species such as those of *Acacia*, they tend to leave *Fagonia indica* and related spiny shrub species alone, and nibble at them only when grazing is poor. Cattle graze and browse on most palatable species and grasses, especially in the monsoon affected mountains of southern Arabia.

Native grazers and browsers such as gazelle and ibex are patchily distributed throughout Arabia, but do not appear to be a threat to vegetation in any way. In enclosures with native plants they are reported to eat *Calotropis procera*, *Convolvulus glomeratus*, *Ephedra foliata*, *Farsetia* spp., *Grewia erythraea*, *Hibiscus micranthus* and *Periploca aphylla*, amongst grasses, *Cymbopogon commutatus*, *Hyparrhenia hirta* and *Stipagrostis* spp., and amongst the annuals, *Argyrolobium* sp. and *Rhynchosia memnonia* (Robertson 1995). Gazelles are also recorded to eat lichens in the central desert of Oman (Hawksworth *et al.* 1984). Where the Arabian Oryx (*Oryx leucoryx*) have been reintroduced, such as in the central desert plains of Oman and in Saudi Arabia, their diet consists of the grasses *Chrysopogon* spp., *Cymbopogon schoenanthus, Lasiurus scindicus, Ochtochloa colona, Stipagrostis plumosa* and *S. sokotrana*, and shrubs and trees such as *Acacia tortilis* and *A. ehrenbergiana* (Spalton 1995).

11.8 Summary

The use of plants for the benefit of man has declined sharply in Arabia in recent years. Plastic and metal, synthetic dyes and perfumes and modern synthetic medicines have largely taken over from the traditional uses of plants. Only a few species, such as the date-palm, are still widely used for making utilitarian objects, and only a few plant species are still used medicinally. The main uses of medicinal plants are for curing colds and coughs, and for those diseases and conditions where modern treatment is prolonged and unsatisfactory. Despite the widespread availability of cheap synthetic perfumes, the use of traditional fragrances such as frankincense and rose water is still popular. Pasturing is still the most common form of animal husbandry, and favourite fodder plants are lopped regularly. The extensive and indiscriminate utilization of rangelands and fodder plants has resulted in widespread rangeland degradation.

CHAPTER 12

Diversity and Conservation

Martin Fisher, Shahina A Ghazanfar,
Shaukat A Chaudhary[1], *Philip J Seddon*[1], *E Fay Robertson*[1],
Samira Omar[2], *Jameel A Abbas*[3] *&Benno Böer*[4]

"To sum up, I have not collected more than 250 species over the whole of the Immamat of Muscat. The local people say that very shortly after the rains (which occur once or twice a year) the land is covered with flowers. In any case I am convinced that in this country, the most barren in the world, it would be difficult to find more than 500 species."
Aucher-Éloy, in 1838, after his botanical excursion to Oman (Jaubert 1843).

12.1 Introduction

Not all of the early travellers and plant collectors to Arabia shared Aucher-Éloy's pessimistic musings about the barrenness of the land and its impoverished diversity, although his estimate of the number of species in northern Oman was not far wrong (Ghazanfar 1996a). Theophrastus, probably using accounts brought home by Greek sailors, wrote of the Yemeni highlands in 295 BC that "... the mountains, they say, are lofty, forest-covered and subject to snow, and rivers from

Sections [1]12.4.5 Saudi Arabia, [2]12.4.2 Kuwait, [3]12.4.1 Bahrain,
[4]12.4.6 United Arab Emirates

S. A. Ghazanfar and M. Fisher (eds.), Vegetation of the Arabian Peninsula, 265-302.

them flow down into the plain..." (quoted in Groom 1981, from Hort 1916). Niebuhr (in the introduction to Forsskål 1775) wrote of the valley of Surdud in Yemen that "... surrounded by mountains, and profiting from cool climate and abundant water, [it] was exuberantly rich in plants... ". In addition to these pleasant impressions, the great trade between southern Arabia and Europe in frankincense, myrrh and spices gained the Arabs of southern Arabia a reputation for great wealth, and their country became known as *Arabia Felix* (Groom 1981: pp. 9-11), providing the title for Bertram Thomas' well-known work (1932) of the same name.

These earlier travellers were of course correct in their time and for the corners of Arabia about which they were writing, but the landscape of the Peninsula has greatly changed during the twentieth century. The Greek sailors, Niebuhr, Forsskål and Aucher-Éloy might not recognise the same places were they able to go back today. The role of this chapter is to describe the diversity of plants in the Arabian Peninsula as we see it in the twilight of the twentieth century, and to summarise current efforts to preserve both plants and landscapes. To this end, we provide an overview of the patterns of species richness and endemism in Arabia, outline the current threats to diversity and vegetation cover, and then provide an overview of the protected areas, relevant legislation and organisations in each country of the Peninsula.

12.2 Species Richness and Endemism

Although the flora of the Arabian Peninsula is still under explored, it has received increased attention during the last two decades, and a number of regional floras and checklists have now been published (Table 12.1). From these works it is now apparent that the flora of Arabia consists of a rich and diverse grouping of plants. This is a result of past plant migrations from the Ethiopian, Palaearctic and Oriental regions, possibly during more pluvial climatic conditions, combined with the climatic isolation of the montane floras following the development of the present arid regime (see section 4.2).

Both the greatest number of species and the majority of the Peninsula's endemic species occur in the mountains, which presumably provided a climatic refuge during the alternating arid and pluvial phases of the Pleistocene for those species requiring mesic conditions. The isolation of these populations by the intervening desert plains has led to vicariant evolution. Thus a number of palaeo-African and palaeo-Indo-Malesian relict species have disjunct distributions, and there are several relatively small areas with concentrations of endemic species in the western and southwestern mountains of Saudi Arabia and Yemen and in the northern and southern mountains of Oman (Ghazanfar 1994b, Miller & Nyberg 1991, see also section 4.2.3).

Table 12.1 Regional floras and checklists for the Arabian Peninsula in English.

Region/Country	Reference
Arabian Penisula	Cope (1985, grasses), Miller and Cope (1996, vol. 1, vascular cryptogams, gymnosperms, angiosperms, Myricaceae to Neuradaceae, classification follows Engler & Prantl), Schwartz (1939)
Bahrain	Cornes and Cornes (1989), Phillips (1988)
Kuwait	Al-Rawi (1987), Boulos and Al Dosari (1994), Daoud (1985), Shuaib (1995)
Oman	Ghazanfar (1992a), Mandaville (1977, 1985, northern Oman), Miller and Morris (1988, Dhofar), Radcliffe-Smith (1980, Dhofar)
Qatar	Batanouny (1981), El Amin (1983)
Saudi Arabia	Chaudhary (1989a, grasses), Chaudhary and Cope (1983, grasses), Collenette (1985), Mandaville (1990, eastern Saudi Arabia), Migahid (1988a, 1988b)
United Arab Emirates	Western (1989), Western, Jongbloed and Böer (unpubl.)
Yemen	Balfour (1882a, 1882b, 1884a, 1884b, 1888, Socotra), Blatter (1914-1916), Boulos (1988), Gabali and Al-Gifri (1990, S Yemen), Gabali and Al-Gifri (1991, Hadhramaut), Hepper and Friis (1994), Wood (1983b, 1997)

12.2.1 Species Richness

The flora of the Arabian Peninsula consists of *c*. 3,500 species of vascular plants. Of these, there are *c*. 3,440 species of angiosperms, 12 species of gymnosperms and *c*. 56 vascular cryptogams (Table 12.2). The vascular cryptogams consist of a single species of *Psilotum*, (Psilotophyta, whisk ferns), 4 species of *Selaginella* (Microphyllophyta, spike mosses), a single species of *Equisetum* (Arthrophyta, horsetails) and *c*. 50 species of ferns (Pterophyta). Only two genera of gymnosperms occur, *Juniperus* and *Ephedra*, distributed mainly in Saudi Arabia, Oman and Yemen. *Ephedra foliata* is the commonest gymnosperm species, distributed widely in the desert plains. The highest number of angiosperm species is found in Yemen (including Socotra), followed by Saudi Arabia, Oman, the UAE, Kuwait, Qatar and Bahrain (Table 12.3). Since a unified checklist for the recently united Yemen is not yet available, statistics for both species richness and endemism still follow the former north-south division.

Table 12.2. Number of families, genera and species in the flora of the Arabian Peninsula (adapted from Miller & Cope 1996).

	Families	Genera	Species
Vascular cryptogams	18	29	56
Gymnosperms	2	4	12
Angiosperms	*c*. 144	*c*. 1,100	*c*. 3,440

Table 12.3. Species richness in the Arabian Peninsula by country.

Country	Area (km^2)	Approximate number of species
Bahrain	660	195
Kuwait	17,818	374
Oman	270,000	1,174
Qatar	11,000	220
Saudi Arabia	2,400,000	1,800
United Arab Emirates	75,000	600
Yemen, North	190,000	1,370
Yemen, South	287,000	960

The majority of species are distributed in the families Poaceae (*c.* 450 spp.), Fabaceae (*c.* 320 spp.), Asteraceae (*c.* 300 spp.), Lamiaceae (*c.* 120 spp.), Euphorbiaceae (*c.* 120 spp.), Boraginaceae (*c.* 120 spp.), Brassicaceae (*c.* 115 spp.), Asclepiadaceae (*c.* 110 spp.), Acanthaceae (*c.* 110 spp.), and Scrophulariaceae (*c.* 100 spp.), and the largest genera are ***Euphorbia*** (67 spp.), ***Heliotropium*** (46 spp.), ***Cyperus*** (41 spp.), ***Indigofera*** (37 spp.), ***Caralluma*** (30 spp.) and ***Convolvulus*** (30 spp.) (Miller & Cope 1996). Most of the genera are monospecific and almost half the number of families are monogeneric. An analysis of species frequency per genus for the flora of Oman (Figure 12.1) is more or less applicable to the entire Arabian flora.

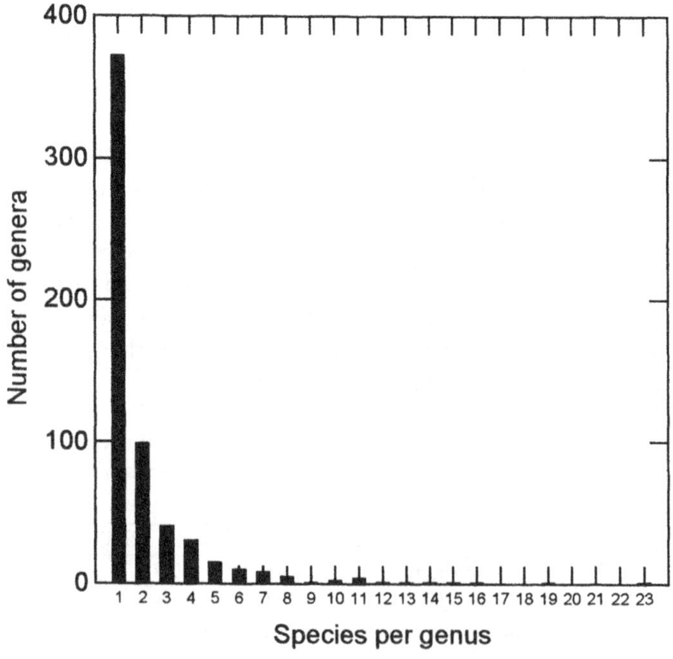

Figure 12.1 Species frequency per genus for the flora of Oman (after Ghazanfar 1992b).

Table 12.4 Endemism in the countries of the Arabian Peninsula (after Miller & Nyberg 1991). The table does not include regional endemics; i.e. those species which are found in more than one country.

Country	Endemic species (% of total species)	Endemic genera
Yemen, N	60 (4%)	Asteraceae: *Centaurothamnus*
Yemen, S	80 (8%)	Amaranthaceae: *Saltia* Lamiaceae: *Isoleucas*
Yemen, Socotra	240 (30%)	Acanthaceae: *Ballochia*, *Tricocalyx* Asclepiadaceae: *Duvaliandra*, *Socotranthus* Brassicaceae: *Lachnocapsa* Caryophyllaceae: *Haya* Cucurbitaceae: *Dendrosicyos* Rubiaceae: *Placoda* Umbelliferae: *Nirarathamnos*
Oman	75 (7%)	Asclepiadaceae: *Cibirhiza* Capparaceae: *Dhofaria* Caryophyllaceae: *Xerotia*
Saudi Arabia	35 (2%)	Asteraceae: *Centaurothamnus* Brassicaceae: *Dolichorhynchus*

12.2.2 ENDEMISM

12.2.2.1 Endemic Genera and Species

There are approximately 600 endemic species in the Arabian Peninsula, with *c.* 360 species on the mainland, of which *c.* 110 are regional endemics, and *c.* 240 species in Socotra (Table 12.4). Endemism is highest in the archipelago of Socotra, followed by the monsoon-affected escarpment mountains of southeast Yemen and Dhofar, and the mountains of the western escarpments of Saudi Arabia and northern Yemen. There is a lesser degree of endemism in the northern mountains of Oman. There are no endemic species in Bahrain, Kuwait, Qatar, or the UAE, but the Emirates shares three regionally endemic species with Oman, distributed in the foothills and wadis of the Hajar mountains.

70% of the endemic species of the Arabian Peninsula and Socotra are distributed in 15 families (Table 12.5). Many of the endemic species are succulents and form characteristic components of the vegetation of southern Arabia (see also Chapter 4). The family Asclepiadaceae is richest in endemics, with most distributed in five succulent genera: *Caralluma* (30 spp.) with 21 endemics, *Duvalia* (4 spp.) with 2 endemics, *Echidnopsis* (4 spp.) with 2 endemics, *Huernia* (6 spp.), all endemics, and *Rhytidocaulon* (4 spp.), all endemics. Similarly in the family Euphorbiaceae there are a large number of succulent endemics in the genus *Euphorbia*. Endemism is also high in the succulent family Aloaceae.

Table 12.5. Plant families in Arabia with the highest endemism (adapted from Miller & Nyberg 1991)

Families	Approximate total number of species	Number of endemic species (% of total)
Asclepiadaceae	103	55 (53%)
Acanthaceae	104	45 (43%)
Liliaceae (incl. Aloaceae)	84	31 (37%)
Scrophulariaceae	96	35 (36%)
Lamiaceae	122	43 (35%)
Boraginaceae	119	37 (31%)
Euphorbiaceae	121	37 (31%)
Asteraceae	297	57 (19%)
Caryophyllaceae	86	11 (13%)
Leguminosae	313	28 (9%)
Brassicaceae	114	9 (8%)
Chenopodiaceae	77	5 (6%)
Poaceae	457	18 (4%)

12.2.2.2 Centres of Diversity and Endemism

Although the Arabian Peninsula does not qualify as a regional centre of endemism, the Somalia-Masai regional centre of endemism extends from eastern Africa into southern and southwestern Arabia, and the Afromontane regional centre of endemism extends from the highlands of eastern Africa into the mountains of southwestern Arabia (Figure 4.3). In addition, two subzones of the Saharo-Sindian regional zone, the Arabian regional subzone and the Nubo-Sindian local centre of endemism, are found in central Arabia and northern Oman respectively (White & Léonard 1991). In the Arabian Peninsula there are four local centres of diversity and endemism of the Somalia-Masai, two of the Afromontane and four of the Saharo-Sindian phytochoria (Boulos *et al.* 1994, Ghazanfar 1992b, in press-b, Miller & Nyberg 1991), the location, vegetation type, species richness and endemism of which are given below.

Somalia-Masai Regional Centre of Endemism

The escarpment mountains of Dhofar and the Mahra region (Dhofar Fog Oasis)
Location: Dhofar mountains (Jebel Qamr, Jebel Qara, Jebel Samhan), eastern part of the Mahra region, the surrounding Nejd desert.
Altitude: 0-2,100 m.
Typical vegetation: ***Anogeissus-Acacia-Commiphora*** deciduous woodland and semi-deciduous shrubs (***Olea europaea***, ***Dodonaea viscosa***, ***Carissa edulis***, ***Rhus somalensis***), succulents (***Aloe***, ***Caralluma***, ***Euphorbia***, ***Adenium***, ***Cissus***) and semi-desert grassland.
Species richness: *c.* 850 spp.
Endemism: *c.* 90 spp., 2 endemic genera, 10.5% endemism.

Hadhramaut (Yemen)

Location: Southern Yemen, Hadramaut and western part of the Mahra region.
Altitude: 1,000-2,200 m.
Typical vegetation: *Acacia-Commiphora* deciduous woodland and evergreen and semi-deciduous shrubs (*Olea europaea*, *Dodonaea viscosa*, *Carissa edulis*, *Euphorbia*), succulents (*Aloe*, *Caralluma*, *Euphorbia*, *Adenium*, *Cissus*).
Species richness: *c.* 700 spp.
Endemism: *c.* 40 spp., 5.7% endemism.

Jebel 'Ureys (Yemen)

Location: Southern coast of Yemen (c. 150 km east of Aden).
Altitude: 0-1,700 m.
Typical vegetation: Deciduous shrubland, succulent shrubland (dominated by *Euphorbia balsamifera* on the seaward facing slopes), and semi-desert grassland, succulents (*Kleinia deflersii*).
Species richness: *c.* 500 spp.
Endemism: *c.* 15 spp., c. 3% endemism.

Socotra Archipelago (Yemen)

Location: Northern part of the Indian Ocean, east of Somalia.
Altitude: Socotra, 0-1,519 m; 'Abd al Kuri, 0-850 m; Semhan, 0-779 m; Darsa 0-357 m.
Typical vegetation: Open deciduous shrubland on the coastal plains (*Croton socotranus*, *Euphorbia arbuscula*, *Dendrosicyos socotranus*, *Ziziphus spina-christi*) and lower slopes of the mountains (*Croton socotranus*, *Jatropha unicostata*), submontane, semi-deciduous thicket (*Rhus thyrsiflora*, *Buxus hildebrandtii*, *Carphalea obovata* and *Croton* spp.), and grassland and rock vegetation (*Rhus thyrsiflora*, *Cephalocroton socotranus*, *Allophylus rhoidiphyllus*, *Dracaena cinnabari*).
Species richness: 790 spp.
Endemism: *c.* 240 spp, 9 genera, 1 near-endemic family (Dichramaceae), 30 % endemism.

Afromontane Regional Centre of Endemism

Highlands of southwestern Arabia (Saudi Arabia and Yemen)

Location: Taif southwards to Jebel Sawdah and east to Mukayras.
Altitude: 200-3,760 m.
Typical vegetation: *Acacia-Commiphora* deciduous shrubland, evergreen shrubland (*Barbeya oleoides*, *Carissa edulis*, *Dodonaea viscosa*, *Euclea schimperi*, arborescent *Euphorbia*) and *Juniperus procera* woodland, *Juniperus procera-Erica arborea* woodland (*Kniphofia sumarae*, *Helichrysum arwae*).
Species richness: *c.* 2,000 spp.
Endemism: *c.* 170 spp., 2 genera, 8.5% endemism.

Hajayrah mountains (Yemen)

Location: Mountains south of Taiz.
Altitude: 2,000-3,000 m.
Typical vegetation: Similar to highlands of southwestern Arabia.
Species richness: Unknown, probably *c.* 600 spp.
Endemism: 99 spp., incl. 8 restricted to this area, *c.* 12% endemism.

Saharo-Sindian Regional Zone

Northern Hijaz (Saudi Arabia)

Location: Jebel Lawz, Jebel Dibbagh, Jebel Radhwa, Jebel Shada and the southern fringe of Harrat ar Rahah (northwestern Saudi Arabia).
Altitude: 600-2,300 m.
Typical vegetation: Luxuriant vegetation along wadis with perennial water (*Hyphaene thebaica, Nerium oleander, Phragmites australis*). Open *Acacia* shrubland at lower altitudes (*A. raddiana, A. tortilis, Retama raetam*). Dwarf shrubland above *c.* 800 m (*Artemisia sieberi, Ferula* spp., *Pistacia khinjuk*, *Prunus korschinskyi*). Semi-evergreen *Juniperus phoenicea* bushland above 1,400 m on some isolated peaks. Rich in Mediterranean annual and perennial herbs and shrubs.
Species richness: Unknown; one of the two richest areas in extra-tropical Arabian endemics.
Endemism: >5 spp., including one endemic genus. The endemics are mostly high altitude species. % endemism unknown.

Harrat al Harrah

Location: Northern Saudi Arabia.
Altitude: 500-1,000 m.
Typical vegetation: Dwarf shrubs (*Haloxylon salicornicum, Salsola* spp., *Artemisia* spp., *Achillea fragrantissima*, *Zilla spinosa*), sand inhabiting shrubs (*Haloxylon persicum, Calligonum comosum*) and perennial grasses (*Stipagrostis* spp.).
Species richness: >290 spp.
Endemism: None.

Jiddat al Harasees (Oman)

Location: Central Oman.
Altitude: *c.* 150 m
Typical vegetation: Open *Acacia* scrub (*Acacia tortilis*, *A. ehrenbergiana*), drought- deciduous shrubland (*Ochradenus harsusiticus*, *Fagonia*, *Convolvulus oppositifolia*, *Nannorhops ritcheana*, *Hyoscyamus gallgheri*).
Species richness: *c.* 200 spp.
Endemism: 12 spp., 6% endemism.

Hajar and Musandam mountains (northern Oman)

Location: Western and Eastern mountains of northern Oman (Jebel al Akhdar, Jebel Aswad, Jebel Bani Jabir), Musandam mountains (Jebel Harim).
Altitude: 300-3,009 m.
Typical vegetation: Western Hajar: Evergreen *Olea-Monotheca-Dodonaea* shrubland, (with *Ebenus stellatus*, *Euryops arabicus*, *Helianthemum*, *Teucrium*) open *Juniperus* woodland (*Juniperus excelsa* subsp. *polycarpos*, *Daphne mucronata*, *Berberis*, *Lonicera aucheri*). Eastern Hajar: Very open *Ceratonia oreothauma-Ziziphus hajarensis* woodland (with *Acacia* ssp., *Prunus arabica*), semi-deciduous scrub (*Ebenus stellatus*, *Euryops arabicus*, *Helianthemum* spp.). Musandam: *Artemisia* scrub and open semi-deciduous bushland (*Acacia tortilis*, *Moringa peregrina, Prunus arabica, Ziziphyus spina-christi*)
Species richness: *c.* 600 spp.
Endemism: *c.* 25 spp. (one of the two richest areas in extra-tropical Arabian endemics), 4.2% endemism.

12.3 Threats to Diversity and Vegetation Cover

Threats to plant species diversity and vegetation cover and biomass in Arabia are inextricably linked with graziers and the fate of the rangelands. Of all the threats to diversity and vegetation cover, that of 'overgrazing' is undoubtedly the greatest, and the degradation of rangelands at the mouths of livestock is a problem that needs to be addressed by pastoralists and conservationists alike. How severe is the situation? A detailed survey of the rangelands of Oman in 1981 concluded that "The rangelands of northern and central Oman are in a highly degraded state.." and "In southern Oman, the condition of the rangelands, particularly in Jebel Qara, is critical" (Anon. 1982). This is a statement that probably applied to a major part of the Peninsula in 1982, and certainly applies to an even greater degree as we write in 1997. In the 1970's 85% of the rangelands of Saudi Arabia were already estimated to be in a severely degraded state (Kingery 1971), and in 1985 it was estimated that 75% of the whole country was seriously eroded due to the impoverishment of the natural vegetation (Anon. 1985b).

Detailed, incontrovertible evidence for these points of view is not available, since there have been no long-term studies of rangeland quality in the region, but the evidence is there for all to see: the supplementation of the needs of large herds of domestic livestock by the use of artificial feed and water tankers during droughts, increasing numbers of feral donkeys released from toil by the acquisition of motorised transport, the dominance of plant communities by unpalatable species (see section 11.7), widespread ageing of woodlands with few if any signs of regeneration in many areas, and the surface crust of the desert plains permanently broken by ungulate hooves.

A number of recent comparative surveys indicate the general effect of protecting rangelands from grazing. After 11 years of protection, an exclosure in Abu Dhabi with mixed gravel plain and sand dune substrata had increased plant cover on both substrata compared to adjacent grazed areas, with the fodder grass *Stipogrostis plumosa* showing the greatest increase, and with decreased cover of unpalatable species such as *Zygophyllum mandavillei* (Oatham *et al.* 1995). In another exclosure in Abu Dhabi, measurable increases in cover compared to adjacent grazed areas occurred after only two years, again with increases in fodder species (Böer & Norton 1996). Both plant biomass and diversity was greater in an area of the Asir highlands following five years of protection from grazing, compared with adjacent areas estimated to receive three visits by an average of one hundred sheep and goats per week (Abulfatih *et al.* 1989). In another area of the Asir highlands, the protected areas of Hima Sabihah had a greater cover of palatable grass species and lower cover of unpalatable species such as *Asphodelus fistulosus* (Hajar 1993). An exclosure plot, designed to exclude camels but permit entry by gazelle, on sandy substratum in a sheep- and goat-free reserve in northern Saudi Arabia showed significant increase in perennial plant cover, particularly for palatable grasses, within three years of protection (van Heezik & Seddon 1995). However, exclosure plots on packed silt and on clay substrata in the same reserve showed fewer differences compared with adjacent grazed areas (van Heezik & Seddon 1996).

Taking a wider perspective, what might we expect the overall effect of grazing on vegetation and soils to be? A global overview based on 236 studies, including arid and semi-arid sites in Africa and the Middle East (Milchunas & Lauenroth 1993), found that changes in species composition induced by grazing are proportionally greater on lands with higher productivity, with longer histories of grazing and with higher levels of grazing, in that order of importance. The relationships between grazing parameters and changes in soil organic matter and nitrogen are more complex, with no predictable relationship with changes in species composition. In addition, there is evidence in areas of low herbivore consumption for the controversial (Dyer *et al.* 1993, Painter & Belsky 1993) concepts of overcompensation by grazed plants and herbivore optimisation of plant productivity (i.e. grazing is beneficial to plants and increases community productivity). This influence is not great, though higher in sites with longer histories of grazing and low productivity. The main lessons for Arabia are (1) that the greatest effect of grazing on species composition is likely to be seen in areas of greater productivity, such as the mountains, (2) that increases in grazing pressure will decrease species diversity, and (3) that changes in species diversity cannot necessarily be used to indicate effects of grazing on soil quality. With these points in mind, it is perhaps not a coincidence that one of the most extreme affects of grazing pressure in Arabia can be seen in the mountains of Dhofar in Oman, where serious overstocking with camels, cattle and goats (Anon. 1982) has completely denuded the landscape of what, ironically, is one of the most important 'hot spots' of diversity in Arabia.

Whilst there appears to be no politically or socially acceptable answer at the moment to the 'overgrazing problem', it must be emphasized that it is not the complete elimination, but rather the reduction of grazing pressure that is required. Ungulate grazing has undoubtedly been part of the evolution of desert rangelands, but it is the degree of the current grazing pressure that is overwhelming them. At least part of the answer might lie with the traditional *hima* or *mahjur* system of communal property, whereby areas were lain aside locally as fallow lands and as areas for specific uses such as beekeeping, grass cutting or woodland protection (Draz 1969, 1978, Ghanem & Eighmy 1984, Kessler 1995, Lancaster & Lancaster 1990, Shoup 1990). Several recent conservation reports for specific areas have drawn up plans for the combination of the traditional *hima* system with modern planning for conservation areas (Abualjadayel *et al.* 1991, Ady *et al.* 1995, Ady *et al.* 1994, Kamel *et al.* 1989). In a further extension of this idea a recent land use study in Jebel Dhofar (Anon. 1995) recommended an integrated and holistic approach for managing the rangelands. In this proposed scheme local, self-defined groups of pastoralists would undertake to manage their stock and rangeland so that it recovers ecologically, and in return, the government would undertake to provided legal, technical and physical support, with initial short-term financial assistance.

Extreme grazing pressure may be the most ubiquitous and pernicious threat to diversity and vegetation cover in Arabia, but it is not the only one. Accounts for individual countries (see section 12.4) identify a number of recurring threats to nature reserves that are representative of the threats facing plant life in all areas of the Peninsula: impact of recreational activities, cutting of firewood, off-road driving, invasive alien plant species, military activities, mining, and in coastal regions, oil pollution and sewage effluent. In a sense both the flora and fauna of Arabia are suffering from 'development' in all of its forms, including urban expansion, road building and general habitat degradation. These problems came with the rapid economic expansion that began in Saudi Arabia and UAE around the middle of the century and in Oman in the early 1970s, and were exacerbated by changes in traditional grazing practices and the abandonment of many *hima* areas, in the case of Saudi Arabia by a Royal Decree which opened lands to free grazing.

Finally, the hostilities of the Gulf War had a significant environmental impact on the ecosystems of Kuwait (Al-Houty *et al.* 1993, Anon. 1992: pp. 455-463, Omar *et al.* 1995). Jal az Zor National Park and the proposed Wadi al Battin National Park and other designated conservation areas were badly damaged by troop emplacements, military manoeuvres, tank activities and mines. In addition, the formation of oil lakes destroyed the vegetation in some areas and may potentially have future impacts on the soil.

12.4 Conservation

The pace of the establishment of nature conservation areas in the Arabian Peninsula is variable from country to country, with most of the protected areas having been established within the last 15 years, and with all countries still actively considering the establishment of new reserves. The largest countries with active conservation policies, Oman and Saudi Arabia, have both produced detailed study documents containing proposals for systems of nature conservation areas (Child & Grainger 1990, Clarke 1986).

Most reserves have been created for the protection and/or reintroduction of focal animal species rather than for the protection of natural vegetation *per se*. Nevertheless, all of the established reserves and protected areas afford some sort of protection to the flora, though to varying degrees, largely dependent on whether the areas are fenced and/or adequately patrolled to control grazing incursions. Some of the areas with the highest plant diversity and endemism lie within the mountains (see section 12.2 above), but unfortunately these areas are the ones with the least amount of formal protection. The only established nature conservation areas within these regions are the Asir National Park System and Raydah in Saudi Arabia, and the Jebel Samhan Sanctuary in the Dhofar mountains, the latter established in 1997. However, the Asir National Park System lies within a highly populated area that precludes formal protection (see section 12.4.5) and Raydah has an area of only 9 km^2.

Table 12.6 Brief descriptions of the 1994 IUCN protected area management categories (Anon. 1994b).

Category	Description
I	Strict Nature Reserve/Wilderness Area: protected area managed for science or wilderness protection
II	National Park: protected area managed mainly for ecosystem protection and recreation
III	Natural Monument: protected area managed mainly for conservation of specific natural features
IV	Habitat/Species Management Area: protected area managed mainly for conservation through management intervention
V	Protected Landscape/Seascape: protected area managed mainly for landscape/seascape conservation and recreation
VI	Managed Resource Protected Area: protected area managed mainly for the sustainable use of natural ecosystems

A list of known protected areas in the Arabian Peninsula is given below, in alphabetical order by country (see also Figure 12.2). For each area the IUCN protected area management category, the phytochorion in which the area lies (see Chapter 4 and Figure 4.3), the area and altitude, status, reasons for protection, physical features, climate (see Chapter 2), principal vegetation and floral features, and threats to vegetation are given. We have attempted to classify each protected area using the 1994 IUCN protected area management categories (Anon. 1994b, see Table 12.6), which are applied to sites of over 1,000 hectares (10 km^2) or to offshore or oceanic islands of at least 100 hectares (1 km^2) where the whole island is protected. Although the protected area systems designed within the various countries of the Peninsula do not readily map across to the IUCN categories, the use of this classification enables us to determine the total number of conservation areas of different types for the whole Peninsula (Section 12.4.8 and Table 12.7), and thus provides a comparative overview. For climate, a reference is given to the climagram for the nearest meteorological station, if available in Figure 2.6, and if representative of the area. In other cases the climate description is based on the climatic contour maps in Figures 2.7-2.11 and other sources as noted. The definition of the seasons follows that established in Chapter 2; i.e. Winter is December-March, Spring is April-May, Summer is June-September, and Autumn is October-November.

There is some unevenness in the details of the accounts both between reserves within a country and between countries, and this is simply a reflection of the incompleteness of the current botanical knowledge of our protected areas, and in some cases the difficulties involved in obtaining current information. For each country we also provide an outline of the main administrative organisations involved with vegetation matters and conservation areas, and a brief summary of the main items of relevant legislation. These are not intended to be comprehensive, and neither do they represent any 'official' view nor, in the case of legislation, any 'officially' sanctioned translations or any implication as to the current status

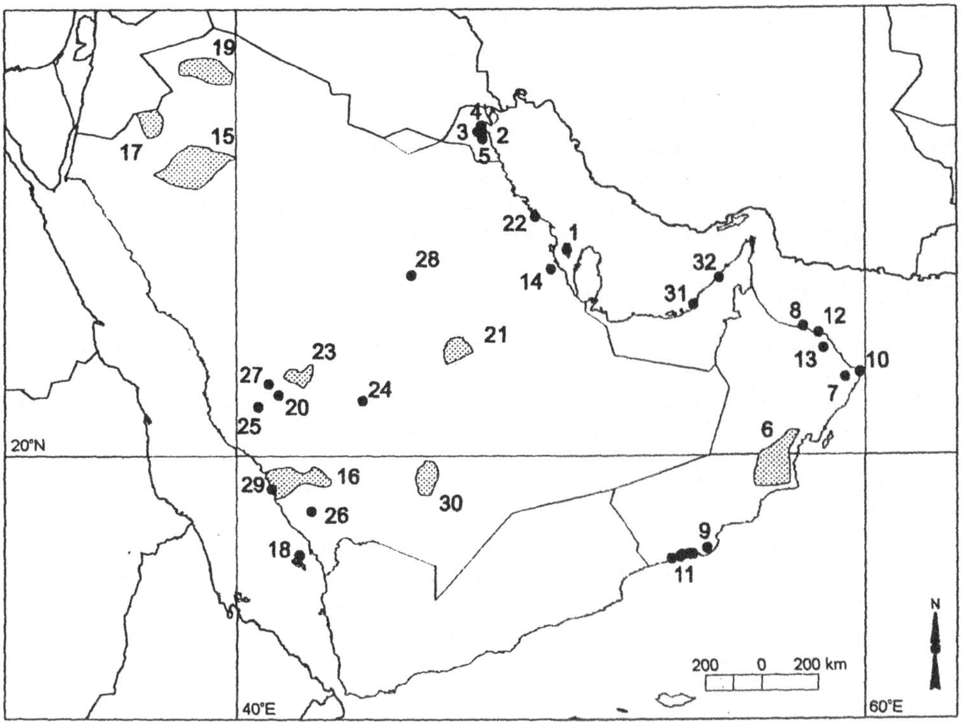

Figure 12.2 Conservation areas in the countries of the Arabian Peninsula. *Bahrain*: 1, al Areen Reserve. *Kuwait*: 2, Doha Reserve, 3, Jahra Reserve, 4, National Park of Kuwait, 5, Sulaibiya Research Station. *Oman*: 6, Arabian Oryx Sanctuary, 7, As Saleel Natural Park, 8, Dimaaniyat Islands, 9, Jebel Samhan Sanctuary, 10, Ra's al Hadd Turtle Reserve, 11, Salalah coast Nature Sanctuaries at Khawr al Mughsayl, al Baleed, Soly (Sawli), al Dahriz, Taqa, Rori (Rawri), Auqad, al Qurom al Saghir and al Qurom al Kabir, 12, Sultan Qaboos Park and Nature Reserve, 13, Wadi Sareen. *Saudi Arabia*: 14, al Hassa National Park, 15, al Khunfa, 16, Asir National Park System, 17, At Tubayq, 18, Farasan Islands, 19, Harrat al Harrah, 20, Hima Saysad, 21, Ibex Reserve, 22, Jubail Marine Wildlife Sanctuary, 23, Mahazat as Sayd, 24, Majami al Hadb, 25, National Wildlife Research Centre, 26, Raydah, 27, Sahal Rukbah, 28, Thumamah Nature Park, 29, Umm al Qamari, 30, 'Uruq Bani Ma'arid. *United Arab Emirates*: 31, Abu Dhabi Nature Reserve, 32, Khawr Dubai.

of enforcement. Rather, our intention is to provide an indication of the main ground covered within the different countries of Arabia in matters of conservation, environmental protection and rangeland management, inasmuch as they relate to plant diversity, conservation and vegetation in general.

Information on the protected areas of each country has been collated both from a variety of published and unpublished literature and from our field work. It has therefore generally not been possible to cite references in the usual way, and citations of relevant sources for each country are, where appropriate, given at the end of each country section. To the best of our knowledge, the information in this section is correct as at October 1997.

12.4.1 BAHRAIN

12.4.1.1 Relevant Legislation

Amiri Decree 20/1983. Establishment of laws for agricultural quarantine.
Amiri Decree 21/1983. Prohibits the cutting or damaging of date palms.
Amiri Decree (2/1995). Concerned with the protection of wildlife, including the formation of the National Committee for the Protection of Wildlife.

12.4.1.2 Government Agencies

Environmental Affairs (EA) of the Ministry of Housing, Municipalities and Environment. By Amiri Decree 21/1996 EA replaces the former Environmental Protection Committee. The role of EA is to achieve sustainable development through appropriate legislation, environmental and pollution monitoring, assessment of the environmental impact of development projects, and the promotion of public awareness.
Ministry of Works and Agriculture. Responsible for the implementation of Amiri Decrees 20/1983 and 21/1983 (see above).
National Committee for the Protection of Wildlife. Established by Amiri Decree 2/1995. Deals with all issues related to biodiversity, including international conventions, the protection of endangered species, and proposals for relevant laws and regulations.

12.4.1.3 Protected Areas

Al Areen Reserve (Figure 12.2: 1)
IUCN Management Category: Unassigned.
Phytochorion: Nubo-Sindian local centre of endemism of the Saharo-Sindian regional zone.
Area and altitude: 8 km^2, completely fenced; 3-45 m.
Status: Established in 1976 as both a park and a reserve.
Reasons for protection: To provide a refuge for endangered wildlife and a wildlife education centre. Divided into two sections, the park is open to the public and contains a representative collection of ungulate species from Arabia, Africa and Asia, whilst the reserve (4 km^2) contains only native Arabian wildlife and has restricted access.
Physical features: The western section at *c.* 3 m above sea level consists of salt flats with aeolian sand and unconsolidated sediments in the form of small dunes. A gentle slope extends from the western edge to 45 m at the highest point to the east. This area is dissected by runnels and small wadis.
Climate: Mean annual precipitation 86 mm, mostly falling in winter-early spring; mean annual temperature 26.5°C (see climagram for Bahrain, Figure 2.6 and Table 2.2).
Vegetation: 53 recorded spp., of which 31 are perennials and 22 annuals. Most are xerophytic, a few halophytic. The site was heavily grazed prior to the erection of the fence, but the vegetation is now thriving. Seven community types have been recognized within the reserve: (1) The *Zygophyllum qatarense* community occurs mainly in areas close to the coast which have small dunes. (2) The *Heliotropium crispum* community also occurs in the coastal areas, but where soil is relativley coarser. (3) The *Panicum turgidum* community occurs in the higher middle and eastern parts of the reserve on relatively deep soil. (4) The *Leptadenia pyrotechnica* community is mainly confined to areas with large accumulations of sand. (5) The *Sporobolus arabicus* community occurs in shallow saline depressions. (6) The *Pennisetum divisum* community occurs in long runnels and depressions. (7) The *Helianthemum kahiricum* community occurs in a small area with shallow stony-sandy soil.

Threats to vegetation: A large number of exotic trees have been planted in the reserve, changing the character of the natural vegetation.

12.4.1.4 Literature Sources

See Abbas *et al.* (1991a, 1991b), Anon. (1992: pp. 399-402) and Samour *et al.* (1989).

12.4.2 KUWAIT

12.4.2.1 Relevant Legislation

Law no. 15, 1972. Gives Kuwait Municipality the legal right to develop and protect all lands in Kuwait.

Law no. 112, 1976. Regulations concerning the import of plants and plant parts, living or dead.

Law no. 62, 1980. Gives responsibility for the protection of the environment to the Environment Protection Council. Amongst other things, the decree includes policies for protection of the environment from pollution, conservation of nature and natural resources, and the establishment of protected areas.

Decision No. 9, 1980, Ministry of Public Works. Prohibition of the uprooting or destruction of *Rhanterium epapposum.*

Law no. 9, 1988. Amending the provisions of Law no. 94 of 1983. PAAFR (see section 12.4.2.2 below) given responsibility for rangeland conservation, and the establishment and supervision of national parks.

Law no. 41, 1988. Empowering PAAFR to ban grazing in certain areas.

Decision no. 244, 1989. Stipulates the conditions for issuing grazing permits and their duration.

12.4.2.2 Government Agencies

Environment Protection Authority (EPA). Established by Law no. 21, 1995, EPA (formerly the Environment Protection Council) is in charge of environmental issues and *in situ* biological conservation.

Kuwait Environment Protection Society (KEPS). KEPS is a local non-governmental organization responsible for environmental awareness and the development of education programmes. The Wildlife Protection Committee of KEPS (established in 1994) is responsible for enhancing the public awareness of wildlife protection and conservation.

Kuwait Institute for Scientific Research (KISR). Established by Law no. 28, 1981, KISR conducts agricultural and environmental research under three divisions: Environmental and Earth Science, Water Resources and Food and Biological Resources. Conservation projects are the responsibility of the Aridland Agriculture Department in the Division of Food and Biological Resources.

Ministry of Defense (MOD). The MOD has recently become involved in *in situ* conservation by taking over the protection of Kuwait National Park.

National Committee for Wildlife Conservation. Established by the EPA in 1996. Responsible for reviewing international conventions on wildlife protection and conservation, and for recommending legislation and other measures for wildlife protection.

Public Authority for Agriculture and Fish Resources (PAAFR). Established by Law no. 94, 1983, PAAFR is responsible for agricultural development, including the management of wildlife, livestock and fisheries. The overall role of the Authority is to supervise the use of land and water for agricultural and fishery purposes so as to ensure their proper use and conservation. PAAFR was formerly a department of the Ministry of Public Works.

12.4.2.3 Protected Areas

Doha Reserve (Figure 12.2, 2)

IUCN Management Category: Unassigned.
Phytochorion: Within the Nubo-Sindian regional centre of endemism of the Saharo-Sindian regional zone, but with partly azonal vegetation.
Area and altitude: 4.5 km^2; sea-level.
Status: Established, but without endorsement by law.
Reasons for protection: Bird breeding and nesting.
Physical features: Sandy sabkha associated with coastal mud flats.
Climate: Mean annual precipitation *c*. 110 mm, falling mostly in winter-early spring; mean annual temperature 26°C.
Vegetation: Halophytic vegetation and an important stand of *Phragmites australis*.
Threats to vegetation: None.

Jahra Reserve (Figure 12.2, 3)

IUCN Management Category: Unassigned.
Phytochorion: Within the Nubo-Sindian regional centre of endemism of the Saharo-Sindian regional zone, but with partly azonal vegetation.
Area and altitude: 2.5 km^2; sea-level.
Status: Established, but without endorsement by law.
Reasons for protection: Resident and migratory birds.
Physical features: A man-made pond covered by secondary treated sewage effluent flowing across sandy sabkha to the sea, as well as a coastal zone with tidal area, mud flats, shore line and reed beds.
Climate: As for Doha Reserve, above.
Vegetation: Halophytic chenopods, with an important stand of *Phragmites australis*.
Threats to vegetation: The alteration of natural surface runoff by the construction of drainage systems causes occasional draining of the pond.

National Park of Kuwait (Figure 12.2, 4)

IUCN Management Category: VI (Managed Resource Protected Area).
Phytochorion: Within the Nubo-Sindian regional centre of endemism of the Saharo-Sindian regional zone, but also contains elements of the Arabian regional subzone of the Saharo-Sindian regional zone.
Area and altitude: Approximately 330 km^2; 0-150 m.
Status: Designated by Kuwait Municipality Council, decision no. 87/1988, and established by PAAFR. Responsibility for protecting the area was given in 1995 to the MOD, since when the area has been completely fenced.
Reasons for protection: Allocation of an area of physical and biological importance for the conservation and management of natural renewable resources.
Physical features: Generally flat with minor undulations. The most significant physical feature is the Jal az Zor escarpment running parallel to the seashore. The escarpment forms a natural watershed. The deepest depression is Wadi Umm-Arimam which covers *c*. 11 km^2 of the northwestern corner of the park.
Climate: As for Doha Reserve, above.
Vegetation: The status of vegetation in the area is poor due to soil erosion and intensive grazing prior to establishment. Important habitats are dominated by the following plant communities: *Rhanterium*, *Haloxylon*, *Halocnemum*, *Seidlitzia*, *Nitraria* and *Zygophyllum*. Important species are *Ochradenus baccatus*, *Calligonum crinitum*, *Gynandriris sisyrinchium* and *Convolvulus oxyphyllus*.

Threats to vegetation: Occasional grazing due to poor protection and fence damage.

Sulaibiya Research Station (Figure 12.2, 5)

IUCN Management Category: I (Strict Nature Reserve/Wilderness Area.
Phytochorion: Within the Nubo-Sindian regional centre of endemism of the Saharo-Sindian regional zone, but also contains elements of the Arabian regional subzone of the Saharo-Sindian regional zone.
Area and altitude: *c.* 20 km^2; 75-130 m.
Status: Established in 1978 as a range management project by KISR, fenced.
Reasons for protection: For conducting range management research.
Physical features: With a gentle SW-NE slope. Northern and western boundaries subject to intrusions by mobile sand.
Climate: As for Doha Reserve, above.
Vegetation: 58 species recorded: three annual grasses, five perennial grasses, seven perennial shrubs and 43 annuals. Vegetation belongs to the *Rhanterium/Cyperus* steppe formation, and under grazing conditions is dominated by *Rhanterium epapposum* and *Cyperus conglomeratus*. Most commonly occurring annual and perennial forb species are *Astragalus schimperi, Atractylis carduus, Ifloga spicata, Koelpinia linearis, Lotus halophilus, Plantago boissieri, P. ciliata, Savignya parviflora, Schimpera arabica, Fagonia glutinosa* and *Polycarpaea repens*. After 10 years of protection from grazing *R. epapposum* showed a marked increase and *C. conglomeratus* declined (Omar 1991).
Threats to vegetation: None, since the site is completely protected.

Range Exclosures

There are several exclosures designated to protect specific habitats or plant species namely: Mutla, Rawdatain, Sheikh Zayed Park, Shigaya and Um al Qurayn. The vegetation in the areas are monitored by the PAAFR and is recovering from extensive damage incurred from the invasion of Kuwait during the Gulf War. The exclusion of livestock is providing permanent protection for the vegetation.

12.4.2.4 Literature Sources

See Anon. (1994a), Anon. (1992: pp. 455-463), Omar (1991, 1996) and Omar *et al.* (1994).

12.4.3 Oman

12.4.3.1 Relevant Legislation

Ministerial Decision no. 4/1976. Provides for a total prohibition of the hunting of all birds in coastal and island regions and some species of mammals; superseded by Ministerial Decision no. 207/93 (see below).

Royal Decree 49/77. Establishment of a plant quarantine law for both the import and export of plants

Royal Decree no. 26/1979. Provides authority for the establishment of national parks and nature reserves.

Royal Decree no. 6/1980. Provides authority for the protection of national heritage, including rare groups or samples of flora and fauna.

Royal Decree no. 10/1982. Establishment of laws for the protection of the environment and prevention of pollution. Various sections have been subject to amendment by Royal decrees 63/85, 71/89 and 111/96 (for the latter, see below).

Ministerial Decision no. 128/93. Prohibition of the cutting of live trees and the collection and transport of dead wood except with a permit.

Ministerial Decision no. 207/93. Prohibition of hunting, capture or firing at wild animals including birds (supersedes Ministerial Decision no. 4/1976).

Ministerial Notice no. 25/96. Issued by the Ministry of Commerce and Industry. Bans the export of seeds of the frankincense tree, ***Boswellia sacra***.

Royal Decree no. 111/96. Amends article 30 of Royal Decree no. 10/1982 (see above), prescribing increased penalties for causing damage or harm to nature reserves and the creatures therein.

12.4.3.2 Government Agencies

Directorate General of Nature Reserves (DGNR), Ministry of Regional Municipality and Environment (formerly Ministry of Environment). The DGNR was established in 1991, when it was known as the Directorate-General of Nature Protectorates. It processes the formal designation of all protected areas, preparing and implementing management plans, carrying out or commissioning field studies where necessary. Responsible for nature conservation both inside and outside protected areas, it has established a network of ranger units whose main tasks are to increase public awareness, collect field data and assist in law-enforcement. It chairs advisory specialist groups, including the Plant Group, with members drawn from various ministries. The DGNR also scrutinises all development projects for their impact on wildlife and habitats.

Directorate General of Environmental Affairs (DGEA), Ministry of Regional Municipality and Environment. The DGEA is the authority for issue of all environmental permits for development projects, with specialised sections dealing with air, noise and water pollution. The marine section is responsible, *inter alia*, for marine pollution prevention and implementing the IUCN Coastal Zone Management Project for Oman (Anon. 1986a, 1988, 1989a, 1989b, 1991). DGEA is also responsible for the development of the draft National Conservation Strategy.

Ministry of Commerce and Industry. In the past has been associated with conservation matters through its involvement with the IUCN Coastal Zone Management Project for Oman (Anon. 1986a, 1988, 1989a, 1989b, 1991).

Natural History Museum, Ministry of National Heritage and Culture. Established in 1990. Houses a variety of museum displays on natural history, a large collection of fauna and flora, and the National Herbarium of Oman with approximately 14,000 accessions covering the vegetation of the whole country.

Office of the Adviser for Conservation of the Environment (ACE), Diwan of Royal Court. Established in 1974 with the remit of designing a programme for establishing a system to bring conservation of the environment to the support of the rational development of the Sultanate's natural and human resources. ACE has played a leading role in conservation and related activities in Oman, in particular with regard to the first surveys of the flora and fauna (Harrison 1977, Reade *et al.* 1980) and to the establishment of the White Oryx Project (Stanley-Price 1989). Amongst the many other initiatives and projects with which ACE has been involved, the following have been of particular importance for the study and conservation of the vegetation of Oman: publication in both Arabic and English of books on the wild flowers of northern Oman (Mandaville 1978) and traditional uses of plants in Dhofar (Miller & Morris 1988), IUCN surveys and reports for a system of nature conservation areas (Clarke 1986) and for a Coastal Zone Management Plan (Anon. 1986a, 1988, 1989a, 1989b, 1991), a WWF/IUCN survey of the Wadi Sareen Nature Reserve (Munton 1985), and the Royal Geographical Society's multi-disciplinary study of the ecosystem of Ramlat al Wahibah (Dutton 1988b).

12.4.3.3 Protected Areas

A joint project between the IUCN and the government of Oman resulted in a report on Proposals for a System of Nature Conservation Areas (Clarke 1986), which proposed a total of 91 nature conservation areas dispersed widely thoughout the country, and covering a total area of 119,798 km^2 (*c.* 40% of the land surface of Oman). Three primary classes of nature conservation areas were proposed, viz:

National Nature Reserves (NNRs): The most important type of nature conservation area, to be managed for wildlife, scientific study and the protection of geomorphological phenomena. Approximately equivalent to IUCN category I.

National Scenic Reserves (NSRs): Applied to selected areas of scenic value that have features making them potential choices for NNR status, but that already have permanent settlements within them. Equivalent to IUCN categories II - VI.

National Resource Reserves (NRRs): A temporary status for areas that have potential interest as nature conservation areas, but about which insufficient information is available. No equivalent IUCN category.

Of the 91 proposed nature conservation areas, 59 were NNRs, 20 NSRs and 12 NRRs. In practice, no areas have yet been declared as National Scenic or Resource Reserves. Of the sites recommended by Clarke (1986) three NNRs have now been declared and part or all of three NNRs (Jiddat al Harasees, Janabah Hills and North Jazir) and az Zahr NSR have been included in the Arabian Oryx Sanctuary (see below).

Arabian Oryx Sanctuary (ACE, DGNR) (Figure 12.2, 6)
IUCN Management Category: II (National Park)
Phytochorion: Within the Somalia-Masai regional centre of endemism, but also contains vegetation elements of the Arabian regional subzone of the Saharo-Sindian regional zone.
Area and altitude: At *c.* 34,000 km^2, the largest protected area in the Peninsula; 0-50 m.
Status: Partially protected since 1979 by staff of the White Oryx Project. Declared on 8 January 1994 by Royal Decree No. 4/1994, and inscribed on the World Heritage List in December 1994. The Arabian Oryx Sanctuary is now the responsibility of DGNR, whilst the White Oryx Project remains under ACE.
Reasons for protection: An important site for wildlife and biodiversity in general, including the Arabian oryx, houbara bustard, two species of gazelle, ibex and other threatened wildlife species. The area is also a local centre of plant endemism (see section 12.2.2.2), and has important ecological, geological and wilderness values.
Physical features: Two distinct zones within the sanctuary: the coastline and adjacent areas running up to the Huqf escarpment, and the Jiddat al Harasees plateau above and to the west of the Huqf. The Jiddah is a flat or undulating Tertiary limestone plateau at 100-150 m, bounded to the west by the sands of the Rub' al Khali, to the north by the wadis of the northern mountains, to the south by wadis that run to the coastal plain, and to the east by the Huqf escarpment and depression. There are no major drainage features on the plateau, and the main surface features are shallow sandy depressions known as *haylah*, which collect water after rain. These are the main vegetated areas. Areas between *haylah* are typically stony and highly weathered, with relatively sparse vegetation. The Huqf depression contains outcrops of sedimentary rocks of nearly all geological periods, dating back to the Proterozoic; small dunes and large areas of sabkha are interspersed between rocky outcrops.
Climate: Mean annual precipitation 39 mm, falling mainly in Spring, though inter-annual variability of rainfall very high; mean annual temperature 26.6°C (see climagram for Ja'aluuni,

Figure 2.6 and Table 2.2). A fog desert, receiving an average of 54 fog days per year (see Figure 2.13).
Vegetation: The Jiddah is recognised as a local centre of endemism (see section 12.2.2.2 above), with 12 endemic (including 1 regionally endemic) species. Although a hyperarid area, frequent heavy fogs and dews enhance water availability, with growth of many perennials supported entirely by this source during droughts. The *haylah* vegetation consists of an open *Acacia* scrub with *A. tortilis*, *A. ehrenbergiana* and *Prosopis cineraria* as the dominant structural components. *Rhazya stricta* and the endemic *Ochradenus harsusiticus* form a major component of the shrubby vegetation, and the ground cover includes, amongst other grasses, *Stipagrostis sokotrana*, which is an important food source for Arabian oryx. Other notable species include *Convolvulus oppositifolia* (endemic), *Caralluma flava*, *Nannorhops ritcheana*, *Pulicharia undulata* and *P. pulvinata* (endemic). On the coastal limestone hills the endemic shrub *Hyoscyamus gallagheri* is dominant. A detailed vegetation survey has not been carried out, but *c.* 200 spp. are recorded from the Jiddah, and with the inclusion of the Huqf depression and coastal areas there are *c.* 250 spp.
Threats to vegetation: Unrestricted grazing by camels, goats and feral donkeys; off-road driving.

As Saleel Natural Park (DGNR) (Figure 12.2, 7)

IUCN Management Category: Unassigned.
Phytochorion: Saharo-Sindian subzone.
Area and altitude: 220 km^2; 100-150 m.
Status: Declared in 1997 by Royal Decree no. 50/1997.
Reasons for protection: Area with gazelle and other wildlife, and an important fossilised reef hill. It is planned that the Arabian oryx will be introduced in the Park and then some areas opened to the public.
Physical features: Flat desert plain with sand and gravel, covered with an open *Acacia tortilis* woodland.
Climate: Nearest representative station is the coastal station of Sur with mean annual precipitation of 92 mm, falling mainly in winter and early spring; mean annual temperature 29.3°C (Figure 2.6 and Table 2.2). However, being inland, as Saleel will have a greater temperature range than Sur, and greater maximum and lower minimum temperatures.
Vegetation: *Acacia tortilis* woodland with associated shrubs.
Threats to vegetation: Overgrazing by goats and camels.

Dimaaniyat Islands (DGNR) (Figure 12.2, 8)

IUCN Management Category: IV (Habitat/Species Management Area).
Phytochorion: Nubo-Sindian regional centre of endemism of the Saharo-Sindian regional zone.
Area and altitude: 200 km^2 of sea and seabed, including nine islands, and various rocks, reefs and offshore shoals situated 18-20 km off the Batinah coast; 0-50 m.
Status: Declared in 1996 by Royal Decree no. 23/1996.
Reasons for protection: Host to a high density of nesting seabirds and the only known osprey nesting site in the Capital Area. Known to have the largest nesting population of hawksbill turtles (*Eretmochelys imbricata*) in Oman. Green turtles (*Chelonia mydas*) and sooty falcons also nest, and there is a variety of reefs with a high diversity of corals and molluscs. The islands are free of mainland predators such as foxes, dogs, cats and rats.
Physical features: Consists largely of bare sedimentary limestone and skeletal sandy soils.
Climate: Nearest representative station is Seeb with mean annual precipitation of 86 mm, falling mainly in winter-early spring, with the occasional summer storm; mean annual temperature 28.7°C (Figure 2.6 and Table 2.2).
Vegetation: The perennial vegetation consists entirely of halophytic species, with *c.* 13 spp. recorded. The dominant halophytic shrub is *Suaeda monoica* which occurs in monospecific

stands on the vegetated islands. Other common species are *Suaeda aegyptiaca*, *Cyperus conglomeratus*, *Sporobolus virginicus* and *Halopyrum mucronatum*.
Threats to vegetation: Uninhabited apart from a small ranger presence on one island; there are no domestic livestock. Only potential source of threat comes from visits by tourist parties and fishermen, especially those that involve overnight camps.

Jebel Samhan Sanctuary (DGNR) (Figure 12.2, 9)

IUCN Management Category: Unassigned
Phytochorion: Somalia-Masai regional centre of endemism.
Area and altitude: 4,500 km^2; 0-1,800 m.
Status: Declared in June 1997 by Royal Decree no. 48/1997.
Reasons for protection: An area of wilderness and an important site for wildlife and biodiversity. Contains the last remnants of a relatively well-preserved deciduous tropical woodland, and is also perhaps the last refuge for the Arabian leopard. Several other large mammals including the nubian ibex, hyaena, gazelle and wolf live in the mountains and the green and loggerhead turtles nest on the beaches. The area is also part of a local centre of diversity and endemism (see section 12.2.2.2).
Physical features: The area is a barren mountain massif with elevated limestone highlands rising steeply from the coast overlooking the Zalawt foothills and Mirbat coastal plains. Deep wadis and gorges intersperse the high peaks. Several water pools are present including the Andhur pools, which are of both archaeological and ecological importance.
Climate: Except for Jebel Habrer, the Sanctuary is largely outside the influence of the southwest monsoon. Salalah (20 m) and Qairoon Hariti (878 m) are the nearest meteorological stations (Figure 2.6, Table 2.2), though neither are particularly representative, both being within the influence of the monsoon. Mean annual precipitation at the lower altitudes is probably *c.* 50-100 mm, falling mainly in winter-early spring, with rain in some years falling in the summer, and mean annual temperature *c.* 26°C. Rainfall of 100-200 mm may be expected at the higher altitudes.
Vegetation: Jebel Samhan is recognised as a part of the Dhofar mountain local centre of endemism (see section 12.2.2.2), with *c.* 20 endemic species. The main vegetation is an open *Commiphora-Acacia* woodland with succulents, *Aloe* spp. and *Cissus quadangularis,* and low shrubs. Species of special interest include those of *Caralluma*, *Maytenus*, *Launaea* and *Dracaena*. Endemic species include *Anogeissus dhofarica*, *Lavandula hasikensis, Maytenus dhofarensis* and *Salvia hillcoatiae*. The only Arabian location of *Pappea capensis* occurs on Jebel Habrer. A detailed vegetation survey has not been carried out, but >500 spp. are expected to be present, similar to the number of species recorded for the Dhofar mountains.
Threats to vegetation: Unrestricted grazing by camels, goats and feral donkeys; potential unplanned development; off-road driving.

Ra's al Hadd Turtle Reserve (DGNR) (Figure 12.2, 10)

IUCN Management Category: IV (Habitat/Species Management Area).
Phytochorion: Nubo-Sindian regional centre of endemism of the Saharo-Sindian regional zone.
Area and altitude: 80 km^2; 0-100 m.
Status: Declared in 1996 by Royal Decree no. 25/1996. IUCN category IV.
Reasons for protection: The Ras al Hadd peninsula forms part of a complex of turtle nesting beaches which extend south beyond Ras al Jinz to Ras al Khabbah. The beaches attract 6,000-13,000 nesting green turtles each year, and are of global importance for the species. The reserve also contains important archaeological sites.
Physical features: The northern and western parts of the reserve are principally low-lying limestone hills, dissected by wadis, and ending as low cliffs at the sea. The remainder of the Ras al Hadd peninsula is low-lying and sandy, with the generally flat relief broken by mounds that

indicate the sites of shell middens and other archaeological deposits. There are two large bays within the reserve: Khawr al Jaramah and Khawr al Hajar. The southern and western shores of both bays are backed by sabkha.
Climate: Nearest representative station is Sur, with a mean annual precipitation 92 mm, falling mainly in winter-early spring, with the occasional summer storm; mean annual temperature of 29.3°C (Figure 2.6 and Table 2.2).
Vegetation: Typical coastal and foothill vegetation. The coastal vegetation consists of a *Limonium stocksii-Sphaerocoma aucheri-Cornulaca monacantha* community, with *Cyperus conglomeratus* and *Suaeda* spp. as common associates. There are several small stands of mangrove (*Avicennia marina*) along the shore of Khawr al Jaramah. Vegetation of the foothills consists typically of *Acacia-Commiphora-Euphorbia* scrub with species such as *Lycium shawii, Pulicaria glutinosa* and *Ochradenus arabicus* forming the main components of the shrubby vegetation. No thorough plant survey carried out, but the estimated number of species is *c.* 60.
Threats to vegetation: Off-road driving, grazing of domestic livestock, camping parties.

Salalah coast Nature Sanctuaries at Khawr al Mughsayl, al Baleed, Soly (Sawli), al Dahriz, Taqa, Rori (Rawri), Auqad, al Qurom al Saghir and al Qurom al Kabir (DGNR) (Figure 12.2, 11)

IUCN Management Category: Unassigned.
Phytochorion: Somalia-Masai regional centre of endemism.
Area and altitude: Range of size of each individual ***khawr*** is 0.017 km^2 (al Qurom al Saghir) to 0.58 km^2 (Rawri); sea-level.
Status: Declared in June 1997 by Royal Decree no. 49/1997.
Reasons for protection: Permanent brackish water bodies maintained by a unique system of fresh over- and underground water from the escarpment mountains, and seepages of seawater, utilized for fishing, livestock watering and grazing, and irrigation; presence of mangroves; wintering grounds for *c.* 300 species of birds; natural scenic beauty.
Physical features: The ***khawrs*** (coastal lagoons) are separated from the sea by sand bars and contain brackish water of varying salinities (see sections 10.2.3 & 10.2.4). Of the nine khawrs, two (al Qurom al Saghir and al Qurom al Kabir) are not fed by an underground fresh water system and are therefore highly saline.
Climate: Nearest meteorological stations is Salalah, with mean annual precipitation of 85 mm, falling mainly in spring and summer; mean annual temperature 26.4°C (Figure 2.6, and Table 2.2).
Vegetation: The ***khawrs*** are surrounded by the Salalah coastal plains which is characterised by mounds of semi-desert, salt tolerant plants (*Cressa cretica*, *Limonium axillare* and *Urochondra setulosa*) and rocky outcrops with small shrubs and succulents (*Cassia italica* and *Cissus quadangularis*). The coastal dunes are colonized by characteristic grasses and sedges (*Cyperus conglomeratus, Halopyrum mucronatum, Sporobolus spicatus* and *S. iocladus*), satlbushes (*Atriplex farinosum*) and the trailing shrub *Ipomoea pes-caprae*. The margins of ***khawrs*** are lined with the grasses *Paspalum vaginatum* and *Sporobolus virginicus* and reed beds with different species (including *Juncus*, *Phragmites*, *Schoenoplectus* and *Typha*) occur at the water edge reflecting the water salinity of the ***khawr***. ***Khawrs*** with high salinity have the mangrove *Avicennia marina*. The water vegetation consists of the submerged species *Potamogeton pectinatus*, *Najas marina*, *Ceratophyllum demersum* and *Ruppia maritima* and the macroalgae *Chara* and *Enteromorpha*.
Threats to vegetation: Unrestricted grazing by camels, especially mangroves; soil erosion and trampling by off-road driving and livestock leading to environmental degradation and reduction of plant cover and species diversity.

Sultan Qaboos Park and Nature Reserve (Figure 12.2, 12)

IUCN Management Category: Unassigned

Phytochorion: Nubo-Sindian regional centre of endemism of the Saharo-Sindian regional zone.
Area and altitude: *c*. 1 km^2; 0-10 m.
Status: Declared by Royal Decree no. 19/81. Part of the area functions as a public park, and the remainder consists of dense mangrove stands.
Reasons for protection: An important green area, being the only khawr that lies within the immediate Capital Area.
Physical features: Flat tidal khawr, fed with underground freshwater from the landward side by Wadi Adai.
Climate: Nearest representative station is Seeb, with a mean annual precipitation of 86 mm, falling mainly in winter-early spring, with the occasional summer storm, and a mean annual temperature 28.7°C (Figure 2.6, Table 2.2)
Vegetation: The natural vegetation has been vastly altered by the construction of the park and its associated features. Where it is still preserved, the vegetation consists of halophytic shrubs of ***Suaeda aegyptiaca*** and ***Indigofera oblongifolia*** on dry calcareous sands, and ***Halopeplis perfoliata-Aeluropus lagopoides*** communities on sandy-silty soils. The important remaining natural vegetation are the mangroves, ***Avicennia marina***, which form a dense fringe on the edges of the ***khawr***, reaching up to 4 m in height.
Threats to vegetation: Probably none within the ***khawr*** itself, though any further expansion of the area as a public recreation facility would threaten the mangroves.

Wadi Sareen (ACE) (Figure 12.2, 13)
IUCN Management Category: Unassigned.
Phytochorion: Nubo-Sindian regional centre of endemism of the Saharo-Sindian regional zone.
Area and altitude: Neither area nor boundaries specifically delineated as yet, though the general area patrolled is *c*. 1,800 km^2; 250-2,000 m.
Status: Commonly known as Wadi Sereen Nature Reserve or the Arabian Tahr Reserve, though it has not yet been officially declared as a conservation area. The area currently patrolled is part of what was proposed as the Jebel Aswad NNR by Clarke (1986). Protection is provided by rangers employed by the Diwan of Royal Court, who live within the area.
Reasons for protection: Primarily for the conservation of the regionally endemic Arabian tahr.
Physical features: Limestone mountain and wadi terrain of 300 - 2,000 m. Dominated by cliffs, steep mountain and ridge slopes and steeply incised drainage lines. At lower elevations there are scree slopes and minor hills and wadi outwash fans. At higher altitudes there is much bare rock with skeletal soils in cracks, and at lower elevations there are gravelly sands, lithosols and coarse alluvials, with alluvial gravels in the wadis.
Climate: Seeb and Sur are the nearest meteorological stations (Figure 2.6, Table 2.2), though they are not particularly representative, being at sea level. Mean annual precipitation at the lower altitudes is probably *c*. 100-200 mm, falling mainly in winter-early spring, and mean annual temperature *c*. 26°C.
Vegetation: *c*. 10 endemic spp., including the trees ***Ceratonia oreothauma*** subsp. ***oreothauma*** and ***Ziziphus hajarensis***, the shrubs ***Dionysia mira***, ***Lavandula subnuda*** and ***Rhus aucheri***, and the annual ***Pycnocycla prostrata***. There are several species rich sites around water seepages, where rare species such as the orchid ***Epipactis veratrifolia***, ***Lindenbergia*** sp. and various ferns are found. No full vegetation survey, but the total number of species is similar to that of the northern mountains in general; i.e. *c*. 300 species.
Threats to vegetation: Overgrazing due to domestic livestock, though less of a problem than elsewhere since there is an agreement between the residents and ACE that grazing is restricted to certain areas.

12.4.3.4 Literature Sources

See Anon. (1982), Anon. (1986a, 1988, 1989a, 1989b, 1991), Anon. (1992: pp. 491-500), Anon. (1994a), Clarke (1986), Ghazanfar (1991a, 1992a, 1993b, in press-b), Ghazanfar *et al.* (1995), Hillcoat *et al.* (1980), Kürschner (1986c), Mandaville (1977), Munton (1985), Salm (1986, 1989) and Whitcombe (1995).

12.4.4 QATAR

We have been unable to obtain current information on the status of plant conservation in Qatar, and what information we have comes mostly from Anon. (1992: pp. 501-504). The 1993 United Nations List of National Parks and protected areas (Anon. 1994a) lists one site, Ras Asharij Gazelles Conservation Farm, an IUCN category IV site, without location information. There do not appear to be any other specific protected areas or any protected area legislation, although game reserves for wildlife breeding have been set up by the state authorities and the ruling family. No information is available on the vegetation status of these areas, though, since they are large natural areas often over 100 ha, they may be providing incidental protection for plants. Batanouny (1981) provides a basic description of the plant communities and flora. There are no plant species endemic to Qatar. The main responsibility for conservation matters appears to lie with the Environment Protection Committee, established under Law No. 4 of 1981, and the Ministry of Industry and Agriculture. Recommendations for protected area designation have been made (Anon. 1986b). Threats to the environment, and hence to the vegetation include grazing, the spread of cultivation, oil exploration and the opening up of the country through the construction of desert roads. The potential damaging effects of development in wilderness areas are obviously exacerbated in a relatively small country such as Qatar.

12.4.5 SAUDI ARABIA

12.4.5.1 Government Agencies

Meteorological and Environmental Protection Administration (MEPA). Established in 1980, MEPA has responsibility for marine conservation, environmental protection, and meteorology. Responsibility for the identification of protected areas passed to NCWCD in 1986. Part of MEPA's work, in conjunction with MAW and NCWCD, is the assessment of rangeland degradation. The Environmental Support of Nomads (ESON) Project is a programme of rangeland resource mapping and modelling for advising pastoral nomads on suitable grazing regimes. The project's pilot study focuses on four sites totalling 70,000 km^2 and encompassing both natural vegetation as well as traditional, semi-settled and settled nomadic lifestyles. ESON aims to make recommendations about where, which type and how many animals, and for how long to graze a particular area. Study sites encompass areas in the Qassim-Hail, Taysiyah-Jandaliyah, Nafud al Ariq and Mahazat as Sayd areas. The latter three sites include proposed and current NCWCD protected areas.

Ministry of Agriculture and Water (MAW). MAW was the principal agency legally responsible for the management and monitoring of forests, rangelands and wildlife before responsibility for protected areas was given to NCWCD (see below) in 1986. MAW has been active in the area of legislation for vegetation conservation, and the establishment of water-harvesting areas and re-seeding programmes for rangeland improvement. A number of dams have been built to improve the recharge of aquifers and to prevent erosion and flooding, and several large areas of rangeland have been fenced to serve as *in situ* seed reservoirs. A separate

Department of National Parks was created in 1983 with the mandate to establish national parks both for vegetation conservation and recreation. These were based on the system of National Parks in the USA. There is currently no mechanism for the total protection of rangelands, and MAW therefore concentrates its efforts on 32 rangeland sites, totalling approximately 200 km^2, all of which are fenced. Two such sites, Sahal Rukbah and Hima Saysad, are listed in the protected area summary.

National Commission for Wildlife Conservation and Development (NCWCD). The NCWCD was created by Royal Decree no. M/22 of 1986 to "Develop and implement plans to preserve wildlife in its natural ecology and to propose the establishment of proper protected areas and reserves for wildlife in the Kingdom.." (Article 3(4) of Royal Decree No. M/22). The term wildlife covers all indigenous wild plants and animals and their habitats under natural or semi-natural conditions on land and in the sea. NCWCD's first projects focussed on the protection and restoration of high profile animal species, by which it has been able to gain popular support for other, less spectacular, but equally important conservation programmes. The foundation of the NCWCD approach has been the creation of a large network of protected areas and the management of these areas in such a way as to fulfill its mandate to conserve and develop the nation's wildlife.

12.4.5.2 Relevant legislation

1966, Land Development Act. Gives MAW responsibility over the regulation of land development activities, putting a limit on the maximum acreage of land that may be utilized by individuals for agriculture.

1975, Agricultural and Veterinary Regulations. Controls the introduction of plant and animal species into Saudi Arabia and regulates the issue of health certificates.

1977, Forests and Pastures Act. Commits MAW to conserve pastures, public and urban forests, and to regulate their use. Cutting trees and shrubs for private or commercial use without a permit is prohibited, grazing is restricted to allocated sites, and no building may be erected on agricultural land.

1977, National Hunting Decree Law No. 457. Details areas with permanent hunting bans and has provided the basis for incidental protection of vegetation at sites such as the island of Umm al Qamari. Administered by the Ministry of Interior on advice from NCWCD.

1978, Royal Forest Decree Law No. 1392 (incorporating National Park Law of 1977). Lays down regulations for the protection of forests and wildlife. Administered by MAW.

1979, Water Resources Conservation Act. Governs the control and use of water resources, with priority use granted to human and animal needs and to agricultural and industrial purposes. Enforcement is the responsibility of MAW.

1986, Royal Decree No. M/22. Commitment to the creation of a system of protected areas, with responsibility for this given to NCWCD.

1995, Decree No. 128. Lays down the regulations governing a 'Wildlife Protected Areas System', including selection, establishment and management of wildlife protected areas.

12.4.5.3 Protected Areas

The NCWCD System Plan for Protected Areas (Child & Grainger 1990) draws on information from earlier surveys by other government agencies to list a total of 103 candidate protected areas, covering a total of over 170,000 km^2, (8.1% of the Kingdom). By 1995 a total of 12 protected areas had been formally created, with protection from grazing by sheep and goats for over 15,000 km^2. Child and Grainger (1990) proposed four Protected Area types for Saudi Arabia:

Special Natural Reserve (SNR): Sites of high biological excellence. Virtually all forms of settlement and agriculture to be excluded, though some controlled forms of use, such as scientific research and low-impact recreation, may be permissible. Approximately equivalent to IUCN category I.

Natural Reserve (NR): Areas of high natural excellence managed to permit greater public access than for SNRs. Approximately equivalent to IUCN categories II or IV.

Biological Reserve (BR): Small areas of biological importance, such as water catchments, isolated stands of rare trees or localized breeding sites for key species, fully protected from man-made disturbance and administered by local authorities. BRs act as an extension of the *hima* concept. Approximately equivalent to IUCN category III.

Resource Use Reserve (RUR): Relatively large areas within which the emphasis is on resource management rather than conservation. RURs can be created and managed by local communities, with NCWCD assistance in the early stages. Approximately equivalent to IUCN category VI.

A fifth category of protected area is administered by the Ministry of Agriculture and Water:

National Park (NP): Areas of natural beauty managed primarily to provide recreational opportunities with minimal impact on natural features, but within which habitation, development and unregulated traditional resource use is permitted. Approximately equivalent to IUCN category V.

The IUCN equivalent categories are only approximate equivalents of the NCWCD and MAW protected area types, since the two systems do not map directly across in all cases, especially for those reserves which contain areas of differing designation.

Al Hassa National Park (MAW) (Figure 12.2, 14)

IUCN Management Category: V (Protected Landscape).
Phytochorion: Arabian regional subzone of the Saharo-Sindian regional zone.
Area and altitude: 2,400 km^2; 0-120 m.
Status: NP, recreational and regulated use zone extending for *c*. 70 km from Hofuf to the Gulf.
Reasons for protection: Incorporates the large al Hassa oasis and extensive areas of sandy desert and saline plains.
Physical features: Flat and undulating sandy desert and sabkha with areas of marsh, cut by numerous drainage lines.
Climate: For 1970-1986 mean annual precipitation at Hofuf is 78 mm falling mainly in winter-early spring; mean annual temperature 25.3°C (Al-Jerash 1989).
Vegetation: The al Hassa oasis encompasses *c*. 100 km^2 of marshes and drainage channels, vast date-palm groves, and sand dunes which have been stabilised by shelter-belt trees, mostly *Tamarix* and date-palms. The area between the oasis and the Gulf is mostly sandy desert with *Calligonum comosum* and *Haloxylon salicornicum* communities in deep sand. Species list incomplete.
Threats to vegetation: Moderate to heavy unregulated recreational activities, overgrazing in non-protected areas.

Al Khunfa (NCWCD) (Figure 12.2, 15)

IUCN Management Category: IV (Habitat/Species Management Area).
Phytochorion: Arabian regional subzone of the Saharo-Sindian regional zone.
Area and altitude: 20,450 km^2, including a core protected area (Ghurrub sector) of 2,875 km^2; 700-950 m.
Status: Core area SNR, other zones BR/RUR, declared in 1988.

Reasons for protection: Contains the largest remaining wild population of reem gazelle in Saudi Arabia and representative areas of the northern plains and sandstone hills on the edge of the Great Nafud.
Physical features: Gravel plateaux with isolated sandstone hills. Some deeply incised wadis and occasional salt flats.
Climate: Nearest representative meteorological stations is Jouf, with a mean annual precipitation of 63 mm falling mostly in winter-spring, occasionally in autumn; mean annual temperature 21.5°C, with occasional frosts (Figure 2.6, Table 2.2).
Vegetation: The plateaux and hills support few plants, and vegetation is largely confined to wadis and waterlines. There are rare *Acacia* trees and occasional *Tamarix* and *Atriplex leucoclada* bushes. Dwarf shrubland, dominated by chenopods including *Traganum nudatum* and *Suaeda* spp., composites, including *Pulicaria undulata* and *Artemisia* spp., and perennial grasses including *Centropodia fragilis* and *Stipagrostis plumosa* grow in major wadis. >50 species recorded, though the list is incomplete.
Threats to vegetation: Unrestricted grazing of camels. Encroachment of centre-pivot irrigation, and of sheep and goat flocks into the core area.

Asir National Park System (MAW) (Figure 12.2, 16)

IUCN Management Category: V (Protected Landscape).
Phytochorion: Spans the Somalia-Masai and the Afromontane archipelago-like regional centres of endemism, and contains vegetation elements from both.
Area and altitude: 4,500 km^2; 0-3,200 m.
Status: NP, declared in 1981.
Reasons for Protection: Vast recreational and plant conservation zone encompassing high altitude escarpment regions down to coastal plains.
Physical features: Encompassing marine and coastal areas of the flat Tihamah coastal plain and rising to granite and gneiss foothills and the Asir escarpment cut by deep wadi canyons.
Climate: Varies across the large altitudinal range of the park, with a mean annual precipitation >250 mm falling over two periods, late winter-early spring and summer, and a mean annual temperature of *c.* 19°C at the higher altitudes (see Abha, Figure 2.6 and Table 2.2), and at the coast a mean annual precipitation of *c.* 100 mm falling mainly in winter and with a mean annual temperature of *c.* 29°C.
Vegetation: The Asir National Park System consists of several units extending from the Asir escarpment near Abha to the Red Sea. High altitude units such as the al Sawdah National Park and the al Qarra National Park lie in the *Juniperus procera* zone, and encompass some of the best stands of this tree species in the Kingdom. The junipers form stands with *Acacia origena, Olea europaea, Nuxia oppositofolia, Maesa lanceolata* and *Teclea nobilis* at successively lower elevations on western slopes. The extreme lower altitudinal limit for *Juniperus procera* here is *c.* 1,600 m. On the western foothills and plains *Acacia* communities are prominent; *Acacia etbaica*, *A. asak*, *A. hamulosa*, *A. mellifera*, *A. johnwoodii*, *A. tortilis*, *A. seyal* and *A. oerfota* form a mosaic of communities. *c.* 350 species recorded from al Sawdah, but lists for other units incomplete.
Threats to vegetation: Conflicting land use claims in areas adjacent to human habitation and in areas with high densities of goats. The large area of the park system precludes formal protection.

At Tubayq (NCWCD) (Figure 12.2, 17)

IUCN Management Category: IV (Protected Landscape).
Phytochorion: Arabian regional subzone of the Saharo-Sindian regional zone.
Area and altitude: Core area *c.* 3,000 km^2; 750-1,200 m.
Status: Core area SNR, other areas NR/RUR, declared in 1989.

Reasons for protection: Contains the most northerly population of ibex in Saudi Arabia, lies within the historical range of ostrich and includes a representative area of the northern plain.
Physical features: A high sandstone plateau dropping abruptly in a steep, deeply incised escarpment, to a gravel plain with conical hills and sand drifts.
Climate: One of the nearest representative meteorological stations is Guriat (Figure 2.6, Table 2.2), with a mean annual precipitation of 53 mm falling mainly in winter-early spring, and a mean annual temperature of 19.8°C, with absolute minimum temperatures falling to *c*. -8°C, and with occasional frosts and snow.
Vegetation: Dwarf shrubs, including a succulent chenopod *Traganum nudatum* and a salt-accumulating shrublet *Reaumeria hirtella*, grow on the plateau, where they are most numerous along shallow runnels. The annual grass *Stipa capensis* is common on the plateau pavements. Hills and the escarpment slopes are barren, as is much of the gravel plain, except along watercourses and where shallow sand drifts support annual plants following rain. Perennial grasses such as *Centropodia fragilis* and *Stipagrostis drarii* grow on the deeper sand drifts. Canyon wadis have the greatest plant diversity, including a few *Acacia pachyceras*, numerous *Haloxylon persicum*, and dwarf shrubs such as *Haloxylon salicornicum, Artemisia judaica, Centaurea scoparia* and *Deverra triradiata* and, following rain, many annual herbs. >141 spp., none endangered or rare, but generally with far fewer of the 'weedy' species which are so common over large areas of the Peninsula.
Threats to vegetation: Unrestricted grazing by sheep, goats and camels.

Farasan Islands (NCWCD) (Figure 12.2, 18)
IUCN Management Category: V (Protected Landscape/Seascape).
Phytochorion: Within the Somalia-Masai regional centre of endemism, but with partly azonal vegetation.
Area and altitude: Approximately 600 km^2; 0-10 m.
Status: Multiple use area (SNR/NR/RUR) with some regulation of use of marine and terrestrial resources.
Reasons for protection: The islands contain high concentrations of nesting seabirds and ospreys, turtle nesting beaches, a large population of idmi gazelle, and representative examples of unusual vegetation types, including mangrove communities.
Physical features: A group of low-lying islands in the Red Sea formed from uplifted fossil coral reef. In some places the underlying salt substratum has become leached causing the overlying coral to collapse and form long narrow ravines which, together with erosion clefts, hold fine silty clay soils.
Climate: Nearest meteorological station is on the coast at Gizan (Figure 2.6, Table 2.2), with a mean annual precipitation of 129 mm, which, being on the boundary of the winter-early spring and summer rainfall systems, receives rain in both periods; mean annual temperature 30.6°C.
Vegetation: The northwestern plateau and the westward facing shores are exposed to strong prevailing winds and almost devoid of vegetation. Dwarf shrubland is most diverse and abundant in the deep ravines which cut through areas of fossil coral. In larger wadi beds there are trees and shrubs including *Commiphora gileadensis*, *Ziziphus spina-christi* and *Maerua oblongifolia*. There are two groves of *Acacia ehrenbergiana* woodland on the largest island, Farasan Kabir. Two species of mangroves (*Rhizophora mucronata* and *Avicennia marina*) grow on the coastal flats. Sabkha areas support a succulent dwarf shrub community dominated by *Limonium* spp. Endemic or near endemic species are *Glossonema* sp. aff. *boveanum*, *Cucumis* sp. aff. *prophetarum* and *Dipcadi* sp. Within the area of the proposed Naval Base on Farasan Kabir are four rare species found nowhere else in Saudi Arabia (*Micrococca mercurialis*, *Nothosaerva brachiata*, *Basilicium polystachion* and *Vahlia digyna*). Three other rare plants *Cleome noena*

subsp. *brachystyla*, *Ipomoea hochstetteri* and *Brockmannia somalensis* are found north of the port. 179 recorded spp.
Threats to vegetation: The larger islands are inhabited and the SNR is threatened by expansion of cultivation and excessive grazing. Coastal development, including a proposed naval base, threatens the mangroves. Phosphate extraction is a possibility.

Harrat al Harrah (NCWCD) (Figure 12.2, 19)

IUCN Management Category: II (National Park).
Phytochorion: Arabian regional subzone of the Saharo-Sindian regional zone.
Area and altitude: 12,150 km^2; 500-1,000 m.
Status: SNR/NR, declared in 1987.
Reasons for protection: Contains breeding houbara bustards, small populations of reem (*Gazella subgutturosa*) and idmi (*Gazella gazella*), and a representative portion of the most extensive basalt lava field in the Arabian Peninsula.
Physical features: Limestone overlain by basalt lava. An undulating plain at *c.* 850 m with occasional tall volcanic hills. Lower hills and plateaux are covered in limestone or basalt boulders and rock fragments. Small dry watercourses and larger wadis, which consist of a streambed and riverine terraces, drain the plateaux. Deep sands are deposited on the slopes of some hills.
Climate: Nearest representative stations are Turaif and Jouf (Figure 2.6, Table 2.2), with mean annual precipitation of 82 and 63 mm, respectively, falling mainly in winter-early spring, and mean annual temperatures of 18.7 and 21.5°C respectively.
Vegetation: Sparse, patchy dwarf shrub vegetation, concentrated in small drainage lines, wadis and silty depressions. There are no trees, except for the occasional *Tamarix arborea* along large stream beds. Dominant perennial shrub species include *Achillea fragrantissima*, *Artemisia monosperma*, *Artemisia sieberi*, *Astragalus spinosus*, *Calligonum comosum*, *Haloxylon salicornicum*, *Salsola* spp., *Traganum nudatum* and *Zilla spinosa*. *Capparis spinosa* is abundant on the margins of silty depressions. Perennial grasses, which include several *Stipagrostis* spp., are sparsely distributed and poorly represented, probably due to intensive camel grazing in some areas. Chert/gravel plains and rocky basalt slopes (*harrah*) and hills support few perennial plants. A community of *Haloxylon persicum* and *Calligonum comosum* grows on mobile sand drifts. In years with good rainfall there is a flush of annuals in all habitats. In *harrah* areas the edible fungus *Terfezia claveryi (faga)*, a local delicacy, grows in association with *Helianthemum lippii*. *Faga* is highly sought after and fetches high prices in markets throughout the Kingdom. >290 recorded spp.
Threats to vegetation: Unrestricted grazing of camels; encroachment of sheep and goat flocks on the reserve's borders; unregulated collection of *faga*.

Hima Saysad (MAW) (Figure 12.2, 20)

IUCN Management Category: Unassigned.
Phytochorion: Arabian regional subzone of the Saharo-Sindian regional zone.
Area and altitude: 7 km^2, fenced; *c.* 1,000 m.
Status: RUR, non-grazing reserve created in 1985.
Reasons for protection: Protection from livestock grazing of a small area surrounding an old silted up dam.
Physical features: Undulating sand and gravel plain.
Climate: Nearest representative station is Hima Saysad, with a mean annual precipitation of 182 mm falling mainly in spring and late summer, and a mean annual temperature of 21.4°C, for 1970-1986 (Al-Jerash 1989).
Vegetation: Dominated by dense stands of *Acacia tortilis*. Species list incomplete.
Threats to vegetation: Unregulated encroachment by graziers.

Ibex Reserve (NCWCD) (Figure 12.2, 21)

IUCN Management Category: II (National Park).
Phytochorion: Arabian regional subzone of the Saharo-Sindian regional zone.
Area and altitude: *c*. 2,000 km^2; 600-1100 m.
Status: Multiple use site, including SNR, NR and RUR designations, declared in 1988.
Reasons for protection: Supports a relict population of ibex, is a suitable re-introduction site for idmi gazelle and includes representative examples of the plateau and wadis of the central Tuwayq cuesta.
Physical features: Gently undulating, predominantly limestone plateau, dissected by several deeply incised canyon wadi systems. Aeolian sand has accumulated at the mouths of the larger wadis.
Climate: Nearest stations are at Aflaj, Kharj and Yabrin, with mean annual precipitation of 48-77 mm falling in winter-early spring and mean annual temperatures of *c*. 25 °C for 1970-1986 (Al-Jerash 1989).
Vegetation: Little rain infiltrates the fine textured soils and rock pavements of the plateau, which is barren except for a few perennial shrubs and grasses where water accumulates in small depressions and along shallow waterlines. In the canyon wadis, rainfall runoff from the plateau provides a year-round supply of ground water to perennial plants growing along the stream beds. On lower terraces and alluvial fans, *Acacia tortilis* dominates dwarf shrub communities. Perennial species include many composites and crucifers, and grasses such as *Panicum turgidum* and *Cenchrus ciliaris*. Upper wadi terraces are sparsely vegetated with small perennial grasses, including *Stipagrostis*, *Oropetium* and *Tripogon* spp. *Haloxylon salicornicum* dominates dwarf shrub communities on drifts of aeolian sand. The standing crop of plants is considerably higher behind three fences which have been constructed to exclude domestic livestock from small areas of the wadis. Endemic or near endemic species include *Capparis spinosa* var. *mucronifolia* and *Ochradenus arabicus*. Species which are rare in Saudi Arabia include *Heteroderis pusilla*, *Salsola lachnantha* and an unusual form of *Kickxia acerbiana* with sagittate leaves. 262 recorded species.
Threats to vegetation: Quarrying and uncontrolled grazing by domestic livestock.

Jubail Marine Wildlife Sanctuary, (NCWCD) (Figure 12.2, 22)

IUCN Management Category: II (National Park).
Phytochorion: Within the Nubo-Sindian local centre of endemism of the Saharo-Sindian regional zone, but with partly azonal vegetation.
Area and altitude: Terrestrial and intertidal zone approximately 1,200 km^2; marine reserve approximately 1,300 km^2; 0-54 m.
Status: NR/RUR, gazetted in 1995.
Reasons for protection: Contains representative examples of all of the main habitat types found along the western Arabian Gulf coast.
Physical features: Central coastal lowlands consisting of relatively flat white sand dunes, sand sheets, sabkha and salt marshes. There is a gentle slope from east to west, and extensive areas of sand dunes blown into ridges lying north-south. The subtidal zone includes sand, mud and rock substrates and coral reefs.
Climate: Nearest representative station is on the coast at Dhahran, with a mean annual precipitation of 91 mm falling in winter-early spring; mean annual temperature 26.5°C (Figure 2.6, Table 2.2).
Vegetation: The dunes and deep sandsheets are covered in shrubland communities dominated by *Calligonum comosum* or *Leptadenia pyrotechnica*, several dwarf shrubland communities dominated by *Haloxylon salicornicum*, *Rhanterium epapposum* or *Lycium shawii* and *Panicum turgidum* grassland. Annual plants, especially *Plantago* spp., are abundant during spring. *Zygophyllum qatarense* dominates a succulent dwarf shrub community which grows on the

shallower sandsheets that overlay saline soils. Several salt-tolerant perennial grasses are associated with this community. Annual plants are never abundant. Sabkha generally barren, but may support a succulent chenopod, *Halocnemum strobilaceum*. Salt marshes are covered in zoned communities of succulent dwarf shrubs including *Suaeda* spp., *Halocnemum strobilaceum* and *Salicornia europaea*. The muds that fringe a few protected bays support a dwarf form of the black mangrove, *Avicennia marina*. Seagrass beds of *Halodule uninervis, Halophila stipulacea* and *H. ovalis* grow on sandy substrates at depths of 1-20 m. *Zygophyllum mandavillei* is endemic and *Mesembryanthemum nodiflorum* rare. 179 recorded spp.
Threats to vegetation: Grazing by sheep, goats and camels. Military activity, particularly tank exercises. The coastline north of Abu Ali Island has not recovered from the oil spills of the 1991 Gulf War.

Mahazat as Sayd (NCWCD) (Figure 12.2, 23)

IUCN Management Category: I (Strict Nature Reserve/Wilderness Area).
Phytochorion: Arabian regional subzone of the Saharo-Sindian regional zone.
Area and altitude: 2,244 km^2, completely fenced; 900-1,100 m.
Status: SNR, declared in 1988.
Reasons for protection: Reintroduction site for Arabian oryx, reem gazelle, ostrich and houbara; representative area of the Nejd pediplain physiographic region.
Physical features: Gently undulating sand and gravel plain with a few low basalt hills.
Climate: Nearest representative station at Turabah, with a mean annual precipitation of 97 mm falling in winter-spring, and a mean annual temperature of 23.1°C for 1970-1986 (Al-Jerash 1989).
Vegetation: Rapidly recovering, extensive though still patchy vegetation cover, mostly dwarf shrubland with emergent small trees of *Acacia tortilis* and other *Acacia* spp., and *Maerua crassifolia*. Robust perennial grasses, including *Panicum turgidum*, *Lasiurus scindicus* and *Octhochloa compressa* are abundant on deeper sand and on low-lying ground, while *Stipagrostis* spp. are more abundant in rocky areas. Many perennial shrubs and forbs grow among the perennial grasses. *Haloxylon salicornicum* dominates on alkaline soils. *Ajuga arabica* is endemic. The number of plant species recorded in the reserve has increased from 56 at the time of fence completion in 1989, to 112 in 1990 and 156 in 1996.
Threats to vegetation: Intermittent and illegal small-scale harvesting of grasses.

Majami al Hadb (NCWCD) (Figure 12.2, 24)

IUCN Management Category: VI (Managed Resource Protected Area).
Phytochorion: Arabian regional subzone of the Saharo-Sindian regional zone.
Area and altitude: *c.* 2,000 km^2; 900-1,200 m.
Status: SNR/NR/RUR, protected since 1994.
Reasons for protection: Spectacular landscape of granite domes. Forms part of the former range of Arabian oryx, ostrich and two or three species of gazelle. Grazing rights are under dispute and the area has been given to NCWCD to manage for the public benefit.
Physical features: Junction of several major wadis among granite domes and basalt hills.
Climate: Climate contour maps (Figures 2.7-2.11) indicate a mean annual precipitation of 50-100 mm falling mainly in winter and spring, and a mean annual temperature *c.* 24-25°C.
Vegetation: Rainwater runs off a large area of rocky slopes into the wadis, where fresh water is close to the surface. The granite domes are barren, and the basalt hills support a few small perennial plants including *Stipagrostis* spp. Dwarf shrubland, often dominated by the grass *Cymbopogon schoenanthus*, grows in the waterlines which run off the basalt hills. As the waterlines merge into narrow wadis, trees of *Acacia tortilis*, *A. raddiana*, *A. ehrenbergiana* and *Maerua crassifolia* emerge above the shrubland. Shrubs including *Lycium shawii*, robust perennial grasses including *Cenchrus ciliaris*, *Panicum turgidum* and *Octhochloa compressa* and

many dwarf shrubs grow along the riverbanks and on slightly raised terraces. Where the wadi mouths open into wide sandy plains, trees are less numerous. *Rhazya stricta*, which is not grazed by livestock, is abundant on silty areas in the major wadis. 57 recorded spp., but surveys incomplete.
Threats to vegetation: Grazing by sheep, goats, camels and feral donkeys. Poisoning and cutting of trees for firewood.

National Wildlife Research Centre, NWRC (NCWCD) (Figure 12.2, 25)

IUCN Management Category: II (National Park).
Phytochorion: Somalia-Masai regional centre of endemism.
Area and altitude: Fenced stock exclusion zone of *c*. 15 km^2; 1,400 m.
Status: NR, declared in 1986, with stock exclusion extension created in 1993.
Reasons for protection: Research and breeding centre for Arabian oryx, houbara bustard, and ostrich. Representative portion of Asir highland rangelands.
Physical features: Lies within the narrow belt of high country known as the Asir highlands, on a gently undulating gravel and sandy plain cut by numerous shallow wadis. Isolated fragments of the crystalline rocks of the Arabian Shield protrude from the plain.
Climate: Nearest representative station is Taif, with a mean annual precipitation of 204 mm falling mainly in spring and late summer-autumn; mean annual temperature 22.9°C (Figure 2.6 and Table 2.2).
Vegetation: Typical plant communities of the inland Asir plateau, including associations of *Acacia tortilis*, *A. ehrenbergiana*, *Fagonia indica*, *Zygophyllum simplex*, *Salsola spinescens*, *Salvia aegyptiaca*, *Ajuga arabica*, *Panicum turgidum*, *Launaea mucronata*, *Indigofera spinosa* and *Polycarpaea repens*, with *Citrullus colcynthis* and *Cucumis prophetarum* commonly present in depressions. *Ochradenus baccatus* increased in abundance following the removal of livestock. After spring rains there is a luxuriant growth of *Stipagrostis plumosa* and annuals, including, *Astragalus tribuloides*, *Eragrostis ciliaris*, *Malva parviflora* and *Aizoon canariense*. 222 recorded spp.
Threats to vegetation: Any future expansion of ungulate breeding programmes may encroach on botanical reserve areas; some short-term encroachment by small herds of goats into the stock exclusion extension.

Raydah (NCWCD) (Figure 12.2: 26)

IUCN Management Category: Unassigned.
Phytochorion: Lies at the border of the Somalia-Masai and Afromontane archipelago-like regional centres of endemism and contains elements of both zones.
Area and altitude: 9 km^2; 1,460-2,900 m.
Status: SNR designation, but RUR in practice, declared in 1989.
Reasons for protection: Representative example of the middle and upper altitudinal range of the Asir escarpment. Provides habitat for the nine endemic bird species found in Saudi Arabia.
Physical features: Steep granite and gneiss escarpment.
Climate: Nearest available meteorological station is at 2,093 m at Abha with a mean annual precipitation of 253 mm falling at two times, in late winter-early spring and summer, and with a mean annual temperature18.6°C (Figure 2.6, Table 2.2); at the higher elevations rainfall probably >300 mm. Frequent cloud and fog at higher altitudes and rare snow on the peaks.
Vegetation: Relatively undisturbed, mature juniper woodland dominated by *Juniperus procera* extends from the summit down to *c*. 1,700 m. Trees of tropical African origin, including *Nuxia oppositifolia*, *Maesa lanceolata* and *Celtis africana*, grow in well-watered gullies. *J. procera* is replaced by *Teclea nobilis* and *Tarchonanthus camphoratus* below 1,700 m. Below 1,550 m the slopes are covered in sclerophyllous scrub with trees such as *Ficus* spp., *Buddleja polystachya* and *Ziziphus spina-christi* confined to gullies. *Aloe sabaea* and other rare aloes dominate an unusual

succulent community on rocky south-facing slopes. The lowest slopes are occupied by *Acacia etbaica* xeromorphic woodland, among which grow many rare succulents, including *Caralluma* and *Ceropegia* spp. Endemic and near endemic species include *Aloe abhaica, Aloe rivieri* var. *gracilis*, *Ceropegia aristolochiodes* subsp. *deflersiana*, *Huernia saudi-arabica*, *Sarcostemma* sp., *Senecio asirensis* and *Albuca pendula*. *Centaurothamnus maximus*, which also occurs in Yemen, belongs to an endemic Arabian genus. *Euphorbia agowensis* used to grow in the reserve, but is now believed to be extinct in Saudi Arabia. 25 other species occur which are rare in Saudi Arabia including *Rhyncosia buramensis*, *Pittosporum viridiflorum* subsp. *arabicum*, *Dombeya torrida*, *Halleria lucida, Hybanthus enneaspermus* and *Rhoicissus tridentata*. 347 recorded spp.
Threats to vegetation: Unregulated grazing of cattle and goats. Livestock trampling and runoff along the unsealed road that runs through the reserve are causing soil erosion. An alien invasive, *Opuntia ficus-indica,* is spreading on the lower slopes.

Sahal Rukbah (MAW) (Figure 12.2, 27)

IUCN Management Category: Unassigned.
Phytochorion: Arabian regional subzone of the Saharo-Sindian regional zone.
Area and altitude: 16 km^2, fenced; *c.* 1,000 m.
Status: RUR, non-grazing reserve created in 1985.
Reasons for protection: Investigation of the effects of removal of livestock on the regeneration of the western central desert rangeland.
Physical features: Undulating sand and gravel plain.
Climate: Nearest representative station is Hima Saysad, see Hima Saysad above.
Vegetation: The major plant community consists of *Acacia tortilis* with *Lycium shawii, Salsola spinescens* and *Indigofera spinosa*. 64 recorded spp.
Threats to vegetation: Absence of patrols allows fence-cutting and encroachment by graziers.

Thumamah Nature Park (RDA) (Figure 12.2, 28)

IUCN Management Category: II (National Park).
Phytochorion: Arabian regional subzone of the Saharo-Sindian regional zone.
Area and altitude: *c.* 150 km^2, of which about 6 km^2 is fenced to exclude livestock; 500-900 m.
Status: NR, under partial protection since the time of the late King Khaled.
Reasons for protection: Protection of plants and animals in their natural environment and recreational use. There are four public campsites in the Park.
Physical features: Two spectacular limestone escarpments, separated by a plateau, dominate a gravel plain which is crossed by several shallow wadis.
Climate: Nearest representative station is Riyadh with a mean annual precipitation of 126 mm falling in winter-spring; mean annual temperature 24.8°C (Figure 2.6, Table 2.2).
Vegetation: Woodlands dominated by *Acacia pachyceras* and *Acacia ehrenbergiana* grow in the main channels of the major wadis. There are many shrubs, dwarf shrubs and perennial grasses in the undergrowth along these wadis and a profuse growth of annual plants following good rain. Dwarf shrublands, usually dominated by *Haloxylon salicornicum*, cover level ground in much of the Park. *Rhanterium epapposum* dominates a dwarf shrub community on windblown sand over limestone. Grasslands dominated by perennials such as *Panicum turgidum* and *Lasiurus scindicus* grow along wadis and waterlines and in depressions. Some gravel plains support sparse tufts of small perennial grasses, mainly *Oropetium* and *Tripogon* spp., while others are largely barren, as are the steep slopes and the escarpment. Endemic or near endemic species include *Acacia pachyceras* var. *najdensis*. 203 recorded spp.
Threats to vegetation: Grazing by camels and sheep, except in the fenced area which has been protected for almost ten years, occasional tree cutting and off-road driving.

Umm al Qamari (NCWCD) (Figure 12.2, 29)

IUCN Management Category: Unassigned.
Phytochorion: Within the Somalia-Masai regional centre of endemism, but consists largely of azonal vegetation.
Area and altitude: <0.25 km^2; sea-level.
Status: SNR, protected against hunting since 1977.
Reasons for protection: Supports thousands of resident doves and other birds, including a large egret/heron colony.
Physical features: A coral atoll consisting of two small islands surrounded by a coral shelf with shallow sea. The main body of the islands are fossil coral but with some surface sand accumulation and soil formation.
Climate: Climate contour maps (Figures 2.7-2.11) and data for 1970-1986 for the station at 50 m at Kiyat (Al-Jerash 1989) indicate a mean annual precipitation of 50-100 mm falling in two periods, winter-early spring and summer (similar to the climate at Gizan, Figure 2.6, Table 2.2), and a mean annual temperature of *c*. 30°C.
Vegetation: Consists of two categories of plants, one dew-dependent, the other succulent. The first group forms dense thickets up to 3 m tall and is dominated by *Salvadora persica*, with *Atriplex farinosa*, *Cadaba rotundifolia*, *Suaeda monoica* and *Suaeda fruticosa*. The *S. persica* is denser and taller towards the edges of the vegetated zone, becoming smaller and less frequent toward the middle of the island where the two species of *Suaeda* are more abundant. The second group is composed of scattered herbs including the succulents *Suaeda vermiculata*, *Zygophyllum album, Euphorbia fractiflexa* and the grass-like *Cyperus conglomeratus*. Vegetation surveys incomplete, number of species unknown.
Threats to vegetation: none.

'Uruq Bani Ma'arid (NCWCD) (Figure 12.2, 30)

IUCN Management Category: II (National Park).
Phytochorion: Arabian regional subzone of the Saharo-Sindian regional zone.
Area and altitude: c. 5,000 km^2, including a 2,400 km^2 central area from which camps, sheep, goats and most camels are excluded; 640-1,062 m.
Status: NR/BR/RUR, declared in 1994.
Reasons for protection: Suitable habitat for reintroduction of Arabian oryx, reem and idmi gazelle, and ostrich. Representative of the exposed limestone escarpment, parallel dunes and interdune corridors of the western Rub' al Khali.
Physical features: The west-facing Tuwayq escarpment terminates a narrow limestone plateau that is dissected by numerous incised wadis draining eastwards into interdune corridors. Parallel linear dunes, up to 150 m and covered in mobile red sand, separate the interdune corridors (*shiqqat*) which are filled with sand, silt or gravels.
Climate: Climate contour maps (Figures 2.7-2.11) indicate a mean annual precipitation of *c*. 50 mm falling mainly in winter-spring, and a mean annual temperature *c*. 26°C.
Vegetation: The limestone plateau is largely barren but the incised wadis support a diverse dwarf shrub community with several perennial legumes, grasses and *Acacia* spp. Perennial grasses and dwarf shrubs, including *Fagonia indica*, grow on gravels in the interdune corridors. Dwarf shrub communities on sands are dominated by *Tribulus arabicus*. A very diffuse community dominated by a woody shrub, *Calligonum crinitum* subsp. *arabicum,* together with a sedge *Cyperus conglomeratus* and the perennial grasses *Stipagrostis drarii* and *Centropodia fragilis* grow on the dunes. Annual plants are relatively unimportant, and the increase in biomass following rain is due to the germination of perennials. Endemic or near endemic species are *Calligonum crinitum* subsp. *arabicum*, *Cornulaca arabica*, *Limuem arabicum* and *Crotalaria* sp. aff. *leptocarpa*. A rare, unusual variety (var. *divaricatus*) of *Ziziphus spina-christi* occurs. 106 recorded spp.

Threats to vegetation: Seasonally intermittent encroachment of camels into the non-grazing central area of the reserve.

12.4.5.4 Literature Sources

See Abdul-Baqi (1996), Ahmed (undated), Al-Farhan *et al.* (1994), Alwelaie (1989), Alwelaie *et al.* (1993), Anon. (1985b), Anon. (1985a), Anon. (1987), Anon. (1992: pp. 505-515), Anon. (1994a), Barth (1995), Böer (1994), Chaudhary (1980, 1995) Chaudhary and Al-Juwaid (in press), Chaudhary *et al.* (1988), Child and Grainger (1990), Collenette (1985, 1996), Draz (1969), El-Sheikh and Chaudhary (1988), Fisher (1997), Gillet and Launay (1990), Habibi (1989), Jones *et al.* (1995), Kingery (1971), König (1986), Llewellyn (1987), Mandaville (1990), Robertson (1993, 1995, 1996), Robertson and Collenette (1996), Robertson *et al.* (1996), Robertson *et al.* (1997), Seddon (1996a, 1996b), Seddon *et al.* (1994), Seddon and Khan (1996), Seddon and van Heezik (1996), Shorbagi (1996), Thouless and al Bassri (1989), Thouless and Tatwany (1989) and van Heezik and Seddon (1995, 1996).

12.4.6 United Arab Emirates

12.4.6.1 Relevant Legislation

Each municipality of the seven Emirates has its own environmental regulations, though there is a "Draft Proposal for a Federal Law concerning Preservation and Development of the Environment in the United Arab Emirates". It has not been possible to obtain details of the relevant legislation of each Emirate, though in general environmental and wildlife protection appears to be less developed than that of several of the other countries of the Peninsula.

12.4.6.2 Government Agencies

Arabian Leopard Trust (ALT). A non-governmental organisation mainly concerned with the preservation of the Arabian leopard, but also concerned with conservation in general, including that of plants.

Environmental Research and Wildlife Development Agency (ERWDA). Established by Law no. 4, 1996, amended by Law no. 1, 1997, the overall aim of ERWDA is to enhance the sustainable development of the environment of the Emirate of Abu Dhabi. ERWDA includes a Terrestrial Environmental Research Institute, a Marine Research Institute and a Wildlife Veterinary Research Institute. As we write in October 1997 the precise role of the Agency is not yet clear, but one of its objectives is the creation of protected areas.

National Avian Research Centre (NARC). Established by Royal Decree in 1989, and now absorbed into ERWDA, the main goal of NARC is to secure a population of the houbara bustard in Abu Dhabi that can be hunted in the traditional, sustainable way by Arab falconers. As part of this work a number of ecological projects have been undertaken on grazing, habitat degradation and restoration and vegetation surveys.

12.4.6.3 Protected Areas

Abu Dhabi Nature Reserve (Figure 12.2: 31)

IUCN Management Category: V (Protected landscape).
Phytochorion: Nubo-Sindian local centre of endemism of the Saharo-Sindian regional zone.
Area and altitude: *c.* 20 km^2, located at the eastern lagoon of Abu Dhabi island; sea-level.
Status: Nature Reserve, declared in 1987. Controlled by the police with no public access.

Reasons for protection: Important grounds for crustaceans, as a breeding site for birds and as a hatching ground for fish.
Physical features: Intertidal mudflat with both natural and artificial drainage channels.
Climate: Nearest representative station is Abu Dhabi, with a mean annual precipitation of 80 mm falling mainly in late winter-early spring; mean annual temperature 27.1°C (Figure 2.6, Table 2.2).
Vegetation: Three perennial species recorded: ***Avicennia marina*** and ***Arthrocnemum macrostachyum*** with ***Cistanche tubulosa*** parasitic on the roots of ***A. macrostachyum***. Some unrecorded annuals probably also present.
Threats to vegetation: Prior to the proclamation of the reserve there was damage to young trees by off-road vehicles and honey collectors. Seawater channels dredged in 1987 and 1988 have caused some soil erosion and may be affecting the hydrological features of the ecosystem.

Khawr Dubai (Figure 12.2: 32)
IUCN Management Category: Unassigned.
Phytochorion: Nubo-Sindian local centre of endemism of the Saharo-Sindian regional zone.
Area and altitude: 450 hectares.
Status: Nature Reserve by Royal Decree (Dubai).
Reasons for protection: Conservation of bird life.
Physical features: Intertidal mudflats.
Climate: Nearest representative station is Dubai, with a mean annual precipitation of 116 mm falling mainly in winter-early spring; mean annual temperature 27.2°C (Figure 2.6, Table 2.2).
Vegetation: Three perennial species recorded: ***Avicennia marina***, which is largely planted, ***Arthrocnemum macrostachyum*** and ***Halopeplis perfoliata***. Some unrecorded annuals probably also present.
Threats to vegetation: Sewage effluent and the removal of mangrove litter by the Municpality.

12.4.6.4 Literature Sources

See Anon. (1992: pp. 547-550), Böer (1991, 1995, 1996a, 1996b), Böer and Griggs (1994) and Hellyer (1988).

12.4.7 Yemen

The majority of our information on the status of conservation efforts in the country comes from Anon. (1992: pp. 551-556). This document indicated that, as of 1991, there were no protected areas in Yemen, nor any specific protected area legislation, and as far as we have been able to ascertain, this situation has not yet changed. The 1993 United Nations List of National Parks and protected areas (Anon. 1994a) has no entries for Yemen. However, despite the lack of designated areas, Yemen does appear to have an extant tradition of protected rangeland areas, known as ***mahjur*** in Yemen and as ***hima*** elsewhere, that is perhaps stronger than in other countries of the Peninsula (Anon. 1992, and references therein, Kessler 1995). The use of ***mahjur*** areas in the rangelands of the Dhamar montane plains in northern Yemen is integrated into the local agro-pastoral land-use system, their main importance being as forage reserves in the dry season (Kessler 1995). In comparison with adjacent communal rangelands, ***mahjur*** areas are characterized by a dense cover of perennial grasses, whereas the communal rangelands are dominated by dwarf shrubs. Although no information is available on the incidental role of these ***mahjur*** areas in protection of threatened or endemic plant species, they could clearly play an important role in vegetation conservation.

There are areas of both regional and global significance in Yemen, and it is ironic that the three most important areas of plant diversity in Arabia (see section 12.2) all lie either totally or partly within a country that has no protected areas or appropriate legislation. However, we understand that efforts to document and conserve the flora of Socotra are being jointly undertaken by the Government of Yemen and the Royal Botanic Garden, Edinburgh, and that an ethnoflora of the archipelago of Socotra is in preparation.

Table 12.7 Number of protected areas in the six IUCN protected area management categories (Anon. 1994b, see Table 12.6) and number of unassigned protected areas (see text for details), and whether or not the Convention on International Trade in Endangered Species (CITES), the UNESCO World Heritage Convention, and the Biodiversity Convention have been signed, with year of ratification, for each of the countries of the Arabian Peninsula.

Country	IUCN Categories						Unassigned areas	CITES	World Heritage	Biodiversity Convention
	I	II	III	IV	V	VI				
Bahrain	0	0	0	0	0	0	1	no	1991	1996
Kuwait	1	0	0	0	0	1	2	no	no	no
Oman	0	1	0	2	0	0	5	no	1981	1995
Qatar	?	?	?	?	?	?	?	no	1984	1996
Saudi Arabia	1	6	0	2	3	1	4	1996	1978	no
UAE	0	0	0	0	1	0	1	1990	no	no
Yemen	0	0	0	0	0	0	0	no	1980	1996
Total	2	7	0	4	4	2	13			

12.4.8 Protected Areas Summary

Table 12.7 summarises the number of officially declared reserves in each country of the Peninsula using the 1994 IUCN Protected Area Management Categories (Anon. 1994b, see Table 12.6 for a description of the categories). In the Peninsula as a whole there are currently 19 protected areas in assigned categories and 13 unassigned, the latter consisting of areas <10,000 hectares, one protected area that is not yet officially designated (Wadi Sareen in Oman), and three areas in Oman that were declared in 1997. The table also indicates which of the seven countries of the region have signed the 1973 Convention on International Trade in Endangered Species of Wild Flora and Fauna (CITES), the 1972 UNESCO World Heritage Convention and the 1992 Biodiversity Convention. Currently the only World Heritage protected area site in the Peninsula is the Category II site in Oman, the Arabian Oryx Sanctuary.

12.5 Summary

The varied climates and topography of Arabia, in combination with vicariant evolution and biogeographic influences from Africa, the Mediterranean and Asia, have all contributed to the diversity of the present-day flora of the Peninsula. Of the *c.* 3,500 species of vascular plants there is a high degree of endemism in the flora of Socotra (30%) and 2-8% endemism in other regions, notably in the monsoon-affected Dhofar/eastern Mahra region of Oman and Yemen and in the southwestern highlands of Saudi Arabia and Yemen.

Besides general threats associated with 'development', the main threat to species diversity and vegetation cover comes from overstocking the rangelands with domestic camels, cattle, goats and sheep, and in some areas from unchecked breeding of feral donkeys. Although the 'overgrazing problem', as it is commonly known, is being addressed to some extent by active conservation policies and the creation of new conservation areas, the problem is still greater than the current solutions. Besides the need for governments and conservation organisations to address the over-stocking problem directly by enforcing grazing system plans and controlling stocking density, much could be gained locally by the extension of the traditional protected area or *hima* concept.

Most of the conservation areas in the Peninsula have been declared in the last 15 years, and active conservation area policies are still being pursued. However, some of the areas of high diversity and endemism which urgently require conservation (in particular on Socotra and in mainland Yemen, the southwestern and northwestern highlands of Saudi Arabia and in the southern region of Oman) currently receive little or no formal protection.

References

Abbas, J.A. (1995) Monthly variation of chloride accumulation of *Zygophyllum qatarense* from saline and non-saline habitats in Bahrain. Pp. 183-189 in *Biology of Salt Tolerant Plants*. M.A. Khan & I.A. Ungar, eds. Karachi University Press, Karachi.

Abbas, J.A. & El-Oqlah, A.A. (1992) Distribution and communities of halophytic plants in Bahrain. *Journal of Arid Environments* 22: 205-218.

Abbas, J.A. & El-Oqlah, A.A. (1996) Distribution of plant species along a transect across Bahrain Island. *Journal of Arid Environments* 34: 37-46.

Abbas, J.A., Saleh, M.A. & Mohammed, S.A. (1991a) Edaphic factors and plant species distribution in a protected area in the desert of Bahrain Island. *Vegetatio* 95: 87-93.

Abbas, J.A., Saleh, M.A. & Mohammed, S.A. (1991b) Plant communities of a protected area in the desert of Bahrain island. *Journal of Arid Environments* 20: 21-42.

Abd El-Ghani, M.M. (1994) Vegetation along a transect in the Hijaz mountains (Saudi Arabia). *Feddes Repertorium* 105: 517-530.

Abdel-Razik, M.S. (1991) Population structure and ecological performance of the mangrove *Avicennia marina* (Forssk.) Vierh. on the Arbian Gulf coast of Qatar. *Journal of Arid Environments* 20: 331-338.

Abdel-Razik, M.S. & Ismail, A.M. (1990) Vegetation composition of a maritime salt marsh in Qatar in relation to edaphic factors. *Journal of Vegetation Science* 1: 85-88.

Abdul-Baqi, H. (1996) *The National Parks in Saudi Arabia*. Unpublished report. Ministry of Agriculture and Water, Riyadh, Saudi Arabia.

Abualjadayel, H., Badahdah, A., Baghdadi, M., Romaih, S., Taibah, A., Aktar, O., Ady, J. & Adas, A. (1991) *At-Taif Escarpment Nature Reserve and Recreation Area, Report and Master Plan with Guidelines*. Department of Landscape Architecture, School of Environmental Design, Faculty of Engineering, King Abdulaziz University, Jeddah, for National Commission for Wildlife Conservation and Development, Saudi Arabia.

Abulfatih, H.A. (1984) Elevationaly restricted floral elements of the Asir mountains, Saudi Arabia. *Journal of Arid Environments* 6: 247-252.

Abulfatih, H.A. (1992) Vegetation zonation along an altitudinal gradient between sea level and 3000 metres in south-western Saudi Arabia. *Journal King Saud University (Science)* 4: 57-59.

Abulfatih, H.A., Emara, H.A. & El Hashish, A. (1989) The influence of grazing on vegetation and soil of Asir Highlands in south western Saudi Arabia. *Arab Gulf Journal of Scientific Research* 7(1): 69-78.

Abuzinada, A.H., Hawksworth, D.L. & Bokhary, H.A. (1986) The lichens of Saudi Arabia, with a key to the species reported. *Arab Gulf Journal of Scientific Research* 2: 1-51.

Adams, R.P., Demeke, T. & Abulfatih, H.A. (1993) RAPD DNA fingerprints and terpenoids: clues to past migrations of *Juniperus* in Arabia and East Africa. *Theoretical and Applied Genetics* 87: 22-26.

Ady, J., Aktar, O., Joma, A., Mattar, M., Nakshbandi, A., Al-Qahtani, T., Al-Rehaily, Z. & Samarkandi, R. (1995) *The Farasan Islands: Conservation and the Development of Eco-tourism in the Southern Red Sea. Nature Reserves and Recreation Area, Report and Master Plan with Guidelines*. Department of Landscape Architecture, School of Environmental Design, Faculty of Engineering, King Abdulaziz University, Jeddah, for National Commission for Wildlife Conservation and Development, Saudi Arabia.

Ady, J., Bawayan, H., Dagestani, F., Al-Janadi, A., Al-Jehani, S., Al-Qawasmi, Z., Rashwan, A., Al-Shahrani, W., Syed, M.M. & Joma, H. (1994) *Jibal Qaraqir and Jibal Ad-Dubbagh:*

Conservation and the Development of Eco-tourism in the Land of Madyan. Nature Reserves and Recreation Area, Report and Master Plan with Guidelines. Department of Landscape Architecture, School of Environmental Design, Faculty of Engineering, King Abdulaziz University, Jeddah, for National Commission for Wildlife Conservation and Development, Saudi Arabia.

Ahmed, S.S. (undated) *Phyto-ecology of Zifaf Island.* Unpublished report. National Commission for Wildlife Conservation and Development, Riyadh, Saudi Arabia.

Al-Belushi, J.D., Glennie, K.W. & Williams, B.P.J. (1996) Permo-Carboniferous glaciogenic al Khlata Formation, Oman: a new hypothesis for origin of its glaciation. *GeoArabia* 1: 389-404.

Albers, F. (1983) Cytotaxonomic studies in African Asclepiadaceae. *Bothalia* 14: 795-798.

Aleem, A.A. (1978) Comparison of intertidal zonation at Al-Ghardaga (Egypt) and at Obhor (Saudi Arabia), Red Sea. *Proceedings of the Saudi Biological Society* 2: 71-82.

Aleem, A.A. (1979) A contribution to the study of seagrasses along the Red Sea coast of Saudi Arabia. *Proceedings of the Saudi Biological Society* 3: 113-136.

Al-Farhan, A., Al-Johani, A., Al-Sadoon, Q. & Sulayem, M. (1994) *Possibility of Releasing Gazelles in Mjami al Hadb.* Unpublished report (in Arabic). National Commission for Wildlife Conservation and Development, Riyadh, Saudi Arabia.

Al-Gifri, A.N. & Al-Subai, M.J. (1994) Vegetation between Abyan and Modia (Abyan Governorate, Jemen). *Feddes Repertorium* 105: 229-234.

Al-Gifri, A.N. & Hussein, M.A. (1993) Plant communities between Abyan and Modia (Abyan Governorate, Jemen). *Feddes Repertorium* 104: 267-270.

Al-Gifri, A.N. & Kürschner, H. (1996) First records of bryophytes from the Hadramaut and Abyan Governorate, Southern Yemen. Studies in Arabian bryophytes 20. *Nova Hedwigia* 62: 137-146.

Al-Gifri, A.N., Kürschner, H. & Mies, B. (1995) New records, additions and a new species, *Sematophyllum socotrense* Buck (Sematophyllaceae, Musci) to the bryophytre flora of Socotra (Yemen). Studies in Arabian bryophytes 19. *Nova Hedwigia* 61: 467-480.

Al Hassan, A.Y. & Hill, D.R. (1986) *Islamic Technology, an Illustrated History.* Cambridge University Press.

Al-Houty, W., Abdal, M. & Zaman, S. (1993) Preliminary assessment of the Gulf War on Kuwaiti desert ecosystems. *Journal of Environmental Science and Health* 28: 1705-1726.

Al-Hubaishi, A. & Müller-Hohenstein, K. (1984) *An Introduction to the Vegetation of Yemen. Ecological Basis, Floristic Composition, Human Influence.* Deutsche Gesellschaft für Technische Zusammenarbeit, Eschborn.

Al-Jerash, M.A. (1989) *Data for Climatic Water Balance in Saudi Arabia 1970-1986 A.D.* Scientific Publishing Centre, King Abdulaziz University, Jeddah, Saudi Arabia.

Al-Laboun, A.A. (1993) *Lexicon of the Paleozoic and Lower Mesozoic of Saudi Arabia.* Al-Hudud Publishers, Riyadh, Saudi Arabia.

Al-Rawi, A. (1987) *Flora of Kuwait. Volume 2: Compositae and Monocotyledonae.* Kuwait University, Kuwait.

Alwelaie, A.N. (1989) Factors contributing to the degradation of the environment in central, eastern and northern Saudi Arabia. Pp. 37-61 in *Wildlife Conservation and Development in Saudi Arabia. Proceedings of the First Symposium.* National Commission for Wildlife Conservation and Development, Riyadh, Saudi Arabia.

Alwelaie, A.N., Chaudhary, S.A. & Alwetaid, Y. (1993) Vegetation of some Red Sea islands of the Kingdom of Saudi Arabia. *Journal of Arid Environments* 24: 287-296.

Anon. (1982) *Range and Livestock Survey.* GRM International Pty. Ltd., for the Ministry of Agriculture and Fisheries, Oman.

Anon. (1983) *SEPASAT Newsletter No. 1.* Royal Botanic Gardens, Kew.

Anon. (1985a) *Development of the Thumamah Nature Park. Ecological Survey (Task 9).* Unpublished report. Riyadh Development Authority, Riyadh, Saudi Arabia.

Anon. (1985b) *The State of the Environment in the Kingdom of Saudi Arabia. Part II. The Terrestrial Environments*. Unpublished report. Meteorological and Environmental Protection Administration, Jeddah, Saudi Arabia.

Anon. (1985c) *The Use of Medicinal Plants at the Primary Health Care Level*. Unpublished report. Kuwait Institute of Scientific Research, Kuwait.

Anon. (1986a) *Oman Coastal Zone Management Plan: Greater Capital Area*. IUCN, Gland, for the Ministry of Commerce and Industry, Oman.

Anon. (1986b) *Water and Solid Waste Assessment Report; Environmental Management Strategy*. Report prepared for the Environment Protection Committee of the State of Qatar by Environmental Resources Ltd., London.

Anon. (1987) *Report on the NCWCD Extensive Field Survey of the Al 'Arid/Uruq bani-Mu'Arid, December 8th-15th, 1987. Protected Area Report no. 2*. National Commission for Wildlife Conservation and Development, Riyadh, Saudi Arabia.

Anon. (1988) *Oman Coastal Zone Management Plan: Quriyat to Ras al Hadd*. IUCN, Gland, for the Ministry of Commerce and Industry, Oman.

Anon. (1989a) *Oman Coastal Zone Management Plan: Dhofar. Volume 1. Action Plan*. IUCN, Gland, for the Ministry of Commerce and Industry, Oman.

Anon. (1989b) *Oman Coastal Zone Management Plan: Dhofar. Volume 2. Resource Atlas*. IUCN, Gland, for the Ministry of Commerce and Industry, Oman.

Anon. (1990) *Saline Agriculture*. Board on Science and Technology for International Development (BOSTID), National Academy Press, Washington, DC.

Anon. (1991) *Oman Coastal Zone Management Plan: Musandam*. IUCN, Gland, for the Ministry of Commerce and Industry, Oman.

Anon. (1992) *Protected Areas of the World: A Review of National Systems. Volume 2: Palaearctic*. IUCN, Gland, Switzerland and Cambridge, UK.

Anon. (1993) *Khawrs and Springs of the Dhofar Governorate. Land Use Proposals*. Unpublished report. Planning Committee for Development and Environment in the Governorate of Dhofar, Oman.

Anon. (1994a) *1993 United Nations List of National Parks and Protected Areas*. IUCN, Gland, Switzerland and Cambridge, U.K.

Anon. (1994b) *Guidelines for Protected Area Management Categories*. IUCN, Gland Switzerland.

Anon. (1995) *Detailed Land Use Study in Jabal Dhofar. Phase 2. Policy Options and Proposals*. Supreme Committee for Town Planning, Oman.

Anon. (1996) *The 1996 Cyclones and Kharif of Southern Dhofar*. Unpublished report. Ministry of Water Resources, Directorate General of Regional Affairs, Salalah District, Oman.

Anon. (undated) *The Khawrs of Dhofar*. Planning Committee for Development and Environment in the Governorate of Dhofar, Oman.

Anton, D. & Vincent, P. (1986) Parabolic dunes of the Jufurah Desert, Eastern Province, Saudi Arabia. *Journal of Arid Environments* 11: 187-198.

Asmodé, J.-F. (1989) Status and trends of the juniper woodland in the Taif area. Pp. 34-47 in *Quarterly Report*. National Wildlife Research Centre, Taif, Saudi Arabia.

Babikir, A.A. (1984) Vegetation and envrironment on the coastal sand, dunes and playas of Khor El-Odaid Area, Qatar. *Geo Journal* 9: 377-385.

Babikir, A.A. (1986) The vegetation of natural depressions in Qatar in relation to climate and soil. *Journal of Arid Environments* 10: 165-173.

Babikir, A.A. & Kürschner, H. (1992) Vegetational patterns within a coastal saline of NE-Qatar. *Arab Gulf Journal of Scientific Research* 10: 61-75.

Bagnold, R.A. (1941) *The Physics of Blown Sand and Desert Dunes*. Chapman & Hall, London.

Baierle, H.U. (1993) *Vegetation und Flora im südwestlichen Jordanien*. Dissertationes Botanicae 200, Stuttgart.

Baierle, H.U. & Frey, W. (1986) A vegetation transect through Central Saudi Arabia (at-Taif - ar-Riyad). Pp. 111-136 in *Contributions to the Vegetation of Southwest Asia*. H. Kürschner, ed.

Beihefte Tübinger Atlas Vorderen Orients, Reihe A (Naturwissenschaften) Nr. 24. Dr Ludwig Reichert Verlag, Wiesbaden.

Baierle, H.U., Frey, W. & El-Sheikh, M. (1985) *Vegetation und Flora im mittleren Saudi Arabien (at-Ta'if - ar-Riyad)*. Beihefte Tübinger Atlas Vorderen Orients, Reihe A (Naturwissenschaften) Nr. 22. Dr Ludwig Reichert Verlag, Wiesbaden.

Baldry, J. (1982) Textiles in Yemen: Historical references to Trade and Commerce in Textiles in Yemen from Antiquity to Modern times. *British Museum Occasional Paper, No. 27.*

Balfour, I.B. (1882a) Diagnoses plantarum novarum et imperfecte descriptorum phanerogamarum Socotrensium. *Proceedings of the Royal Society of Edinburgh* 11: 498-514.

Balfour, I.B. (1882b) Diagnoses plantarum novarum et imperfecte descriptorum phanerogamarum Socotrensium. *Proceedings of the Royal Society of Edinburgh* 11: 834-832.

Balfour, I.B. (1884a) Diagnoses plantarum novarum et imperfecte descriptorum phanerogamarum Socotrensium. *Proceedings of the Royal Society of Edinburgh* 12: 76-98.

Balfour, I.B. (1884b) Diagnoses plantarum novarum et imperfecte descriptorum phanerogamarum Socotrensium. *Proceedings of the Royal Society of Edinburgh* 12: 402-411.

Balfour, I.B. (1888) Botany of Socotra. *Transactions of the Royal Society of Edinburgh* 31: 1-446.

Balinsky, B.I. (1962) Patterns of animal distribution of the African Continent. *Annales of the Cape Province Museum* 2: 299-310.

Barth, H.-J. (1995) An assessment of the impact of overgrazing within the sanctuary and an evaluation of the potential grazing capacity. Pp. 131-145 in *Establishment of a Marine habitat and Wildlife Sanctuary for the Gulf Region. Final Report for Phase III*. CEC/NCWCD, Frankfurt & Jubail.

Barth, H.K. (1976) Natürraumliche Gliederung und Einzellandschaften. Pp. 51-90 in *Saudi Arabien. Natur, Geschichte, Mensch und Wirtschaft*. H.H. Blume, ed. H. Erdmann Verlag.

Basson, P.W., Burchard, J.E., Hardy, J.T. & Price, A.R.G. (1977) *Biotopes of the Western Arabian Gulf*. Aramco, Dhahran, Saudi Arabia.

Batanouny, K.H. (1979) Vegetation along the Jeddah-Mecca road: pattern and process as affected by human impact. *Journal of Arid Environments* 2: 21-30.

Batanouny, K.H. (1981) *Ecology and Flora of Qatar*. Qatar University Press, Qatar.

Batanouny, K.H. & Turki, A.A. (1983) Vegetation of south-western Qatar. *Arab Gulf Journal of Scientific Research* 1: 5-19.

Baum, B.R. (1978) *The Genus* Tamarix. The Israel Academy of Sciences and Humanities, Jerusalem.

Baum, B.R. (1989) Studies in the Flora of Arabia: XXV. *Tamarix* in the Arabian peninsula *Notes from the Royal Botanic Garden Edinburgh* 46: 1-6.

Bazara, M., Guarino, L., Miller, A.G. & Obadi, N. (1990) Observations on an endangered fan palm in Arabia. *Edinburgh Journal of Botany* 47: 375-379.

Bendali, F.C., Floret, C., Le Floc'h, E. & Pontanier, R. (1990) The Dynamics of Vegetation and Sand Mobility in Arid Regions of Tunisia. *Journal of Arid Environments* 18: 21-32.

Berthoud, T. & Cleuziou, S. (1983) Farming communities of the Oman peninsula and the copper of Makkan. *Journal of Oman Studies* 6: 239-246.

Beydoun, Z.R. (1991) *Arabian Plate Hydrocarbon Geology and Potential: a Plate Tectonic Approach*. AAPG Studies in Geology #33. American Association of Petroleum Geologists, Tulsa, Oaklahoma.

Blatter, E. (1914-1916) Flora of Aden. *Records of the Botanical Survey of India* 7: 1-418.

Blunt, A. (1968) *A Pilgrimage to Nejd*. Frank Cass, London.

Böer, B. (1991) *Vegetationskartierung der Inseln Abu Dhabi und Rafiq mit Erfassung wichtiger ökologischer Parameter*. Diplomarbeit, Universität Osnabrück.

Böer, B. (1994) Status, environmental factors and recovery of the intertidal and terrestrial vegetation between Ras as-Zaur and Abu Ali Island after the Gulf War oil spill. Pp. 229-253 in *Establishment of a Marine Habitat and Wildlife Sanctuary for the Gulf Region. Final Report for Phase II*. CEC/NCWCD, Frankfurt & Jubail.

Böer, B. (1995) Improving habitat in Abu Dhabi. Pp. 79-84 in *Topics of the Euro-Arab Cooperation Center Symposium on Environmental Risk Assessment, Legislation and Technology, June 6-8, 1995.*

Böer, B. (1996a) A botanical excursion in the UAE. *Tribulus* 6: 22-25.

Böer, B. (1996b) Impact of a major oil spill off Fujairah. *Fresenius Environmental Bulletin* 5: 7-12.

Böer, B. (1996c) Plants as soil indicators along the Saudi coast of the Arabian Gulf. *Journal of Arid Environments* 33: 417-423.

Böer, B. (1996d) Trial planting of mangroves (*Avicennia marina)* and salt marsh plants (*Salicornia europaea*) in oil-impacted soil in Jubail area, Saudi Arabia. Pp. 186-192 in *A Marine Wildlife Sanctuary for the Arabian Gulf.* F. Krupp, A.A. Abuzinada & J.A. Nader, eds. National Commission for Wildlife Conservation and Development, Riyadh, Saudi Arabia.

Böer, B. (1997) An introduction to the climate of the United Arab Emirates. *Journal of Arid Environments* 35: 3-16.

Böer, B. & Griggs, A. (1994) Impact of the oil spill on intertidal areas along the east coast of the United Arab Emirates in April 1994. *Tribulus* 4: 20-23.

Böer, B. & Norton, J. (1996) Vegetation rehabilitation in an enclosure in Abu Dhabi. *Fresenius Environmental Bulletin* 5: 79-84.

Böer, B. & Warnken, J. (1992) Qualitative analysis of the coastal and inland vegetation of the Dawkat ad-Dafi and Dawkat al-Mussalamiya region. Pp. 81-101 in *Establishment of a Marine Habitat and Wildlife Sanctuary for the Gulf region. Final Report for Phase I.* CEC/NCWCD, Frankfurt & Jubail.

Boissier, E. (1867-1888) *Flora Orientalis, volumes 1-5 and supplement*. Basle and Geneva.

Bokhary, H.A., Parvez, S. & Abu-Zinada, A.H. (1993) Lichen flora from high altitude areas of Saudi Arabia. *Nova Hedwigia* 56: 491-496.

Botschantzev, V.P. (1984) Two new species of the genus *Salsola* from Saudi Arabia. *Botanic Zurnal* 69: 686-688.

Boulos, L. (1978) Materials for the flora of Qatar. *Webbia* 32: 369-396.

Boulos, L. (1987) A contribution to the flora of Kuwait. *Candollea* 42: 263-275.

Boulos, L. (1988) A contribution to the flora of south Yemen (P.D.R.Y.). *Candollea* 43: 549-585.

Boulos, L. (1991) A new species of *Salsola* from Oman. Studies in the Chenopodiaceae of Arabia: 3. *Kew Bulletin* 46: 297-299.

Boulos, L. (1992) Notes on *Agathophora* (Fenzl) Bunge and *Cornulaca* Del. Studies in the Chenopodiaceae of Arabia V. *Kew Bulletin* 47: 283-287.

Boulos, L. (1994) Notes on *Acacia* Mill. Studies in the Leguminosae of Arabia I. *Kew Bulletin* 50: 527-537.

Boulos, L. & Al Dosari, M. (1994) Checklist of the Flora of Kuwait. *Journal of the University of Kuwait (Science)* 21: 203-217.

Boulos, L., Miller, A.G. & Mill, R.R. (1994) Regional Overview: South West Asia and the Middle East. Pp. 293-308 in *Centres of Plant Diversity. A Guide and Strategy for their Conservation. Volume 1 Europe, Africa, South West Asia and the Middle East.* S.D. Davis, V.H. Heywood & A.C. Hamilton, eds. IUCN Publications Unit, Cambridge, U.K.

Boulos, L. & Wood, J.R.I. (1983) A new *Senecio* (Compositae) from southwest Arabia. *Kew Bulletin* 38: 491-492.

Bowers, J.E. (1982) The plant ecology of inland dunes in western North America. *Journal of Arid Environments* 5: 199-220.

Breed, C.S., Fryberger, S.G., Andrews, S., McCauley, C., Lennartz, F., Gebel, D. & Horstman, K. (1979) *Regional Studies on Sand Seas, using LANDSAT (ERTS) imagery*. A Study of Global Sand Seas. US Geological Society. Washington.

Brooks, W.H. & Mandil, K.S.D. (1983) Vegetation dynamics in the Asir woodlands of south-western Saudi Arabia. *Journal of Arid Environments* 6: 357-362.

Bruyns, P.V. (1987a) The genus *Caralluma* R. Brown (Asclepiadaceae) in Israel. *Israel Journal of Botany* 36: 73-86.

Bruyns, P.V. (1987b) Miscellaneous notes on Stapelieae (Asclepiadaceae). *Bradleya* 5: 77-90.

Bruyns, P.V. (1988) *Cibirhiza*, a new genus of Asclepiadaceae from Oman. *Notes from the Royal Botanic Garden Edinburgh* 45: 51-54.

Bruyns, P.V. (1989) Studies in the Flora of Arabia XXIV: The genus *Ceropegia* in Arabia. *Notes from the Royal Botanic Garden Edinburgh* 45: 287-326.

Bruyns, P.V. & Forster, P.I. (1991) Studies in the Flora of Arabia XXVI: A new species of *Sarcostemma* from Arabia. *Edinburgh Journal of Botany* 48: 333-335.

Bruyns, P.V. & Jonkers, H.A. (1994) The genus *Caralluma* R.Br. (Asclepiadaceae) in Oman. *Bradleya* 11: 51-69.

Bussmann, R.W. & Beck, E. (1995) The forests of Mount Kenia (Kenya), a phytosociological synopsis. *Phytocoenologia* 25: 467-560.

Carapico, S. (1985) Yemeni agriculture in transition. Pp. 241-254 in *Agricultural Development in the Middle East*. P. Beaumont & K. McLachlan, eds. Wiley, Chichester.

Carter, S. (1982) New succulent spiny Euphorbias from East Africa. *Hookers Icones Plantarum* 39, Part III.

Carter, S. (1985) New species and taxonomic changes in *Euphorbia* from east and northeast tropical Africa and a new species from Oman. *Kew Bulletin* 40: 809-825.

Chadwick, J. & Mann, W.N. (1950) *The Medical Works of Hippocrates: A New Ttranslation from the Original Greek made especially for English Readers*. Blackwell Scientific Publications. Oxford.

Chapman, R.W. (1978a) Geology. Pp. 4-19 in *Quaternary Period in Saudi Arabia. 1: Sedimentological, Hydrogeological, Hydrochemical, Geomorphological, and Climatological Investigations in Central and Eastern Saudi Arabia*. S.S. Al-Sayari & J.G. Zötl, eds. Springer Verlag, Wien.

Chapman, R.W. (1978b) Geomorphology. Pp. 19-30 in *Quaternary Period in Saudi Arabia. 1: Sedimentological, Hydrogeological, Hydrochemical, Geomorphological, and Climatological Investigations in Central and Eastern Saudi Arabia*. S.S. Al-Sayari & J.G. Zötl, eds. Springer Verlag, Wien.

Chaudhary, S.A. (1980) *Preliminary Report on some Range Communities of Saudi Arabia*. Unpublished report. Ministry of Agriculture and Water Resources, Riyadh, Saudi Arabia.

Chaudhary, S.A. (1983) Vegetation of the Great Nafud. *Journal of the Saudi Arabian Natural History Society* 2: 32-37.

Chaudhary, S.A. (1989a) *Grasses of Saudi Arabia*. Ministry of Agriculture and Water, Riyadh, Saudi Arabia.

Chaudhary, S.A. (1989b). Understanding the desert range plants of Saudi Arabia. Pp. 156-164 in *Wildlife Conservation and Development in Saudi Arabia*. A.H. Abu-Zinada, P. Goriup & I.A. Nader, eds. NCWCD Publication No. 3, Riyadh, Saudi Arabia.

Chaudhary, S.A. (1995) Vegetation Communities. Pp. 1-62 in *The Land Resources of the Kingdom of Saudi Arabia (English version), Annex D*. Ministry of Agriculture and Water Resources, Riyadh, Saudi Arabia.

Chaudhary, S.A. & Al-Juwaid, A.A. (in press) *Vegetation of Saudi Arabia - An Introduction*. Ministry of Agriculture and Water, Riyadh, Saudi Arabia.

Chaudhary, S.A. & Cope, T. (1983) Studies in the Flora of Arabia: VI. A checklist of grasses of Saudi Arabia. *Arab Gulf Journal of Scientific Research* 1: 313-354.

Chaudhary, S.A., El-Sheikh, A.M., Al Farraj, M.M., Al-Farhan, A.A., Al-Wutaid, Y. & Ahmad, S.S. (1988) Vegetation of some high altitude areas of Saudi Arabia. *Proceedings of the Saudi Biological Society* 11: 237-246.

Chaudhary, S.A. & Lavranos, J.J. (1985) Studies in the Flora of Arabia XVI: Two new species of *Ceropegia* from Saudi Arabia. *Notes from the Royal Botanic Garden Edinburgh* 42: 315-319.

Child, G. & Grainger, J. (1990) *A System Plan for Protected Areas for Wildlife Conservation and Sustainable Development in Saudi Arabia.* IUCN, Gland, Switzerland & National Commission for Wildlife Conservation and Development, Riyadh, Saudi Arabia.

Clarke, J.E. (1986) *Sultanate of Oman: Proposals for a System of Nature Conservation Areas.* IUCN, Gland, Switzerland.

Cleuzion & Constantini (1980) Premiers éléments sur l'agriculture protohistorique de l'Arabie Orientale. *Paléorient* 6: 245-251.

Colfer, C.J.P. (1990) Indigenous knowledge of midwives in Oman's interior: a preliminary account. *Paper presented at the III International Congress on Traditional Asian Medicine, Bombay, 4-7 January, 1990.*

Collenette, S. (1985) *An Illustrated Guide to the Flowers of Saudi Arabia.* MEPA Flora Publication No. 1. Meteorological and Environmental Protection Administration, Jeddah, Saudi Arabia.

Collenette, S. (1996) *Updated Plant Species Lists for some NCWCD Protected Areas.* Unpublished lists. National Wildlife Research Centre, Taif, Saudi Arabia.

Cope, T.A. (1984) Some new Arabian grasses. *Kew Bulletin* 39: 833-836.

Cope, T.A. (1985) A Key to the Grasses of the Arabian Peninsula. *Arab Gulf Journal of Scientific Research Special Publication No. 1*: 1-82.

Cope, T.A. (1988) The flora of the Sands. Pp. 305-312 in *The Scientific Results of The Royal Geographical Society's Oman Wahiba Sands Project 1985-1987. The Journal of Oman Studies Special Report No. 3.* R.W. Dutton, ed. Office of the Adviser for Conservation of the Environment, Muscat, Oman.

Cope, T.A. (1992) Some new Arabian grasses II. *Kew Bulletin* 47: 655-665.

Cope, T.A. (1993) A new species of *Chrysopogon* (Gramineae) from Oman. *Kew Bulletin* 49: 533-535.

Cornes, M.D. & Cornes, C.D. (1989) *The Wild Flowering Plants of Bahrain.* IMMEL Publishing, London.

Costa, P.M. (1985) The palm-frond house of the Batinah. *Journal of Oman Studies* 8: 117-120.

Creutzburg, N. & Habbe, K.A. (1964) Klimatypen der Erde (Map 1:50 Mio). In Blüthgen, J., ed. Springer-Verlag, Berlin.

Cribb, P.J. (1979) The orchids of Arabia. *Kew Bulletin* 33: 651-678.

Cribb, P.J. (1987) New records of Orchidaceae for Arabia. *Kew Bulletin* 42: 461-463.

Crowell, J.C. (1995) The ending of the Late Paleozoic Ice Age during the Permian Period. Pp. 62-74 in *The Permian of Northern Pangea. Volume 1.* P.A. Scholle, T.M. Peryt & D.S. Ulmer-Scholle, eds. Springer-Verlag, Berlin.

Danin, A. (1972) Mediterranean elements in rocks of the Negev and Sinai deserts. *Notes from the Royal Botanic Garden Edinburgh* 31: 437-440.

Danin, A. (1983) *Desert Vegetation of Israel and Sinai.* Cana Publishing House, Jerusalem.

Danin, A. (1991) Plant Adaptations in Desert Dunes. *Journal of Arid Environments* 21: 193-212.

Danin, A. (1996a) Adaptations of *Stipagrostis* spp. to desert dunes. *Journal of Arid Environments* 34: 279-311.

Danin, A. (1996b) *Plants of Desert Dunes.* Springer-Verlag, Berlin.

Daoud, H.S. (1985) *Flora of Kuwait. Volume 1: Dicotyledoneae* (revised by A. Al-Rawi). KPI, London, in association with Kuwait University.

Darwin, C. (1839) *Narrative of the Surveying Voyages of His Majesty's Ships Adventure and Beagle, Vol. 3. Journal and Remarks 1832-1936; Voyage of the Beagle.* Henry Colburn, London.

Dawood, N.J., ed. (1967) *The Muqaddimah: Ibn Khaldûn, an Introduction to History*, translated from the Arabic by Franz Rosenthal. Routledge & Kegan Paul, London, in association with Secker & Warburg, London.

De Clerck, O. & Coppejans, E.V. (1994) Status of the macroalgae and seagrass vegetation after the 1991 Gulf war oil spill. *Courier Forschungs Institut Senckenberg* 166: 18-21.

De Winter, B. (1971) Floristic relationships between the northern and southern arid areas in Africa. *Mittelungen Botanische Staatssammlung Müenchen* 10: 424-437.

Deil, U. (1986a) Die Wadivegetation der nördlichen Tihamah und Gebirgstihama der Arabischen Republik Jemen. Pp. 167-199 in *Contributions to the Vegetation of Southwest Asia*. H. Kürschner, ed. Beihefte Tübinger Atlas Vorderen Orients, Reihe A (Naturwissenschaften) Nr. 24. Dr Ludwig Reichert Verlag, Wiesbaden.

Deil, U. (1986b) Vom Menschen gestaltetes Relief in der Arabischen Republik Jemen und Konsequenzen für vegetationskundliches Arbeiten. *Colloques Phytosociologiques* 13: 373-395.

Deil, U. (1988) Primäre und sekundäre Standorte sukkulentenreicher Pflanzengessellschaften in Südwest-Arabien. *Flora* 180: 41-57.

Deil, U. (1989) Fragmenta Phytosociologica Arabiae-Felicis II. Adiantetea-Gesellschaften auf der Arabischen Halbinsel, Coenosyntaxa in dieser Klasse sowie allgemeine Überlegungen zur Phylogenie von Pflanzengesellschaften. *Flora* 182: 247-246.

Deil, U. (1991) Rock communities in tropical Arabia. *Flora et Vegetatio Mundi* 9: 175-187.

Deil, U. (1996) Zur Kenntnis der Adiantetea-Gesellschaften des Mittelmeerraumes und umliegender Gebiete. *Phytocoenologia* 26: 481-536.

Deil, U. & Müller-Hohenstein, K. (1984) Fragmenta Phytosociologica Arabiae-Felicis I - Eine *Euphorbia balsamifera*- Gesellschaft aus dem jemenitischen Hochland und ihre Behiehungen zu makaronesischen Pflanzengesellschaften. *Flora* 175: 307-326.

Deil, U. & Müller-Hohenstein, K. (1985) Beiträge zur Vegetation des Jemen. 1. Pflanzengesellschaften und Ökotopgefüge der Gebirgstihama am Beispiel des Beckens von At Tur (J.A.R.). *Phytocoenologia* 13: 1-102.

Deil, U. & Müller-Hohenstein, K. (1996) An outline of the vegetation of Dubai (UAE). *Verhandlungen der Gesellschaft für Ökologie* 25: 77-95.

Deil, U. & Rappenhöner, D. (1989) Studies in phytosociology and fodder yield in the Yemen Arabic Republic - a comparison of methods and results. *Colloques Phytosociologiques* 16: 429-443.

Dequin, H. (1977) Water harvesting in the Yemen Arab Republic in ancient times and today. *Geo Journal* 1: 94-96.

Di Nola, L., Meyer, A.M. & Heyn, C.C. (1983) Respiration, photosynthesis and drought tolerance in mosses from various habitats in Israel. *Israel Journal of Botany* 32: 189-202.

Diggle, P.J. (1990) *Time Series: a Biostatistical Introduction*. Clarendon Press, Oxford.

Doornkamp, J.C., Brunsden, D. & Jones, D.K.C. (1980) *Geomorphology and Pedology of Bahrain*. Geo Abstracts, Norwich.

Doughty, C.M. (1888) *Travels in Arabia Deserta*. Cambridge University Press, Cambridge.

Dransfield, J. & Uhl, N.W. (1938) *Wissmannia* (Palmae) tranferred to *Livistonia*. *Kew Bulletin* 38: 199-200.

Draz, O. (1969) *The hima system of range reserves in the Arabian Peninsula. Its possibilities, range improvement and conservation projects in the Near East*. FAO/PL: PFC/13.11.

Draz, O. (1978) Revival of the Hema system of range reserves as a basis for the Syrian Range Development Program. Pp. 100-103 in *Proceedings of the 1st International Range Congress*, Denver, Colorado, USA.

Dubaie, A.S., Gifri, A.N. & El-Monayeri, M. (1993) Studies on the flora of Yemen 3. On the flora of Wadi Dahr. *Candollea* 48: 101-109.

During, H.J. (1979) Life strategies in bryophytes: a preliminary review. *Lindbergia* 5: 2-18.

Durozoy, G. (1972) Hydrogéologie des basaltes du Harrat Rahat. *Bulletin Bureau de Recherches Géologiques et Minéres, Orleans, Section III*: 37-50.

Dutton, R.W. (1988a) Introduction, overview and conclusions. Pp. 1-18 in *The Scientific Results of The Royal Geographical Society's Oman Wahiba Sands Project 1985-1987. The Journal of Oman Studies Special Report No. 3*. R.W. Dutton, ed. Office of the Adviser for Conservation of the Environment, Muscat, Oman.

Dutton, R.W., ed. (1988b) *The Scientific Results of The Royal Geographical Society's Oman Wahiba Sands Project 1985-1987. The Journal of Oman Studies Special Report No. 3*. Office of the Adviser for Conservation of the Environment, Diwan of Royal Court, Muscat, Oman.

Dyer, M.I., Turner, C.L. & Seastedt, T.R. (1993) Herbivory and its consequences. *Ecological Applications* 3: 10-16.

Eckhardt, H.C., van Royen, N. & Bredenkamp, G.J. (1995) The grassland communities of the slopes and plains of the north-eastern Orange Free State. *Phytocoenologia* 25: 1-21.

Eig, A. (1938) On the phytogeographical subdivision of Palestine. *Palestine Journal of Botany* 1: 4-12.

El Amin, H.M. (1983) *Wild Plants of Qatar*. Arab Organization for Agricultural Development.

El-Demerdash, M.A. (1996) The vegetation of the Farasan Islands, Red Sea, Saudi Arabia. *Journal of Vegetation Science* 7: 81-88.

El-Demerdash, M.A., Hegazy, A.K. & Zilay, M.A. (1995) Vegetation-soil relationship in Tihamah coastal plains of Jazan region, Saudi Arabia. *Journal of Arid Environments* 30: 161-174.

El-Demerdash, M.A. & Zilay, A.M. (1994) An introduction to: the plant ecology of Tihama plains of Jizan region, Saudi Arabia. *Arab Gulf Journal of Scientific Research* 12: 285-299.

El-Hadidi, M.N. (1977) Two new *Zygophyllum* species from Arabia. *Publication of the Cairo University Herbarium* 7/8: 327-331.

El-Hadidi, M.N. (1980) On the taxonomy of *Zygophyllum* section *Bipartita*. *Kew Bulletin* 35: 335-340.

El-Karemy, Z.A.R. & Zayed, K.M. (1996) Contribution to the vegetation and habitats types of Baha plateau (Saudi Arabia). *Feddes Repertorium* 107: 135-144.

Ellenberg, H. (1989) *Opuntia dillenii* als problematischer Neophyt im Nordjemen. *Flora* 182: 3-12.

El-Monayri, M.O., Alhubaishi, A.A. & Dubaie, A.S. (1991) Habitats and vegetation of Wadi Dahr, Sana'a, Yemen Arab Republic III: The wadi bed ecosystems. *Bulletin of the Faculty of Science, Assiut University* 20: 20-42.

El-Sheikh, A.M. & Youssef, M.M. (1981) Halophytic and xerophytic vegetation near al Kharj springs. *Bulletin of the Faculty of Science King Saud University* 12: 5-21.

El-Sheikh, M.A. & Chaudhary, S.A. (1988) Plants and plant communities of the Al-Harrah protected areas. *Proceedings of the Saudi Biological Society* 11: 211-235.

El-Sheikh, M.A., Mahmoud, A. & El-Tom, M. (1985) Ecology of the inland salt marsh vegetation at Al-Shiggat in Al-Qassim district, Saudi Arabia. *Arab Gulf Journal of Scientific Research* 3: 165-182.

El-Shourbagy, M.N., Baeshin, N.A. & el-Sahhar, K.F. (1986) Studies on the ecology of the western provinces of Saudi Arabia I. Vegetation and soil of Jeddah-Tuwal area. *Feddes Repertorium* 97: 705-712.

El-Shourbagy, M.N., Al-Eidaros, O.H. & Al-Zahrani, H.S. (1987) Distribution of *Halopeplis perfoliata* (Forssk.) Bge. ex Schweinf. in the Red Sea coastal salt marshes: phytosociological relations and respones to soil. *Journal of Coastal Research* 3: 179-187.

El-Tawil, B.A.H. (1983) Chemical constituents of indigenous plants used in native medicine in Saudi Arabia II. *Arab Gulf Journal of Scientific Research* 1: 395-419.

Eskuche, U. (1992) Sinopsis cenosistematica preliminar de los pajonales mesofilos seminaturales des nordeste de la Argentina. *Phytocoenologia* 21: 237-312.

Evans, G. (1995) The Arabian Gulf: a modern carbonate-evaporite factory: a review. *Cuadernos de Geologia Iberica* 19: 61-96.

Farjon, A. (1992) The taxonomy of multiseed Junipers (*Juniperus* sect. *sabina*) in Southwest Asia and East Africa (Taxonomic notes on Cupressaceae I). *Edinburgh Journal of Botany* 49: 251-283.

Fayed, A.A. & Zayed, K. (1989) Vegetation along Makkah-Taif road (Saudi Arabia). *Arab Gulf Journal of Scientific Research* 7: 97-117.

Fennema, F. & Briede, J.W. (1990) The effect of clipping and water stress on the production of selected grass species in the Yemen Arab Republic. *Journal of Arid Environments* 19: 119-124.

Field, D.V. (1981a) A new *Huernia* (Asclepiadaceae) from Saudi Arabia, with notes on related species. *Kew Bulletin* 35: 735-757.

Field, D.V. (1981b) A new species of *Rhytidocaulon* (Asclepiadaceae) from Saudi Arabia. *Kew Bulletin* 36: 51-54.

Field, D.V. & Collenette, S. (1984) *Ceropegia superba* (Asclepiadaceae), a new species form Saudi Arabia. *Kew Bulletin* 39: 639-642.

Fisher, M. (1994) Another look at the variability of desert climates, using examples from Oman. *Global Ecology and Biogeography Letters* 4: 79-87.

Fisher, M. (1995a) *Indexed Bibliography of Natural History and Conservation in Oman*. Backhuys, Leiden, the Netherlands.

Fisher, M. (1995b) Is it possible to construct a tree-ring chronology for *Juniperus excelsa* (M.Bieb.) subsp. *polycarpos* (K. Koch) Takhtajan from the northern mountains of Oman? *Dendrochronologia* 12: 119-127.

Fisher, M. (1997) Decline in the juniper woodlands of Raydah Reserve in southwestern Saudi Arabia: a response to climate changes? *Global Ecology and Biogeography Letters* 6: 379-386.

Fisher, M. & Gardner, A.S. (1995) The status and ecology of a *Juniperus excelsa* subsp. *polycarpos* woodland in the northern mountains of Oman. *Vegetatio* 119: 33-51.

Fisher, M. & Gardner, A.S. (in press) The potential of a montane juniper tree-ring chronology for the reconstruction of recent climate in Arabia. In *Quaternary Deserts and Climatic Change*. A.S. Alsharhan, K.W. Glennie, G.L. Whittle & C.G.S.C. Kendall, eds. Balkema, Rotterdam.

Fontana, J.L. (1996) Los pajonales mesofilos seminaturales de Misiones (Argentina). *Phytocoenologia* 26: 179-271.

Forsskål, P. (1775) *Flora Aegyptiaco-Arabica. Sive Descriptiones Plantarum, quas per Aegyptum Inferiorem et Arabiam Felicem detexit, Illustravit Petrus Forsskål. Post mortem auctoris edidit Carsten Niebuhr.* Accedit tabula Arabiae Felicis Geographixo-Botanica, Copenhagen.

Foster, G.M. & Anderson, B.G. (1978) *Medical Anthropology*. John Wiley & Sons, New York.

Foucault, B.D. (1978) Premières observations phytosociologiques sur le marais de Saint-Louis-Marie-Galante (Guadeloupe). *Documents Phytosociologiques* N.S.2: 181-189.

Fouda, M.M. & Al-Muharrami, M.A. (1996) Significance of mangroves in the arid environment of the Sultanate of Oman. *Oman Journal of Agricultural Sciences* 1: 41-49.

Freitag, H. (1989a) Contributions to the chenopod flora of Egypt. *Flora* 183: 149-173.

Freitag, H. (1989b) *Piptatherum* and *Stipa* (Gramineae) in the Arabian Peninsula and tropical East Africa. Pp. 115-132 in *Plant Taxonomy, Phytogeography and Related Subjects. The Davis and Hedge Festschrift*. K. Tan, ed. Edinburgh University Press, Edinburgh.

Freitag, H. (1991) The distribution of some prominent Chenopodiaceae in SW Asia and their phytogeographical significance. *Flora et Vegetatio Mundi* 9: 281-292.

Frey, W. & Kürschner, H. (1982) The first records of bryophytes from Saudi Arabia. Studies in Arabian bryophytes 1. *Lindbergia* 8: 157-160.

Frey, W. & Kürschner, H. (1986) Masqat Area (Oman), remnants of vegetation in an urban habitat. Pp. 201-221 in *Contributions to the Vegetation of Southwest Asia*. H. Kürschner, ed. Beihefte Tübinger Atlas Vorderen Orients, Reihe A (Naturwissenschaften) Nr. 24. Dr Ludwig Reichert Verlag, Wiesbaden.

Frey, W. & Kürschner, H. (1987) A desert bryophyte synusia from the Jabal Tuwayq mountain systems (central Saudi Arabia) with the description of two new *Crossidium* species (Pottiaceae). Studies in Arabian bryophytes 8. *Nova Hedwigia* 45: 119-136.

Frey, W. & Kürschner, H. (1988) Bryophytes of the Arabian Peninsula and Socotra. Floristics, phytogeography and definition of the xerothermic Pangean element. Studies in Arabian bryophytes 12. *Nova Hedwigia* 46: 37-120.

Frey, W. & Kürschner, H. (1989) *Die Vegetation im Vorderen Orient. Erläuterungen zur Karte A VI 1 Vorderer Orient. Vegetation des Tübinger Atlas des Vorderen Orients*. Beihefte zum Tübinger Atlas des Vorderen Orients, Reihe A (Naturwissenschaften) Nr. 30. Dr Ludwig Reichert Verlag, Wiesbaden.

Frey, W. & Kürschner, H. (1991a) Conspectus Bryophytorum Orientalum et Arabicorum. An annotated catalogue of the bryophytes of Southwest Asia. *Bryophyte Bibliography* 39: 1-181.

Frey, W. & Kürschner, H. (1991b) Das Fossombronio-Gigaspermetum mouretii in der Judäischen Wüste. 2. Okosoziologie und Lebensstrategien. *Cryptogamic Botany* 2: 73-84.

Frey, W. & Kürschner, H. (1991c) Lebensstategien von terrestrischen Bryophyten in der Judäischen Wüste. *Botanica Acta* 104: 172-182.

Frey, W., Kürschner, H., El-Sheikh, A.M. & Migahid, A.M. (1984) Zonation and photosynthetic pathways of halophytes on the Red Sea coast near Tawwal (Saudi Arabia). *Notes from the Royal Botanic Garden Edinburgh* 42: 45-56.

Frey, W., Kürschner, H. & Stichler, W. (1985) Photosynthetic pathways and ecological distribution of halophytes from four littoral salt marshes (Egypt/Sinai, Saudi Arabia, Oman and Iran). *Flora* 177: 127-134.

Friis, I. (1983) The acaulescent and succulent species of *Dorstenia* Sect. Kosaria (Moraceae) from Northeast tropical Africa and Arabia. *Nordic Journal of Botany* 3: 533-538.

Gabali, S.A. & Al-Gifri, A.N. (1990) Flora of South Yemen, Angiosperms. A provisional checklist. *Feddes Repertorium* 101: 378-383.

Gabali, S.A. & Al-Gifri, A.N. (1991) A survey of the flora of the vegetation of Hadramaut, Republic of Yemen. *Fragmenta Floristica et Geobotanica* 36: 127-134.

Gardner, A.S. & Fisher, M. (1994) How the forest lost its trees: Just so stories about juniper in Arabia. *Journal of Arid Environments* 26: 199-201.

Gardner, A.S. & Fisher, M. (1996) The distribution and status of the montane juniper woodlands of Oman. *Journal of Biogeography* 23: 791-803.

Gass, I.G., Ries, A.C., Shackleton, R.M. & Smewing, J.D. (1990) Tectonics, geochronology and geochemistry of the Precambrian rocks of Oman. Pp. 585-599 in *The Geology and Tectonics of the Oman Region*. A.H.F. Robertson, M.P. Searle & A.C. Ries, eds. Geological Society Special Publication 49. The Geological Society, London.

Ghanem, Y. & Eighmy, J. (1984) *Hema* and traditional land use management among arid zone villagers of Saudi Arabia. *Journal of Arid Environments* 7: 287-297.

Ghazanfar, S.A. (1991a) Floristic composition and the analysis of vegetation of the Sultanate of Oman. *Flora et Vegetatio Mundi* 9: 215-227.

Ghazanfar, S.A. (1991b) Vegetation structure and phytogeography of Jabal Shams, an arid mountain in Oman. *Journal of Biogeography* 18: 299-309.

Ghazanfar, S.A. (1992a) *An Annotated Catalogue of the Flora of Oman*. Scripta Botanica Belgica. National Botanic Garden of Belgium, Meise, Belgium.

Ghazanfar, S.A. (1992b) Quantitative and biogeographic analysis of the flora of the Sultanate of Oman. *Global Ecology and Biogeography Letters* 2: 189-195.

Ghazanfar, S.A. (1993a) A new species of *Euphorbia* (Euphorbiaceae) from Masirah Island, Sultanate of Oman. *Novon* 3: 258-260.

Ghazanfar, S.A. (1993b) Vegetation of the Khawrs and Fresh Water Springs of Dhofar. Part E in: *Khawrs and Springs of the Dhofar Governorate. Survey and Monitoring Studies*. Unpublished report. Planning Committee for Development and Environment in the Governorate of Dhofar, Oman.

Ghazanfar, S.A. (1994a) *Handbook of Arabian Medicinal Plants*. CRC Press, Boca Raton.

Ghazanfar, S.A. (1994b) *Novitates* from the flora of the Sultanate of Oman. *Edinburgh Journal of Botany* 51: 59-63.

Ghazanfar, S.A. (1995) Wasm: A traditional method of healing by cauterization. *Journal of Ethnopharmacology* 47: 125-128.

Ghazanfar, S.A. (1996a) Aucher-Éloy's plant specimens from the Immamat of Muscat. *Taxon* 45: 609-626.

Ghazanfar, S.A. (1996b). Traditional health plants of the Arabian Peninsula. *Plants for Food and Medicine*. Paper presented at the joint meeting of the Society for Economic Botany and International Society for Ethnopharmacology, 1-7 July 1996, London.

Ghazanfar, S.A. (1997) The phenology of desert plants: a 3-year study in a gravel desert wadi in northern Oman. *Journal of Arid Environments* 35: 407-417.

Ghazanfar, S.A. (in press-a) Coastal Vegetation of Oman. *Estuarine, Coastal and Shelf Science.*

Ghazanfar, S.A. (in press-b) Status of the flora and conservation in the Sultanate of Oman. *Biological Conservation*.

Ghazanfar, S.A. & Al-Sabahi, A.A. (1993) Medicinal plants of northern and central Oman. *Economic Botany* 47: 89-98.

Ghazanfar, S.A. & Gallagher, M.D. (in press) Remarkable lichens from the Sultanate of Oman. *Nova Hedwigia*.

Ghazanfar, S.A., Miller, A.G., McLeish, I., Cope, T.A., Cribb, P. & Al Rawahi, S.H. (1995) *Plant Conservation in Oman. A Study of the Endemic, Regionally Endemic and Threatened Plants of the Sultanate of Oman*. Unpublished Report, Sultan Qaboos University, Oman.

Ghazanfar, S.A. & Rappenhöner, D. (1994) Vegetation and flora of the Islands of Masirah and Shagaf, Sultanate of Oman. *Arab Gulf Journal of Scientific Research* 12: 509-524.

Ghazanfar, S.A. & Rechinger, B. (1996). Two multi-urpose seed oils form Oman. *Plants for Food and Medicine*. Paper presented at the joint meeting of the Society for Economic Botany and International Society for Ethnopharmacology, 1-7 July 1996, London.

Giacomini, V., Longhitano, N. & Corti, L. (1979) *Cartography of the Vegetation of the Eastern (Hasa) Province of Saudi Arabia*. Pubblicazioni dell'Instituto di Botanica dell'Universiti di Catania. Universitaria Libraria Catanese, Catania.

Gillet, H. & Launay, C. (1990) *Vegetation studies, Mahazat Assaid Reserve, 1990*. Unpublished report. National Wildlife Research Centre, Taif, Saudi Arabia.

Glennie, K.W. (1987) Desert sedimentary environments, present and past: a summary. *Sedimentary Geology* 50: 135-165.

Glennie, K.W. (1995) *The Geology of the Oman Mountains: an Outline of their Origin*. Scientific Press, Beaconsfield.

Glennie, K.W. (1996) Geology of Abu Dhabi. Pp. 16-35 in *Desert Ecology of Abu Dhabi - a Review and Recent Studies*. P.E. Osborne, ed. Pisces Publications, Newbury, UK.

Glennie, K.W. (in press) The desert of southeast Arabia: a product of Quaternary climatic change. In *Quaternary Deserts and Climatic Change*. A.S. Alsharhan, K.W. Glennie, G.L. Whittle & C.G.S.C. Kendall, eds. Balkema, Rotterdam.

Glennie, K.W., Boeuf, M.G.A., Clarke, M.W.H., Moody-Stuart, M., Pilaar, W.F.H. & Reinhardt, B.M. (1974) *Geology of the Oman Mountains. Part One (Text), Part Two (Tables and Illustrations), Part Three (Enclosures)*. Verhandelingen van het Koninklijk Nederlands geologisch mijnbouwkundig Genootschap, Volume 31, The Hague.

Glennie, K.W., Hughes Clarke, M.W., Boeuf, M.G.A., Pilaar, W.F.H. & Reinhardt, B.M. (1990) Inter-relationship of Makran-Oman Mountains belts of convergence. Pp. 773-786 in *The Geology and Tectonics of the Oman Region*. A.H.F. Robertson, M.P. Searle & A.C. Ries, eds. Geological Society Special Publication No. 49. The Geological Society, London.

Groom, N. (1981) *Frankinscence and Myrrh. A Study of the Arabian Incense Trade*. Longman, London & Libraire du Liban, Beirut.

Grosser, L. (1988) *Untersuchungen zu Vegetation, Boden und Landnutzungsmoglichkeiten in der Gebirgsregion Haraz/Arabische Republik Jemen*. Bayreuther Bodenkundlidie Beridile, Bayreuth.

Guest, E. (1966) *Introduction to the Flora of Iraq*. Royal Botanic Gardens, Kew and Baghdad University, Baghdad.

Gutte, P. (1986) Beitrag zur Kenntnis zentralperuanischer Pflanzengesellschaften III Pflanzengesellschaften der subalpine Stufe. *Feddes Repertorium* 97: 319-371.

Gwynne, M.D. (1968) Socotra. *Acta Phtogeography Suecica* 54: 179-185.

Habibi, K. (1989) *Biology of the Farasan Gazelle*. Fourth Farasan report. National Commission for Wildlife Conservation and Development, Riyadh, Saudi Arabia.

Hajar, A.S., Faragalla, A.A. & Al-Ghamdi, K.M. (1991) Impact of biological stress on *Juniperus excelsa* M. Bieb. in south-western Saudi Arabia: insect stress. *Journal of Arid Environments* 21: 327-330.

Hajar, A.S.M. (1993) A comparative ecological study on the vegetation of the protected and grazed parts of Hema Sabihah, in Al-Bahah region, south western Saudi Arabia. *Arab Gulf Journal of Scientific Research* 11: 259-280.

Hale, M.E. (1983) *The Biology of Lichens*. Edward Arnold, London.

Hall, J.B. (1984) *Juniperus excelsa* in Africa: a biogeographial study of an Afromontane tree. *Journal of Biogeography* 11: 47-61.

Halliday, P. (1983) The genus *Kleinia* (Compositae) in Arabia. *Kew Bulletin* 39: 817-827.

Halwagy, R. (1986) On the ecology and vegetation of Kuwait. Pp. 81-109 in *Contributions to the Vegetation of Southwest Asia*. H. Kürschner, ed. Beihefte zum Tübinger Atlas des Vorderen Orients, Reihe A (Naturwissenschaften) Nr. 24. Dr Ludwig Reichert Verlag, Wiesbaden.

Halwagy, R. & Halwagy, M. (1977) Ecological studies on the desert of Kuwait. III. The vegetation of the coastal salt marshes. *Journal of the University of Kuwait (Science)* 4: 33-73.

Halwagy, R., Moustafa, A.F. & Kamal, S. (1982) On the ecology of the desert vegetation in Kuwait. *Journal of Arid Environments* 5: 95-107.

Hamilton, W.R., Whybrow, P.J. & McClure, H.A. (1978) Fauna of fossil mammals from the Miocene of Saudi Arabia. *Nature* 274: 248-249.

Hardy, A.D., Sutherland, H.H., Vaishnav, R. & Worthing, M.A. (1995) A report on the composition of mercurials used in traditional medicines in Oman. *Journal of Ethnopharmacology* 49: 17-22.

Harrison, D.L., ed. (1977) *The Scientific Results of the Oman Flora and Fauna Survey 1975. The Journal of Oman Studies Special Report No. 1*. Ministry of Information and Culture, Muscat, Oman.

Hawksworth, D.L. & Hill, D.J. (1984) *The Lichen-forming Fungi*. Blackie, Glasgow.

Hawksworth, D.L., Lawton, R.M., Martin, P.G. & Stanley-Price, K. (1984) Nutritive value of *Ramalina duriaei* grazed by gazelles in Oman. *The Lichenologist* 15: 93-94.

Heathcote, J.A. & King, S. (in press) Umm as Samim, Oman: a sabkha with evidence for climatic change. In *Quaternary Deserts and Climatic Change*. A.S. Alsharhan, K.W. Glennie, G.L. Whittle & C.G.S.C. Kendall, eds. Balkema, Rotterdam.

Hedberg, O. (1986) Origins of the Afro-Alpine Flora. Pp. 443-468 in *High Altitude Tropical Biogeography*. F. Vuilleumier & M. Monasterio, eds. Oxford University Press, Oxford.

Hedge, I.C. (1982) Some new and interesting species of Labiatae. *Notes from the Royal Botanic Garden Edinburgh* 40: 63-73.

Hedge, I.C. & King, R.A. (1983) The Cruciferae of the Arabian Peninsula: A check-llist of species and a key to genera. Studies in the Flora of Arabia IV. *Arab Gulf Journal of Scientific Research* 1: 41-66.

Hejny, S. (1960) *Ökologische Characteristik der Wasser - und Sumpfpflanzen in den slowakischen Tiefebenen (Donau - und Thiessgebiet)*. Slowakische Akademie der Wissenschaften, Bratsilava.

Hellyer (1988) A brief survey of the Abu Dhabi Nature Reserve. *Emirates Natural History Group Bulletin* 35: 1-5.

Hemming, C.F. (1961) The ecology of the coastal area of northern Eritrea. *Journal of Ecology* 49: 55-78.

Hemming, C.F. & Radcliffe-Smith, A. (1987) A revision of the Somali species of *Jatropha* (Euphorbiaceae). *Kew Bulletin* 42: 103-122.

Hepper, F.N. (1977) Outline of the vegetation of the Yemen Arab Republic. *Publications of the Cairo University Herbarium* 7/8: 307-322.

Hepper, F.N. & Friis, I. (1994) *The Plants of Pehr Forsskål's 'Flora Aegyptiaco-Arabica'*. Royal Botanic Gardens, Kew.

Hepper, F.N. & Wood, J.R.I. (1979) Were there forests in the Yemen? *Proceedings of the Seminar for Arabian Studies* 9: 65-71.

Herzog, T. (1926) *Geographie der Moose*. Gustav Fischer, Jena.

Hillcoat, D., Lewis, G. & Verdcourt, B. (1980) A new species of *Ceratonia* (Leguminosae - Caesalpinioideae) from Arabia and the Somali Republic. *Kew Bulletin* 35: 261-271.

Holm, D.A. (1953) Dome-shaped dunes of central Nejd, Saudi Arabia. Pp. 107-112 in *19th International Geological Congress*, Comptes Rendus, Section 7, Fascicle 7.

Hooper, S.S. (1984) Two montane forest species of *Carex* in Yemen and northeast Africa - new distribuitional records and a new variety. *Kew Bulletin* 39: 747-751.

Hort, A. (1916) *Theophastus: "Enquiry into Plants" and "Concerning Odours" (in one volume)*. Loeb Classical Library, Heinemann and Harvard University Press.

Hötzl, H., Felber, H., Maurin, V. & Zötl, J.G. (1978) Regions of Investigation: Cuesta region of the Tuwayq mountains. Pp. 194-226 in *Quaternary Period in Saudi Arabia. 1: Sedimentological, Hydrogeological, Hydrochemical, Geomorphological, and Climatological Investigations in Central and Eastern Saudi Arabia*. S.S. Al-Sayari & J.G. Zötl, eds. Springer Verlag, Wien.

Hötzl, H., Kramer, F. & Maurin, V. (1978) Quaternary sediments. Pp. 264-301 in *Quaternary Period in Saudi Arabia. 1: Sedimentological, Hydrogeological, Hydrochemical, Geomorphological, and Climatological Investigations in Central and Eastern Saudi Arabia*. S.S. Al-Sayari & J.G. Zotl, eds. Springer Verlag, Wien.

Hötzl, H. & Zötl, J.G. (1984) Middle and Early Pleistocene. Pp. 332-335 in *Quaternary Period in Saudi Arabia. Volume 2: Sedimentological, Hydrogeological, Hydrochemical, Geomorphological, Geochronological and Climatological Investigations in Western Saudi Arabia*. A.R. Jado & J.G. Zötl, eds. Springer Verlag, Wien.

Houghton, J.T., Jenkins, G.J. & Ephraums, J.J., eds. (1990) *Climate Change, the IPCC Scientific Assessment*. Cambridge University Press, Cambridge.

Hunde, A. (1982) Two new species of *Acacia* (Leguminosae-Mimosoideae) from Ethiopia and Yemen. *Nordic Journal of Botany* 2: 337-342.

Hurni, H. (1981) Hochgebirge von Semien, Äthiopien. *Erdkunde* 35: 98-107.

Ismail, A.M.A. (1983) Some factors controlling the water economy of *Zygophyllum qatarense* (Hadidi) growing in Qatar. *Journal of Arid Environments* 6: 239-246.

Ismail, A.M.A. & El-Ghazaly, G.A. (1990) Phenological studies on *Zygophyllum qatarense* Hadidi from contrasting habitats. *Journal of Arid Environments* 18: 195-205.

Jado, A.R. & Zötl, J.G., eds. (1984) *Quaternary Period in Saudi Arabia. Volume 2: Sedimentological, Hydrogeological, Hydrochemical, Geomorphological, Geochronological and Climatological Investigations in Western Saudi Arabia*. Springer Verlag, Wien.

Jahns, H.M. & Ott, S. (1994) Thallic mycelial and cytological characters in Ascomycete Systematics. In *Ascomycete Systematics: Problems and Perspectives in the Nineties*. D.L. Hawksworth, ed. Plenum, New York.

Jaubert, C., ed. (1843) *Relations de Voyages en Orient de 1830 à 1838, par Aucher-Éloy*. Libraire Encyclopédique de Roret, Paris.

Jones, D.A., Fleming, R.M. & At-Tayyeb, H.H. (1995) *Habitats of the Jubail Marine Wildlife Sanctuary. An Introduction and Field Guide*. European Commission, National Commission for Wildlife Conservation and Development, Saudi Arabia and the Seneckenberg Institute.

Jones, G.C. (1989) *Traditional Spinning and Weaving in the Sultante of Oman*. Historical Association of Oman, Oman.

Jones, R.W. & Racey, A. (1994) Cenozoic Stratigraphy of the Arabian Peninsula and Gulf. Pp. 274-307 in *Micropalaeontology and Hydrocarbon Exploration in the Middle East*. M.D. Simmons, ed. Chapman & Hall, London.

Jonsell, B. (1986) A monograph of *Farsetia* (Cruciferae). *Acta Universitatis Upsalensis* 25: 1-107.

Jupp, B.P., Durako, M.J., Kenworthy, W.J., Thayer, G.W. & Schillak, L. (1996) Distribution, abundance and species composition of seagrasses at several sites in Oman. *Aquatic Botany* 53: 199-213.

Jupp, P.B. (1993) Biomass of Aquatic Macrophytes. Part G in: *Khawrs and Springs of the Dhofar Governorate. Survey and Monitoring Studies*. Unpublished report. Planning Committee for Development and Environment in the Governorate of Dhofar, Oman.

Juyal, N., Glennie, K.W. & Singhvi, A.K. (in press) Chronology and paleoenvironmental significance of Quaternary desert sediments in southeast Arabia. In *Quaternary Deserts and Climatic Change*. A.S. Alsharhan, K.W. Glennie, G.L. Whittle & C.G.S.C. Kendall, eds. Balkema, Rotterdam.

Kamel, H.B., Al-Serhani, A. & Zeidan, D. (1989) *Jibal Fiqrah Conservation Study: Report and Master Plan for a Wildlife Reserve and Country Park with Guidelines for Development*. Department of Landscape Architecture, School of Environmental Design, Faculty of Engineering, King Abdulaziz University, Jeddah, Saudi Arabia.

Kassas, M. (1966) Plant life in deserts. Pp. 145-180 in *Arid Lands: A Geographical Appraisal*. E.S. Hills, ed. Methuen, London.

Kassas, M. & Zahran, M.A. (1967) On the ecology of the Red Sea littoral saltmarsh, Egypt. *Ecological Monographs* 37: 297-316.

Kerfoot, O. (1961) *Juniperus procera* Endl. (the African pencil cedar) in Africa and Arabia. 1. Taxonomic affinities and geographical distribution. *East African Forestry Journal* 26: 170-177.

Kessler, J.J. (1987) A rangeland vegetation study of the Dhamar montane plains with an east and west extension. Unpublished Report, *R.L.I.P Communications*, Amersfoort.

Kessler, J.J. (1989) Feed supply and demand of a sheep production system in the montane plains (Yemen Arab Republic). *Tropical Grasslands* 23: 143-152.

Kessler, J.J. (1995) *Mahjur* areas: traditional rangeland reserves in the Dhamar montane plains (Yemen Arab Republic). *Journal of Arid Environments* 29: 395-401.

King, R.A. (1988) *Teucrium* in the Arabian Peninsula and Socotra. *Notes from the Royal Botanic Garden Edinburgh* 45: 21-42.

Kingery, C.E. (1971) *Report to the Government of Saudi Arabia*. Unpublished report. Food and Agricultural Organisation, Rome.

Knapp, R. (1968) Höhere Vegetationseinheiten von Äthiopien, Somalia, Natal, Transvaal, Kapland und einigen Nachbargebieten. *Geobotanische Mitteilungen* 56: 1-36.

König, P. (1986) Zonation of vegetation in the mountainous region of south-western Saudi Arabia ('Asir, Tihama). Pp. 137-166 in *Contributions to the Vegetation of Southwest Asia*. H. Kürschner, ed. Beihefte zum Tübinger Atlas des Vorderen Orients, Reihe A (Naturwissenschaften) Nr. 24. Dr Ludwig Reichert Verlag, Wiesbaden.

König, P. (1987) *Vegetation und Flora im südwestlichen Saudi Arabien (Asir, Tihama)*. Dissertationes Botanicae, Berlin/Stuttgart.

König, P. (1988) Phytogeography of south-western Saudi Arabia. *Die Erde* 119: 75-89.

Kopp, H. (1975) Die räumliche Differenzierung der Agrarlandschaft in der Arabischen Republik Jemen (Nordjemen). *Erdkunde* 29: 59-68.

Kopp, H. (1990) Landnutzungswandel im Nordjemen seit 1970 - Von der autarken Stammesgesellschaft zum marktwirtschaftlich orientierten Nationalstaat. *Erdkunde* 44: 136-148.

Kraft, W., Garling, F. & Matecki, J. (1971) Die hydrogeologischen Verhältnisse in der mittleren Tihama. *Zeitschrift für Angewaudte Geologie* 17: 239-249.

Kukkonen, I. (1991) Problems in *Carex* section Physodae and *Cyperus conglomeratus* within the Flora Iranica area. *Flora et Vegetatio Mundi* 9: 63-73.

Kukkonen, I. (1995) New taxa, new combinations and notes on the treatment of Cyperaceae for Flora Iranica. *Annales Botanici Fennici* 32: 153-164.

Kürschner, H. (1984a) An epilithic bryophyte community in the Asir Mountains (SW Saudi Arabia). Studies in Arabian bryophytes 4. *Nova Hedwigia* 40: 423-434.

Kürschner, H. (1984b) Epiphytic communities of the Asir mountains (SW Saudia Arabia). Studies in Arabian bryophytes 2. *Nova Hedwigia* 39: 177-199.

Kürschner, H. (1986a) Omanisch-makranische Disjunktionen Ein Beitrag zur pflanzengeographischen Stellung und zu den florengenetischen Beziehungen Omans. *Botanische Jahrbücher für Systematik* 106: 541-562.

Kürschner, H. (1986b) A physiognomical-ecological classification of the vegetation of southern Jordan. Pp. 45-80 in *Contributions to the Vegetation of Southwest Asia*. H. Kürschner, ed. Beihefte zum Tübinger Atlas des Vorderen Orients, Reihe A (Naturwissenschaften) Nr. 24. Dr. Ludwig Reichert Verlag, Wiesbaden.

Kürschner, H. (1986c) A study of the vegetation of the Qurm Nature Reserve, Muscat area, Oman. *Arab Gulf Journal of Scientific Research* 4: 23-52.

Kürschner, H. (1996a) Additions to the bryophyte flora of Yemen. New records from the Taizz and Jiblah areas. Studies in Arabian bryophytes 21. *Nova Hedwigia* 62: 233-247.

Kürschner, H. (1996b) Towards a bryophyte flora of the Near and Middle East. New records from Iran, Jordan, Kuwait, Lebanon, Oman, Saudi Arabia, Syria and Turkey. *Nova Hedwigia* 63: 261-271.

Kürschner, H. (1997) *Die aktuelle und naturliche Vegetation Nord-Omans unter besonderer Berücksichtigung der Masqat-Region, mit einer Rekonstruktion der spät-steinzeitlichen Vegetationsverhältnisse*. *The Capital Area of northern Oman, part II*. Beihefte zum Tübinger Atlas des Vorderen Orients, Reihe A (Naturwissenschaften) Nr. 31/2. Dr Ludwig Reichert Verlag, Wiesbaden.

Kürschner, H., Al-Gifri, A.N., Al-Subai, M.Y. & Rowaished, A.K. (in press) Vegetation pattern within coastal salines of southern Yemen. *Feddes Repertorium*.

Lancaster, W. & Lancaster, F. (1990) Desert devices: the pastoral system of the Rwala bedu. Pp. 177-194 in *The World of Pastoralism: Herding Systems in Comparative Perspective*. J.G. Galaty & D.L. Johnson, eds. Belhaven, London.

Lavranos, J.J. (1962) Two new species of *Caralluma* from south-western Arabia. *Journal of South African Botany* 28: 209-213.

Lavranos, J.J. (1963a) Three new species of *Caralluma* from south-west Arabia. *Journal of South African Botany* 29: 103-110.

Lavranos, J.J. (1963b) Two new species of *Huernia* from south-west Arabia. *Journal of South African Botany* 29: 97-99.

Lavranos, J.J. (1965) Notes on the *Aloes* of Arabia with descriptions of six new species. *Journal of South African Botany* 31: 55-81.

Lavranos, J.J. (1967) Notes on the succulent flora of tropical Arabia (with descriptions of new taxa). *Cactus and Succulent Journal* 39: 3-7, 123-127, 167-172.

Lavranos, J.J. (1974) *Sarcostemma vanlessenii*. A new species from Jemen. *Cactus and Succulent Journal* 29: 35.

Lavranos, J.J. (1979) *Caralluma dioscoridis* Lavranos. *Asclepiadaceae* 18: 10-12.

Lavranos, J.J. (1983a) An interesting plant record from the Gulf of Aden area. *Bothalia* 14: 220-221.

Lavranos, J.J. (1983b) Two new Stapelieae from Saudi Arabia. *Cactus and Succulent Journal* 55: 23-36.

Lavranos, J.J. (1985) *Aloe sheilae* Lavranos: A new species from Saudi Arabia. *Cactus and Succulent Journal* 57: 71-72.

Lavranos, J.J. (1993) Two new species of *Echidnopsis* (Asclepiadaceae-Stapelieae) from Socotra. *Cactus and Succulent Journal* 65: 293-295.

Lavranos, J.J. (1995) *Aloe whitcombei* and *A. collenetteae*, two new cliff dwelling species from Oman, Arabia. *Cactus and Succulent Journal* 67: 30-33.

Lavranos, J.J. & Newton, L.E. (1977) Two new species of *Aloe* from Arabia. *Cactus and Succulent Journal* 49: 113-116.

Le Métour, J., Michel, J.C., Béchennec, F., Platel, J.P. & Roger, H. (1995) *Geology and Mineral Wealth of the Sultanate of Oman*. BRGM, France and Ministry of Petroleum and Minerals, Oman.

Lems, K. (1960) Botanical notes on the Canary Islands II - The evolution of plant forms in the Islands: Aeonium. *Ecology* 41: 1-17.

Léonard, J. (1989) *Contribution à l'étude de la flore et de la vegetation des deserts d'Iran (Dasht-e-Kavir, Dasht-e-Lut, Jaz Kurian). Fascicule 9. Considérations phytogéographiques sur les phytochories irano-touranienne, saharo-sindienne et de la Somalie-pays Masai.* Jardin botanique national de Belgique.

Levell, B.K., Braakman, J.H. & Rutten, K.W. (1988) Oil-bearing sediments of Gondwana glaciation in Oman. *American Association of Petroleum Geologists Bulletin* 72: 775-796.

Llewellyn, O.A. (1987) *The Proposed Hima Jibal al-Humrah*. Unpublished report. National Commission for Wildlife Conservation and Development, Meteorological and Environmental Protection Administration and the Hajj Research Center, Saudi Arabia.

Longton, R.E. (1988) Adaptations and strategies of polar bryophytes. *Botanical Journal of the Linnean Society* 98: 253-268.

Lösch, R. (1987) Die Produktionsphysiologie von *Aeonium gorgoneum* und anderer nicht-kanarischer *Aeonien* (Phanerogamae: Crassulaceae). *Courier Forschungs institut Senckenberg* 95: 210-209.

Mahmoud, A. (1984) Germination of caryopses of the halophyte *Aeluropus massauensis* from Saudi Arabia. *Arab Gulf Journal of Scientific Research* 2: 21-36.

Mahmoud, A., El-Sheikh, A.M. & Abdul Baset, S. (1983) Germination of two halophytes: *Halopeplis perfoliata* and *Limonium axillare* from Saudi Arabia. *Journal of Arid Environments* 6: 87-98.

Mahmoud, A., El-Sheikh, A.M. & Isawi, F. (1982) Ecology of the littoral salt marsh vegetation of Rabigh on the Red Sea coast of Saudi Arabia. *Journal of Arid Environments* 5: 35-42.

Mahmoud, A., El-Sheikh, A.M., Youssef, M.M. & Al-Tom, M. (1985) Ecology of the littoral salt marsh vegetation at al-Magawah, on the Gulf of Aqaba, Saudi Arabia. *Arab Gulf Journal of Scientific Research* 3: 145-163.

Mandaville, J.P. (1965) Notes on the vegetation of Wadi as-Sahba, eastern Arabia. *Journal of the Bombay Natural History Society* 62: 330-332.

Mandaville, J.P. (1977) Plants. Pp. 229-267 in *The Scientific Results of the Oman Flora and Fauna Survey 1975. The Journal of Oman Studies Special Report No. 1.* D.L. Harrison, ed. Ministry of Information and Culture, Muscat, Oman.

Mandaville, J.P. (1978) *Wild Flowers of Northern Oman*. Bartholomew Books, London.

Mandaville, J.P. (1984) Studies in the flora of Arabia XI: Some historical and geographical aspects of a principal floristic frontier. *Notes from the Royal Botanic Garden Edinburgh* 42: 1-15.

Mandaville, J.P. (1985) A botanical reconnaissance in the Musandam region of Oman. *The Journal of Oman Studies* 7: 9-28.

Mandaville, J.P. (1986) Plant life in the Rub' al-Khali (the Empty Quarter), south-central Arabia. Pp. 147-157 in *Plant Life of Southwest Asia*. I. Hedge, ed. Edinburgh University Press, Edinburgh.

Mandaville, J.P. (1990) *Flora of Eastern Saudi Arabia*. Kegan Paul, London.

Mathew, B. & Collenette, S. (1994) A curious new species of *Albuca* (Hyacinthaceae) from Saudi Arabia. *Kew Bulletin* 49: 125-148.

Maxwell-Darling (1937) The outbreak areas of the desert locust (*Schistocera gregaria*, Forssk.) in Arabia. *Bulletin of Entomological Research* 28: 605-618.

McClure, H.A. (1978) Ar Rub' al Khali. Pp. 252-262 in *Quaternary Period in Saudi Arabia. 1: Sedimentological, Hydrogeological, Hydrochemical, Geomorphological, and Climatological Investigations in Central and Eastern Saudi Arabia*. S.S. Al-Sayari & J.G. Zötl, eds. Springer Verlag, Wien.

McClure, H.A. (1984) *Late Quaternary Palaeoenvironments of the Rub' al Khali*. PhD thesis, University of London.

McGinnies, W.G. (1985). Climatic and biological classifications of arid lands. A comparison. Pp. 61-68 in *Arid Lands: Today and Tomorrow*. Proceedings of the International Research and Development Conference, Tucson, Arizona. E.E. Whitehead, C.F. Hutchinson, N. Timmermain & R.G. Varady, eds. Westview Press, USA

McKee, E.D. (1966) Structures of dunes at White Sands National Monument, New Mexico (and a comparison with structures of dunes from other selected areas). *Sedimentology* 7: 1-69.

Membery, D.A. (1985) A unique August cyclonic storm crosses Arabia. *Weather* 40: 108-115.

Membery, D.A. (1997) Unusually wet weather across Arabia. *Weather* 52: 166-174.

Membery, D.A. (in press) Famous for fifteen minutes: causes and effects of the tropical storm that struck southern Arabia in June 1996. *Weather*.

Miehe, S. (1988) *Vegetation ecology of the Jebel Marra Massif in the semiarid Sudan*. Dissertationes Botanicae, Berlin/Stuttgart.

Miehe, S. & Miehe, G. (1994) *Ericeous forest and heathlands in the Bale Mountains of South Ethiopia*. Warnke, Reinbeck.

Mies, B. (1994) Checkliste der Gefabpflanzen, Moose und Flechten und botanische Bibliographie der Insel Sokotora und des Sokotrinischen Archipels (Jemens, Indischer Ozean). *Senckenbergiana Biologica* 74: 213-258.

Mies, B. (1995) Die sukkulenten Pflanzenarten der Insel Sokotra und Abd El Kuri (Jemen). *Die anderen Sukkulenten* 13: 4-17.

Mies, B. & Aschan, G. (1995) Radiation regime and temperature conditions in the canopy of the succulent shrub *Euphorbia balsamifera*. *Vieraea* 24: 115-125.

Mies, B., Lumbsch, H.T. & Theler, A. (1995) *Feigeana socotrana*, a new genus and species from Socotra, Yemen (Roccellaceae, Euascomycetidae). *Mycotaxon* 54: 155-162.

Mies, B. & Zimmer, H. (1993) Die Vegetation der Insel Sokotra im Indischen Ozean. *Natur und Museum* 123: 253-264.

Mies, B. & Zimmer, H. (1994) Die Populationen von *Adenium socotranum* (Balfour) Vierhapper und *Dendrosycios socotrana* Balfour bei Ras Hebak (Insel Sokotra, Jemen) und ihre Gefährdung. *Kakteen und andere Sukkulenten* 45: 1-5.

Migahid, A.M. (1988a) *Flora of Saudi Arabia (3rd edition). Volume 1. Cryptogams and Dicotyledons: Equisetaceae to Neuradaceae*. University Libraries, King Saud University, Riyadh.

Migahid, A.M. (1988b) *Flora of Saudi Arabia (3rd edition). Volume 2. Dicotyledons: Leguminosae to Compositae*. University Libraries, King Saud University, Riyadh.

Milchunas, D.G. & Lauenroth, W.K. (1993) Quantitative effects of grazing on vegetation and soils over a global range of environments. *Ecological Monographs* 63: 327-366.

Miles, S.B. (1901) On the border of the great desert: a journey in Oman. *The Geographical Journal*: 405-425.

Miller, A.G. (1980) A revision of *Campylanthus*. *Notes from the Royal Botanic Garden Edinburgh* 38: 373-385.

Miller, A.G. (1984a) *Psilotum nudum*: A new record for Arabia. *Fern Gazette* 12: 361.

Miller, A.G. (1984b) A revision of *Ochradenus*. *Notes from the Royal Botanic Garden Edinburgh* 41: 491-504.

Miller, A.G. (1985) The genus *Lavandula* in Arabia and tropical NE Africa. *Notes from the Royal Botanic Garden Edinburgh* 42: 503-528.

Miller, A.G. (1988) Studies in the Flora of Arabia XXII: *Dhofaria*, a new genus of Capparaceae from Oman. *Notes from the Royal Botanic Garden Edinburgh* 45: 55-60.
Miller, A.G. & Cope, T.A. (1996) *Flora of the Arabian Peninsula and Socotra, Volume 1*. Edinburgh University Press, Edinburgh.
Miller, A.G., Hedge, I.C. & King, R.A. (1982) Studies in the Flora of Arabia I. A botanical bibliography of the Arabian peninsula. *Notes from the Royal Botanic Garden Edinburgh* 40: 43-61.
Miller, A.G. & Morris, M. (1988) *Plants of Dhofar, the Southern Region of Oman: Traditional, Economic and Medicinal Uses*. The Office of the Adviser for Conservation of the Environment, Diwan of Royal Court, Oman.
Miller, A.G. & Nyberg, J.A. (1991) Patterns of endemism in Arabia. *Flora et Vegetatio Mundi* 9: 263-279.
Miller, A.G., Short, M. & Sutton, D.A. (1982) A revision of *Schweinfurthia*. *Notes from the Royal Botanic Garden Edinburgh* 40: 23-40.
Mirreh, M.M. & Al Diran, M.S. (1995) Range damage and recovery in the Widyan of northern Saudi Arabia. Pp. 155-162 in *Range Management in Arid Zones*. S.A.S. Omar, M.A. Razzaque & F. Alsdirawi, eds. Kegan Paul International, London.
Mitten, W. (1888) Muscineae. Pp. 330-336 in *The Botany of Socotra*. J.B. Balfour, ed. Transactions of the Royal Society of Edinburgh.
Monod, T. (1954) Modes contracté et diffus de la vegetation Saharienne. Pp. 35-44 in *Biology of Deserts*. J.L. Cloudsley-Thompson, ed. London.
Monod, T. (1971) Remarques sur les symtéries floristiques des zones sèches Nord et Sud en Afrique. *Mittelungen Botanische Staatssammlungen Müenchen* 10: 375-423.
Morris, E., T. (1984) *Fragrance: the Story of Perfume from Cleopatra to Chanel*. Charles Scribners, New York.
Mossa, J.S., Al-Yahya, M.A., Al-Meshal, I.A. & Tariq, M. (1983) Phytochemical and biological screening of Saudi medicinal plants. *Fitoterapia* 54: 75-80, 147-152.
Müller, C. (1900) *Genera Muscorum Frondosorum*. Leipzig.
Müller, G.K. & Gutte, P. (1985) Beiträge zur Kenntnis der Vegetation der Flußauen, Sümpfe und Gewässer der zentralperuanischen Küstenregion. *Wissenschaftlidie Zeitschrift der Karl-Marx-Universität Leipzig, Mathematisch-Naturwissenschaftlische Reihe* 34: 410-429.
Müller-Hohenstein, K. (1992) Las dunas como ecosistemas en Europa, Arabia y Africa. *Bosque* 13: 9-21.
Müller-Hohenstein, K., Grosser, L., Rappenhöner, D. & Stratz, C. (1987) Applied vegetation studies in the Yemen Arab Republic: Range management and terrace stabilisation. *Catena* 14: 43-50.
Müller-Hohenstein, K. & Rappenhöner, D. (1991) Vegetation mapping under different aspects of basic and applied vegetation science. - Large scale examples from the Yemen Arab Republic. *Flora et Vegetatio Mundi* 9: 199-213.
Munton, P. (1985) The ecology of the Arabian Tahr (*Hemitragus jayakari* Thomas 1894) and a strategy for the conservation of the species. *Journal of Oman Studies* 8: 11-48.
Munton, P. (1988a) An overview of the ecology of the Sands. Pp. 231-240 in *The Scientific Results of The Royal Geographical Society's Oman Wahiba Sands Project 1985-1987. The Journal of Oman Studies Special Report No. 3*. R.W. Dutton, ed. Office of the Adviser for Conservation of the Environment, Muscat, Oman.
Munton, P. (1988b) Vegegtation and forage availability in the Sands. Pp. 241-250 in *The Scientific Results of The Royal Geographical Society's Oman Wahiba Sands Project 1985-1987. Journal of Oman Studies Special Report No. 3*. R.W. Dutton, ed. Office of the Adviser for Conservation of the Environment, Muscat, Oman.
Musallam, M. (1973) *Sex and Society in Islam*. PhD dissertation, Harvard University.
Musil (1927) *Arabia Deserta*. American Geographical Society, New York.

Musselman, L.J. & Hepper, F.N. (1988) The genus *Striga* in Arabia. *Notes from the Royal Botanic Garden Edinburgh* 45: 43-50.

Nasrallah, H.A. & Balling, R.C. (1993) Spatial and temporal analysis of Middle Eastern temperature changes. *Climatic Change* 25: 153-161.

Newton, L.C. & Lavranos, J.J. (1993) *Huernia rosea*: A new species from the Yemen. *Cactus and Succulent Journal* 65: 279-280.

Niebuhr, C. (1772) *Beschreibung von Arabien*. Aus eigenen beobachtungen und im Lande selbst gesammelten nachrichten abgefasset. Copenhagen.

Noy-Meir, I. (1973) Desert Ecosystems: Environment and Producers. *Annual Review of Ecology and Systematics* 4: 25-51.

Oatham, M.P., Nicholls, M.K. & Swingland, I.R. (1995) Manipulation of vegetation communities on the Abu Dhabi rangelands. 1. The effects of irrigation and release from longterm grazing. *Biodiversity and Conservation* 4: 696-709.

Olivier, J. (1995) Spatial distribution of fog in the Namib. *Journal of Arid Environments* 29: 129-138.

Omar, S. (1991) Dynamics of range plants following 10 years of protection in arid rangelands of Kuwait. *Journal of Arid Environments* 21: 99-111.

Omar, S. (1995) Influence of precipitation on vegetation in the rangelands of Kuwait. Pp. 127-138 in *Range Management in Arid Zones*. S.A.S. Omar, M.A. Razzaque & F. Alsdirawi, eds. Kegan Paul International, London.

Omar, S. (1996) *Conservation of Biodiversity in Kuwait*. Unpublished report. Kuwait Institute for Scientific Research, Kuwait.

Omar, S., Abdali, F.K., Baroon, Z. & Al-Mousa, Z. (1995) State of Country Paper: The Gulf Crisis Impact on the Natural, Environmental, Food, and Human Resources of Kuwait. Pp. 47-77 in *TWOWS International Conference: Women's vision of Science and Technology for Development. The Third World Organization for Women in Science*. P. Dennis & M. Cetto, eds.

Omar, S., Alsdirawi, K. & Razzaque, M.A. (1994) *Agricultural and Environmental Laws, Policies and Regulations*. Unpublished report, Kuwait Institute of Scientific Research, Kuwait.

Painter, E.L. & Belsky, A.J. (1993) Application of herbivore optimization theory to rangelands of the western United States. *Ecological Applications* 3: 2-9.

Pedgley, D.E. (1970) The climate of interior Oman. *The Meteorological Magazine* 99: 29-37.

Perez, J.C., Diaz-Gonzalez, T.E., Areces, M.P.F. & Salvo, E. (1989) Contribución al estudio de las comundades rupicolas Cheilanthetalia marantho-maderensis y Androsacetalia vandellii en la Peninsula Iberica. *Acta Botanica Malacitana* 14: 171-191.

Petrusson, L. & Thulin, M. (1966) Taxonomy and biogeography of *Gymnocarpos* (Caryophyllaceae). *Edinburgh Journal of Botany* 53: 1-26.

Philby, H.St.J.B. (1928) *Arabia of the Wahhabis*. Constable, London.

Phillips, D.C. (1988) *Wild Flowers of Bahrain. A Field Guide to Herbs, Shrubs, and Trees*. Privately published, Manama, Bahrain.

Pichi-Sermolli, R.E.G. (1962) On the fern genus "Actinopteris" Link. *Webbia* 17: 1-32.

Pignatti, E. & Pignatti, S. (1995) A survey of the southwestern Australian vegetation classes. *Colloques Phytosociologiques* 23: 47-66.

Plowes, D.C.H. (1995) A reclassification of *Caralluma* R. Brown (Stapelieae: Asclepiadaceae). *Haseltonia* 3: 49-70.

Popov, G.B. (1957) The vegetation of Socotra. *Journal of the Linnean Society of Botany* 55: 706-720.

Poppendiek, H.-H. & Ihlenfeldt, H.-D. (1978) *Delosperma harazianum* (Deflers) Poppendiek & Ihlenfeldt, eine wenig bekannte Mesembryanthemaceae aus dem Südjemen. *Mitteilungen aus dem Institut für Allgemeine Botanik Hamburg* 16: 183-187.

Powers, R.W., Ramirez, L.F., Redmond, C.D. & Elberg, E.L. (1966) *Geology of the Arabian Peninsula. Sedimentary Geology of Saudi Arabia. USGS Professional Paper 560-D*. United States Government Printing Office, Washington.

Price, A.R.G. (1990) Rapid assessment of coastal zone management requirements: case study in the Arabian Gulf. *Ocean and Shoreline Management* 13: 1-19.
Proctor, M.C.F. (1979) Structures and eco-physiological adaptations in bryophytes. Pp. 479-509 in *Bryophyte Systematics*. G.C.S. Clarke & J.G. Duckett, eds. Academic Press, London.
Pursell, R.A. & Kürschner, H. (1987) *Fissidens arabicus* (Fissidentaceae), a new species from the Asir Mountains (Saudi Arabia). *Nova Hedwigia* 44: 101-103.
Qaiser, M. & Lack, H.W. (1985) The genus *Phagnalon* (Asteraceae, Inuleae) in Arabia. *Willdenowia* 15: 3-22.
Raadts, E. (1981) Über zwei arabische *Kalanchoe*-Arten (Crassulaceae). *Willdenowia* 11: 327-331.
Raadts, E. (1995) Über zwei arabische *Kalanchoe*-Arten (Crassulaceae) und eine neue Varietät aus dem Jemen. *Willdenowia* 25: 253-259.
Radcliffe-Smith, A. (1980) The vegetation of Dhofar. Pp. 59-86 in *The Scientific Results of the Oman Flora and Fauna Survey 1977 (Dhofar). The Journal of Oman Studies Special Report No. 2*. S.N.S. Reade, J.B. Sale, M.D. Gallagher & R.H. Daly, eds. Office of the Adviser for Conservation of the Environment, Muscat, Oman.
Rauh, W. (1966) Little known succulents of southern Arabia. *Cactus and Succulent Journal* 38: 165-176, 207-219.
Rauh, W. & Wertel, H.P. (1965) *Caralluma lavarani*, Rauh et Wertel. Eine neue Art aus Südarabien. *Kakteen und andere Sukkulenten* 16: 62-64.
Reade, S.N.S., Sale, J.B., Gallagher, M.D. & Daly, R.H., eds. (1980) *The Scientific Results of the Oman Flora and Fauna Survey, 1977 (Dhofar). The Journal of Oman Studies Special Report No. 2*. Office of the Adviser for Conservation of the Environment, Diwan of Royal Court, Muscat, Oman.
Revri, R. (1983) Catha edulis *Forssk. Geological, Dispersal, Botanical, Ecological and Agronomical Aspects with Special Reference to the Yemen Arab Republic*. Göttingen Beiträge zur Land- und Forstwirtschaft der Tropen und Subtropen 1, 157 pp.
Rivas, G., S. & Esteve, C.F. (1965) Ensayo fitosociologico de la clase Crassi-Euphorbietea macaronesia y estudio de los tabaibales et cardonales de Gran Canaria. *Anales del Instituto Botanico Cavanilles* 22: 220-339.
Robertson, E.F. (1993) *Camel Densities in the Special Ibex Reserve at Hawtah Bani Tamim*. Unpublished report. King Khalid Wildlife Research Centre, Riyadh, Saudi Arabia.
Robertson, E.F. (1995) A land unit classification of Wadi Ghaba in the Ibex Reserve. Pp. 65-76 in *King Khalid Wildlife Research Centre Annual Report 1995*. Riyadh, Saudi Arabia.
Robertson, E.F. (1996) *Protected areas as an effective means for the consevation of plant species*. First Meeeting of the Arabian Plant Specialist Group, June 1996. National Commission for Wildlife Conservation and Development, Riyadh, Saudi Arabia.
Robertson, E.F. & Collenette, S. (1996) *Checklist of Plants in the Ibex Reserve*. Unpublished report, updated June 1996. King Khalid Wildlife Research Centre, Riyadh, Saudi Arabia.
Robertson, E.F., Collenette, S., Strauss, M. & Wacher, T. (1996) *Checklist of Plants in Uruq bani Mu'Arid*. Unpublished report, updated June 1996. King Khalid Wildlife Research Centre, Riyadh, Saudi Arabia.
Robertson, E.F., Dunham, K. & Collenette, S. (1997) *Report on a Visit to the at-Tubayq Protected Area*. Unpublished report. King Khalid Wildlife Research Centre, for the National Commission for Wildlife Conservation and Development, Riyadh, Saudi Arabia.
Robertson, F. (1995b) *Food Plants of Mountain Gazelles and Ibex in the Ibex Reserve*. Unpublished Report. King Khalid Wildlife Research Centre, Riyadh, Saudi Arabia.
Robson, N.B.K. (1987) An Iranian link with Socotra in *Hypericum* (Hypericaceae). *Plant Systematic Evolution* 155: 89-92.
Roger, J., Platel, J.P., Bourdillon-de-Grissac, C. & Cavelier, C. (1994) *Geology of Dhofar*. BRGM, France and Ministry of Petroleum and Minerals, Oman.

Sale, J.B. (1980) The ecology of the mountain region of Dhofar. Pp. 25-54 in *The Scientific Results of the Oman Flora and Fauna Survey 1977 (Dhofar). The Journal of Oman Studies Special Report No. 2*. S.N.S. Reade, J.B. Sale, M.D. Gallagher & R.H. Daly, eds. Office of the Adviser for Conservation of the Environment, Muscat, Oman.

Salm, R.V. (1986) *The Proposed Daymaniyat Islands National Nature Reserve Management Plan*. IUCN Coastal Zone Management Project, Oman.

Salm, R.V. (1989) *A Proposed Management Plan for the Turtle Nesting Beaches in the Ras al Hadd National Scenic Reserve and Ras al Junayz National Nature Reserve*. IUCN Coastal Zone Management Project, Oman.

Samour, J., Irwin-Davies, J., Mohanna, M. & Faraj, E. (1989) Conservation at Al-Areen Wildlife Park, Bahrain. *Oryx* 23: 142-145.

Sanlaville, P. (1992) Changements climatiques dans la péninsule Arabique durant le Pléistocène Supérieur et l'Holocène. *Paleorient* 18: 5-26.

Savulescu, T. (1928) Contributions à la flore d'Arabie. *Bulletin de l'Académie des Sciences de Roumanie* 11: 14-24.

Schandelmeier, H. & Reynolds, P.-O. (1997) *Palaeogeographic-Palaeotectonic Atlas of North-Eastern Africa, Arabia, and Adjacent Areas: Late Neoproterozoic to Holocene*. Balkema, Rotterdam.

Schieferstein, B. & Loris, K. (1992) Ecological investigations on lichen fields of the Central Namib. 1. Distribution patterns and habitat conditions. *Vegetatio* 98: 113-128.

Scholte, P., Al-Khuleidi, A.W. & Kessler, J.J. (1991) *The Vegetation of the Republic of Yemen (Western part)*. DHV Consultants, Amersfoorts.

Scholz, F. (1984) Höhensiedlungen am Jabal Akdar. Tendenzen und Probleme der Entwicklung einer peripheren Region im Oman-Gebirge. *Zeitschrift für Wirtschaftsgeographie* 28: 16-30.

Scholz, H. (1969) Bemerkungen zu einigen *Stipagrostis*-Arten aus Afrika und Arabien. *Österrische Botanische Zeitschrift* 117: 284-292.

Scholz, H. (1983) *Stipagrostis masirahensis* sp. nov. aus Oman (Arabien). *Willdenowia* 13: 389-391.

Scholz, H. & König, P. (1984) Die Gattung *Oropetium* (Gramineae) in Arabien. *Willdenowia* 14: 161-163.

Scholz, H. & König, P. (1988) Eine neue *Aristida* (Gramineae) aus Arabien. *Willdenowia* 17: 111-113.

Schönherr, J. & Ziegler, H. (1975) Hydrophobic cuticular ledges prevent water entering the air pores of liverwort thalli. *Planta* 124: 51-60.

Schönig, H. (1995). *Traditional cosmetics*. Paper presented at the seminar for Arabian Studies 1995, Cambridge, UK.

Schopen, A. (1983) *Traditionelle Heilmittel in Jemen*. Franz Steiner Verlag, Wiesbaden.

Schulz, E. & Whitney, J.W. (1986) Vegetation in north-central Saudi Arabia. *Journal of Arid Environments* 10: 175-186.

Schwartz, O. (1939) Flora des Tropischen Arabien. *Mitteilungen aus dem Institut für Allgemeine Botanik Hamburg* 10: 1-393.

Schweinfurth, G.A. (1884) Allgemeine Betrachtungen über die Flora von Socotra. *Botanische Jahrbüscher* 5: 40-49.

Schweinfurth, G.A. (1891) Über die Florengemeinschaft von Sudarabien und Nordabessinien. *Verhandlungen der Gesellschaft für Erdkunde. Berlin* 9-10: 1-20.

Schweinfurth, G.A. (1894) Sammlung Arabisch-Äthiopischer Pflanzen. Ergebnisse von Reisen in den Jahren 1881, 88, 91, 92 und 94. *Bulletin de l'herbier Boissier* 2: 1-113.

Schweinfurth, G.A. (1896) Sammlung Arabisch-Äthiopischer Pflanzen. Ergebnisse von Reisen in den Jahren 1881, 88, 91, 92 und 94. *Bulletin Herbarium Boissier* 4: 115-266.

Schweinfurth, G.A. (1899) Sammlung Arabisch-Äthiopischer Pflanzen. Ergebnisse von Reisen in den Jahren 1881, 88, 91, 92 und 94. *Bulletin Herbarium Boissier* 7: 267-340.

Schweinfurth, G.A. (1912) *Arabische Pflanzennamen aus Aegypten, Algerien und Jemen*. Dietr Reimar, Berlin.

Schyfsma, E. (1978) Climate. Pp. 31-44 in *Quaternary Period in Saudi Arabia. 1: Sedimentological, Hydrogeological, Hydrochemical, Geomorphological, and Climatological Investigations in Central and Eastern Saudi Arabia*. S.S. Al-Sayari & J.G. Zötl, eds. Springer Verlag, Wien.

Scotese, C.R. & Barret, S.F. (1990) Gondwana's movement over the South Pole during the Palaerozoic: evidence from lithological indicators of climate. Pp. 75-85 in *Palaeozoic Palaeogeography and Biogeography*. W.S. McKerrow & C.R. Scotese, eds. Geological Society Memoir No. 12. The Geological Society, London.

Scotese, C.R. & McKerrow, W.S. (1990) Revised World Maps and Introduction. Pp. 1-21 in *Palaeozoic Palaeogeography and Biogeography*. W.S. McKerrow & C.R. Scotese, eds. Geological Society Memoir No. 12. The Geological Society, London.

Scott, A.J. (1981) A new *Suaeda* (Chenopodiaceae) from Dhofar. *Kew Bulletin* 36: 558.

Scott, G.A.M. (1982) Desert bryophytes. Pp. 105-122 in *Bryophyte Ecology*. A.J.E. Smith, ed. Chapman and Hall, London.

Scott, H. (1942) *In the High Yemen*. Chapman and Hall, London.

Sebsebe, D. (1985) The genus *Maytenus* (Celestraceae) in NE tropical Africa and tropical Arabia. *Symbolae Botanicae Upsalensis* 25: 1-102.

Seddon, P.J. (1996a) *Harrat al-Harrah Protected Area Master Management Plan*. Unpublished report. National Commission for Wildlife Conservation and Development, Riyadh, Saudi Arabia.

Seddon, P.J. (1996b) *Mahazat as-Sayd Protected Area Master Management Plan*. Unpublished report. National Commission for Wildlife Conservation and Development, Riyadh, Saudi Arabia.

Seddon, P.J. & Khan, A.B. (1996) *At-Tubaiq Protected Area Master Management Plan*. Unpublished report. National Commission for Wildlife Conservation and Development, Riyadh, Saudi Arabia.

Seddon, P.J. & van Heezik, Y. (1996) Seasonal changes in houbara bustard *Chlamydotis undulata macqueenii* numbers in Harrat al Harrah, Saudi Arabia: Implications for managing a remnant population. *Biological Conservation* 75: 139-146.

Seddon, P.J., van Heezik, Y. & Khoja, A.R. (1994) *Measuring the Impact of Camel Grazing on the Recovery of Steppe Vegetation: the Conflicts Between Pastoralism and the Conservation of Endangered Wildlife*. Unpublished report presented at the Symposium on Desert Studies, King Saud University, Riyadh, Saudi Arabia.

Sharon, D. (1972) The spottiness of rainfall in a desert area. *Journal of Hydrology* 17: 161-175.

Sheppard, C.R.C., Price, A.R.G. & Roberts, C. (1992) *Marine Ecology of the Arabian Region: Patterns and Processes in Extreme Tropical Environments*. Academic Press, London.

Shinn, E.A. (1973) Sedimentary accretion along the leeward, SE coast of Qatar Peninsula, Persian Gulf. Pp. 199-209 in *The Persian Gulf - Holocene Carbonate Sedimentation and Diagenesis in a Shallow Epicontinental Sea*. B.H. Purser, ed. Springer-Verlag, Berlin.

Shorbagi, A. (1996) *List of MAW Fenced Range Areas in the Kingdom of Saudi Arabia*. Unpublished report. Ministry of Agriculture and Water, Riyadh, Saudi Arabia.

Shoup, J. (1990) Middle Eastern sheep pastoralism and the hima system. Pp. 195-215 in *The World of Pastoralism: Herding Systems in Comparative Perspective*. J.G. Galaty & D.L. Johnson, eds. Belhaven Press, London.

Shuaib, L. (1995) *Wildflowers of Kuwait*. Stacey International, U.K.

Simpson, B.B. & Conner-Ogorzaly, M. (1986) *Economic Botany: Plants in Our World*. McGraw-Hill, New York.

Smith, E.A. (1986a) The structure of the Arabian heat low. Part I: Surface energy budget. *Monthly Weather Review* 114: 1067-1083.

Smith, E.A. (1986b) The structure of the Arabian heat low. Part II: Bulk tropospheric heat budget and implications. *Monthly Weather Review* 114: 1084-1102.

Spalton, J.A. (1995) *Effects of Rainfall on the Reproduction and Mortality of the Arabian oryx,* Oryx leucoryx *Pallas in the Sultanate of Oman*, Ph.D. thesis, University of Aberdeen.

Stanley-Price, M.R. (1989) *Animal Re-introductions: the Arabian Oryx in Oman*. Cambridge Studies in Applied Ecology and Resource Management. Cambridge University Press, Cambridge.

Stanley-Price, M.R., Al-Harthy, A.H. & Whitcombe, R.P. (1988) Fog moisture and its ecological effects in Oman. Pp. 69-88 in *Arid Lands: Today and Tomorrow*. E.E. Whitehead, C.F. Hutchinson, B.N. Timmermann & R.G. Varady, eds. Westview Press, Boulder, Colorado.

Stone, F., ed. (1985) *Studies on the Tihama. The Report of the Tihama Expedition 1982 and Related Papers*. F. Stone, ed. Longman, Harlow.

Täckholm, V. (1974) *Students' Flora of Egypt (2nd edition)*. Beirut.

Thalen, D.C.P. & Kessler, J.J. (1988) The role of rangelands in livestock production. Pp 11-27 in *The Role of Livestock Production in Yemen Agriculture*. Ministry of Agriculture and Fisheries. San'a, Yemen.

Thériot, I., Dixon, H.N. & Buch, H. (1934) Bryophyta nova. *Annals Bryologie* 7: 157-162.

Thesiger, W. (1949) A futher journey across the Emply Quarter. *The Geographical Journal* 113: 21-46.

Thesiger, W. (1959) *Arabian Sands*. Longmans.

Thomas, B. (1932) *Arabia Felix*. Cape, London.

Thouless, C. & al Bassri, K. (1989) *Rheem Gazelles in Al Khunfah Reserve, May 1989*. Unpublished report. King Khalid Wildlife Research Centre, Riyadh, Saudi Arabia.

Thouless, C. & Tatwany, H. (1989) *Sand Gazelles in Al Khunfah, October 1989*. Unpublished report. King Khalid Wildlife Research Centre Riyadh Saudi Arabia.

Thulin, M. & Al-Gifri, A., N, (1995) *Euphorbia applanata* sp nov. (Euphorbiaceae) from Yemen, with a note of *E. quaitensis*. *Nordic Journal of Botany* 15: 193-195.

Torrent, H. & Sauveplan, S. (1977) *Orientation Map for Groundwater Exploration Related to Mineral Investigation in the Arabian Shield*. BRGM, Jeddah.

Townsend, C.C. (1983) A new species of *Peucedanum* (Umbelliferae) from the Yemen Arab Republic. *Kew Bulletin* 38: 53-55.

Townsend, C.C. (1986a) A new *Pimpinella* (Umbelliferae) from the Yemen Arab Republic. *Kew Bulletin* 41: 59-60.

Townsend, C.C. (1986b) *Oreoschimperella* and *Trachyspermum* (Umbelliferae) in East Africa and Arabia. *Kew Bulletin* 41: 453-548.

Troll, C. & Pfaffen, K. (1965) Karte der Jahreszeiten-Klimate der Erde. Erdkunde 18: 5-28.

van Heezik, Y. & Seddon, P.J. (1995) *Exclosure plots in Harrat al-Harrah. Annual report 1995*. Unpublished report. National Wildlife Research Centre, Taif, Saudi Arabia.

van Heezik, Y. & Seddon, P.J. (1996) *Exclosure plots in Harrat al-Harrah. Annual report 1996*. Unpublished report. National Wildlife Research Centre, Taif, Saudi Arabia.

Verdcourt, B. (1989) A new genus *Nogalia* (Boraginaceae-Heliotropioideae) from Somalia and southern Arabia. *Kew Bulletin* 43: 431-435.

Vesey-Fitzgerald, D.F. (1955) Vegetation of the Red Sea coast south of Jedda, Saudi Arabia. *Journal of Ecology* 43: 477-489.

Vesey-Fitzgerald, D.F. (1957a) The vegetation of central and eastern Arabia. *Journal of Ecology* 45: 779-798.

Vesey-Fitzgerald, D.F. (1957b) The vegetation of the Red Sea coast north of Jedda, Saudi Arabia. *Journal of Ecology* 45: 547-562.

Vierhapper, F. (1907) Beiträge zur Kenntnis der Flora Südarabiens und der Inseln Sokotra, Semha und Abdel Kuri I. *Denkschriften der Mathematisch-Naturwissenschaftliden Classe der Kaiserlichen Akademie der Wissenschaften* 71: 1-170.

Villwock, G. (1991) *Beiträge zur physischen Geographie und Landschaftsgliederung des südlichen Jemen (ehemals VDRJ)*. Jemen Studien Band 10. Wiesbaden.

Vogel, H. (1988a) Deterioration of a mountainous agro-system in the third world due to emigration of rural labour. *Mountain Research and Development* 8: 321-329.

Vogel, H. (1988b) Impoundment-type bench terracing with underground conduits in Jibal Haraz, Yemen Arab Republic. *Transactions of the Institute of the British Geographical Society* 13: 29-38.

Volk, O. (1984) Beiträge zur Kenntnis der Marchantiales in Südwest Africa/Namibia IV. Zur Biologie einiger Hepaticae mit besonderer Berücksichtigung der Gattung Riccia. *Nova Hedwigia* 39: 117-143.

Wagenitz, G., Dittrich, M. & Damboldt, J. (1982) *Centaurothamnus*, eine neue Gattung der Compositae-Cardueae aus Arabien. *Candollea* 37: 101-115.

Waisel, Y. (1991) The glands of *Tamarix aphylla*: a system for salt secretion or for carbon concentration? *Physiologia Plantarum* 83: 506-510.

Walker, T.R. (1979) Red Color in Dune Sand. Pp 61-81 in *A Study of Global Sand Seas (Geological Survey Professional Paper 1052)*. E.D. McKee, ed. US Government Printing Office, Washington.

Walter, H. & Lieth, H. (1960) *Klimadiagramm Weltatlas*. VEB Gustav Fischer, Jena.

Warren, A. (1988) A note on sand movement and vegetation in the Wahiba Sands. Pp. 251-255 in *The Scientific Results of The Royal Geographical Society's Oman Wahiba Sands Project 1985-1987. The Journal of Oman Studies Special Report No. 3.* R.W. Dutton, ed. Office of the Adviser for Conservation of the Environment, Muscat, Oman.

Werger, M.J.A. (1978) The Karoo-Namib Region. Pp. 231-299 in *Biogeography and Ecology of Southern Africa*. M.A. Werger, ed. Dr W. Junk, The Hague.

Wessels, D. & Büdel, B. (1989) A rock pool lichen community in northern Transvaal, South Africa: composition and distribuition patterns. *Lichenologist* 21: 259-277.

Western, A.R. (1982) The natural vegetation of Abu Dhabi Islands. *Emirates Natural History Group Bulletin* 17: 18-24.

Western, A.R. (1983) The vegetation of offshore islands in the Gulf. *Emirates Natural History Group Bulletin* 20: 16-23.

Western, A.R. (1987) The coastal vegetation of Fujeirah. *Emirates Natural History Group Bulletin* 31.

Western, A.R. (1989) *The Flora of the United Arab Emirates, an Introduction*. United Arab Emirates University.

Whitcombe, R.P. (1995) *The Arabian Oryx Sanctuary. Preliminary Land Use and Management Plan*. Directorate-General of Nature Reserves, Ministry of Regional Municipalites and Environment, Oman, Office of the Adviser for conservation of the Environment, Diwan of Royal Court, Oman, and The World Heritage Fund, UNESCO.

White, F. (1983) *The Vegetation of Africa. A Descriptive Memoir to Accompany the UNESCO, AETFAT, UNSO Vegetation Map of Africa*. UNESCO, Paris.

White, F. & Léonard, J. (1991) Phytogeographical links between Africa and Southwest Asia. *Flora et Vegetatio Mundi* 9: 229-246.

Whitney, J.W., Faulkender, D.J. & Rubin, M. (1983) *The Environmental History and Present Condition of the Northern Sand Seas of Saudi Arabia*. Ministry of Petroleum and Mineral Resources, Deputy Ministry for Mineral Resources, Jiddah, Saudi Arabia.

Whybrow, P.J. & Hill, A., eds. (in press) *Fossil Vertebrates of Arabia*. Yale University Press.

Whybrow, P.J. & McClure, H.A. (1981) Fossil mangrove roots and palaeoenvironment of the Miocene of the eastern Arabian Peninsula. *Paleogeography, Palaeoclimatology and Palaeoecology* 32: 213-225.

Wickens, G.E. (1982) Studies in the Flora of Arabia: III. A biographical index of plant collectors in the Arabian peninsula (including Socotra). *Notes from the Royal Botanic Garden Edinburgh* 40: 301-330.

Wickens, G.E., Goodin, J.R. & Field, D.V. eds. (1989) *Plants for Arid Lands. Proceedings of the Kew International Conference on Economic Plants for Arid Lands held in the Jodrell Laboratory, Royal Botanic Gardens, Kew, England, 23-27 July 1984*. Unwin Hyman, London.

Widrlechner, M.P. (1981) History and utilization of *Rosa damascena*. *Economic Botany* 35: 42-58.

Wilson, I.G. (1973) Ergs. *Sedimentary Geology* 10: 77-106.

Wilson, R.T.O. (1985) The Tihama from the beginning of the Islamic period to 1800. Pp. 31-36 in *Studies on the Tihama. The Report of the Tihama Expedition 1982 and Related Papers*. F. Stone, ed. Longman, Harlow.

Winstanley, D. (1972) Sharav. *Weather* 27: 146-160.

Wissmann, H. von (1937) Arabien. Pp. 178-211 in *Handbuch der Geographischen Wissenschaft, Band Vorder-und Südasien*. F. Klute, ed. Akademie-Verlag, Potsdam.

Wissmann, H. von (1948) Die pflanzenklimatischen Grenzen der warmen Tropen. Erdkunde 2: 81-92.

Wissmann, H. von (1972) Die *Juniperus*-Gebirgswälder in Arabien. Ihre Stellung zwischen dem borealen und tropisch-afrikanischen Florenreich. *Erdwissenschaftliche Forschuug* 4: 157-176.

Wood, J.R.I. (1983a) The Aloes of the Yemen Arab Republic. *Kew Bulletin* 38: 13-31.

Wood, J.R.I. (1983b) An outline of the vegetation of the Yemen Arab Republic. Pp. 29-89 in *Soil Survey of the Yemen Arab Republic*. J.W. King, ed. Cornell University, Washington.

Wood, J.R.I. (1984) Eight new species and taxonomic notes on the Flora of Yemen. *Kew Bulletin* 39: 123-139.

Wood, J.R.I. (1985) The vegetation of the Tihama. Pp. 14-15 in *Studies on the Tihama. The Report of the Tihama Expedition 1982 and Related Papers*. F. Stone, ed. Longman, Harlow.

Wood, J.R.I. (1997) *A Handbook of the Yemen Flora*. Royal Botanic Gardens, Kew.

Woolf, D.A. (1990) Blood lead levels in Omani children admitted to hospital. *Journal of Tropical Pediatrics* 36: 314-315.

Worthing, M.A. & Sutherland, H.H. (1996) The composition and origin of massicot, litharge (PbO) and a mixed oxide of lead used as a traditional medicine in the Arabian Gulf. *Mineralogical Magazine* 60: 509-513.

Younes, H.A., Zahran, M.A. & El-Qyrashy, E.M. (1983) Vegetation-soil relationships of a sea-landward transect, Red Sea coast, Saudi Arabia. *Journal of Arid Environments* 6: 349-356.

Zahran, M.A. (1975) Biogeography of mangrove vegetation along the Red Sea coast. Pp. 45-51 in *Proceedings of the International Symposium on the Biology and Management of Mangroves*. Gainsville.

Zahran, M.A. (1977) Wet formations of the African Red Sea Coast. Pp. 215-231 in *Wet Coastal Ecosystems*. V.J. Chapman, ed. Elsevier, Amsterdam.

Zahran, M.A. (1993) Dry coastal ecosystems of the Asian Red Sea coast. Pp. 17-29 in *Dry Coastal Ecosystems. Africa, America, Asia, Oceania*. E. van der Maarel, ed. Elsevier, Amsterdam.

Zahran, M.A., Younes, H.A. & Hajrah, H.H. (1983) On the ecology of the mengal vegetation of the Saudi Arabian Red Sea coast. *Journal of the University of Kuwait (Science)* 10: 87-99.

Zohary, D. & Hopf, M. (1994) *Domestication of Plants in the Old World*. Clarendon Press, Oxford.

Zohary, M. (1963) On the geobotanical structure of Iran. *Bulletin of the Research Council Israel* 11, D Supplement: 1-113.

Zohary, M. (1966) *Flora Palestina, Part I*. Jerusalem.

Zohary, M. (1973) *Geobotanical Foundations of the Middle East*. Fischer Verlag, Stuttgart.

Zwemer, S.M. (1986) *Arabia: The Cradle of Islam*. Darf Publishers, London.

Index

Geobotany

1. J.B. Hall and M.D. Swaine (eds.): *Distribution and Ecology of Vascular Plants in a Tropical Rain Forest.* Forest Vegetation in Ghana. 1981 ISBN 90-6193-681-0
2. W. Holzner and M. Numata (eds.): *Biology and Ecology of Weeds.* 1982 ISBN 90-6193-682-9
3. N.J.M. Gremmen: *The Vegetation of the Subantarctic Islands Marion and Prince Edward.* 1982 ISBN 90-6193-683-7
4. R.C. Buckley (ed.): *Ant-Plant Interactions in Australia.* 1982 ISBN 90-6193-684-5
5. W. Holzner, M.J.A. Werger and I. Ikusima (eds.): *Man's Impact on Vegetation.* 1983 ISBN 90-6193-685-3
6. P. Denny (ed.): *The Ecology and Management of African Wetland Vegetation.* 1985 ISBN 90-6193-509-1
7. C. Gómez-Campo (ed.): *Plant Conservation in the Mediterranean Area.* 1985 ISBN 90-6193-523-7
8. J.B. Faliński: *Ecological Studies in Białowieza Forest.* 1986 ISBN 90-6193-534-2
9. G.A. Ellenbroek: *Ecology and Productivity of an African Wetland System.* The Kafue Flats, Zambia. 1987 ISBN 90-6193-638-1
10. J. van Andel, J.P. Bakker and R.W. Snaydon (eds.): *Disturbance in Grasslands.* Causes, Effects and Processes. 1987 ISBN 90-6193-640-3
11. A.H.L. Huiskes, C.W.P.M. Blom and J. Rozema (eds.): *Vegetation Between Land and Sea.* Structure and Processes. 1987 ISBN 90-6193-649-7
12. G. Orshan (ed.): *Plant Pheno-morphological Studies in Mediterranean Type Ecosystems.* 1988 ISBN 90-6193-656-X
13. B. Dell, J.J. Havel and N. Malajczuk (eds.): *The Jarrah Forest.* A Complex Mediterranean Ecosystem. 1988 ISBN 90-6193-658-6
14. J.P. Bakker: *Nature Management by Grazing and Cutting.* 1989 ISBN 0-7923-0068-8
15. J. Osbornová, M. Kovářová, J. Lepš and K. Prach (eds.): *Succession in Abandoned Fields.* Studies in Central Bohemia, Czechoslovakia. 1990 ISBN 0-7923-0401-2
16. B. Gopal (ed.): *Ecology and Management of Aquatic Vegetation in the Indian Subcontinent.* 1990 ISBN 0-7923-0666-X
17. B.A. Roberts and J. Proctor (eds.): *The Ecology of Areas with Serpentinized Rocks.* A World View. 1991. ISBN 0-7923-0922-7
18. J.T.A. Verhoeven (ed.): *Fens and Bogs in the Netherlands.* Vegetation, History, Nutrient Dynamics and Conservation. 1992 ISBN 0-7923-1387-9
19. Woo-seok Kong and D. Watts: *The Plant Geography of Korea.* With an Emphasis on the Alpine Zones. 1993 ISBN 0-7923-2068-9
20. R. Aerts and G.W. Heil (eds.): *Heathlands.* Patterns and Processes in a Changing Environment. 1993 ISBN 0-7923-2094-8
21. W. van Vierssen, M. Hootsmans and J. Vermaat (eds.): *Lake Veluwe, a Macrophyte-dominated System under Eutrophication Stress.* 1994 ISBN 0-7923-2320-3
22. Y. Laumonier: *The Vegetation and Physiography of Sumatra.* 1997 ISBN 0-7923-3761-1
23. C.M. Finlayson and I. von Oertzen (eds.): *Landscape and Vegetation Ecology of the Kakadu Region, Northern Australia.* 1996 ISBN 0-7923-3770-0
24. R. Peters: *Beech Forests.* 1997 ISBN 0-7923-4485-5
25. S.A. Ghazanfar and M. Fisher (eds.): *Vegetation of the Arabian Peninsula.* 1998 ISBN 0-7923-5015-4

KLUWER ACADEMIC PUBLISHERS – DORDRECHT / BOSTON / LONDON

SPRINGER NATURE

GPSR Compliance

The European Union's (EU) General Product Safety Regulation (GPSR) is a set of rules that requires consumer products to be safe and our obligations to ensure this.

If you have any concerns about our products, you can contact us on ProductSafety@springernature.com

In case Publisher is established outside the EU, the EU authorized representative is:

Springer Nature Customer Service Center GmbH
Europaplatz 3
69115 Heidelberg, Germany

The manufacturer's authorised representative in the EU is Springer Nature Customer Service Centre GmbH, Europaplatz 3, 69115 Heidelberg, Germany. If you have any concerns regarding our products, please contact ProductSafety@springernature.com

Printed and bound by CPI Group (UK) Ltd, Croydon, CR0 4YY
30/09/2026
02225436-0019